# PRINCIPLES OF DESIGN AND OPERATION OF CATERING EQUIPMENT

## ELLIS HORWOOD SERIES IN FOOD SCIENCE AND TECHNOLOGY

*This publishing programme provides an organised coverage on food science for professional technologists, research workers, and students in universities and polytechnics.*

**Series Editors:**

**Professor I. D. Morton**, Head of Food Science, Queen Elizabeth College, University of London;

**Dr R. Scott**, formerly of Reading University

**Dr R. Osner**, Bath College of Higher Education

**FUNDAMENTALS OF FOOD CHEMISTRY**
W. Heimann, Director, Institute for Food Chemistry, University of Karlsruhe, Germany

**HYGIENIC DESIGN AND OPERATION OF FOOD PLANT**
Edited by R. Jowitt, National College of Food Technology, University of Reading

**HANDBOOK OF DOMESTIC ENGINEERING**
D. Kirk, Sheffield City Polytechnic and
A. Milson, The Queen's College, Glasgow

**PRINCIPLES OF DESIGN AND OPERATION OF CATERING EQUIPMENT**
A. Milson, The Queen's College, Glasgow and
D. Kirk, Sheffield City Polytechnic

# PRINCIPLES OF DESIGN AND OPERATION OF CATERING EQUIPMENT

**A. MILSON**
M.Sc.Tech., B.Sc., C.Eng., F.I.Gas.D., A.I.F.S.T., A.H.C.I.M.A.
Vice-Principal
The Queen's College, Glasgow

and

**D. KIRK**
B.Sc., A.I.F.S.T., M.Phil
Principal Lecturer
Department of Hotel and Catering Studies and Home Economics
Sheffield City Polytechnic

**ELLIS HORWOOD LIMITED**
Publishers · Chichester
**AVI PUBLISHING COMPANY**
West Port, Connecticut, USA

First published in 1980 by

**ELLIS HORWOOD LIMITED**

Market Cross House, Cooper Street, Chichester, West Sussex, PO19 1EB, England

*The publisher's colophon is reproduced from James Gillison's drawing of the ancient Market Cross, Chichester.*

**Distributors:**

*Australia, New Zealand, South-east Asia:*
Jacaranda-Wiley Ltd., Jacaranda Press,
JOHN WILEY & SONS INC.,
G.P.O. Box 859, Brisbane, Queensland 40001, Australia.

*Europe, Africa:*
JOHN WILEY & SONS LIMITED
Baffins Lane, Chichester, West Sussex, England.

*The Americas, Japan, The Phillipines:*
AVI PUBLISHING COMPANY
250 Post Road East, Westport, Connecticut 06880, U.S.A.

**British Library Cataloguing in Publication Data**
Milson, Anthony
Principles of design and operation of catering equipment. –
(Ellis Horwood series in food science and technology).
1. Food service – Equipment and supplies
I. Title II. Kirk, D
647'.95'028 TX943 79–41556

ISBN 0–85312–132–X (Ellis Horwood Ltd., Publishers, Library Edition)
ISBN 0–87055–357–7 (AVI Publishing Company)

Typeset in Press Roman by Ellis Horwood Ltd.
Printed in Great Britain by Fakenham Press Ltd.

# Table of Contents

**CHAPTER 10 – MASS TRANSFER**



**CHAPTER 11 – ENERGY SOURCES AND CONTROL**

# Authors' Preface

We have written this book because there is no book which deals adequately with the design and operation of catering equipment. Indeed little attempt has been made in the past to understand the processes taking place inside catering equipment. This means that there is no understanding of the limits of existing equipment nor is it possible to manipulate the parameters of equipment design to achieve optimum results. Optimum results involve not only excellence of food quality, but they also involve factors such as speed of cooking, floor space required and the cost of preparation (including capital and operating costs of equipment).

Our major objective is to show how knowledge of process engineering may be applied to the study of equipment used to prepare food in the catering industry. This approach has been achieved with a large measure of success for the food industry but not so far for catering. This situation is now changing and the publication of research papers on catering equipment is increasing. In this book we have attempted to include references to all this literature, to the middle of 1979.

Our aims are to explain the basic concepts on which the heating and other processes of food preparation are based and to show how these concepts apply in the design and operation of catering equipment. Process engineering is, whenever possible, a quantitative approach. We have adopted such an approach, at the same time keeping the mathematics simple. It will become evident as the reader progresses through the book that whilst it is feasible to be quantitative for many processes, a descriptive approach only is possible for others. We have included many illustrative problems for solution in the belief that caterers and others should be encouraged to calculate outcomes. A set of solutions is provided. An extensive guide to the data necessary for making calculations is given in the Apendices.

This book arose out of a course on catering equipment taught by the authors to the Sheffield City Polytechnic's B.Sc. (Catering Systems) course. Teaching is a two-way process and we are grateful for the help our students have given us in clarifying our thoughts on the material presented here. We thank too our friends in the catering-equipment industry and in equipment research for time spent in discussions over a number of years. In particular, we would like to thank Dr. George Wilson of Manchester Polytechnic who commented in detail on the text. We hope that this book will contribute to the understanding of catering equipment.

Mick Weller produced excellent proofs from our difficult manuscript and made the task of proof-reading so much easier than it might have been.

Finally, our gratitude to our long-suffering families whose forebearance made it possible for us to complete the task of writing this book.

# Acknowledgements

We acknowledge permissions to use materials from a number of sources:

Applied Science Publishers Ltd., D. G. Taylor, D. M. Fettes, S. Fischer, N. Bengtsson and G. Glew for data in Tables 2.2, 5.1 and 6.1 and for Figs. 3.21, 3.22, 4.17, 5.19, 6.17 and 8.8, all taken from Glew, G. (ed.), *Catering Equipment and Systems Design*, 1977.

Automatic Vending Association of Britain for Figs. 14.1 and 14.2 taken from the Association's book: *Vending*, 1972.

British Gas for Figs. 4.16, 5.14, 5.15, 5.16, 5.18, 5.21, 5.22, 5.23, 5.25, 5.26, 5.28, 5.29, 5.31, 5.32, 5.33, 6.15, 6.16, 9.1, 9.2, 11.18, 11.21, 11.22, 11.23, 11.24, 11.29, 11.30, 11.31, 11.32, 13.2, 13.3 and 13.4 taken from its publication: *A Student's Guide to Gas Catering*, and for Figs. 5.11 and 11.33 taken from *Watson House Bulletin*, **34** (1970), 130-133 and **40** (1976), 243-244.

G. C. Lowrison for Figs. 12.1, 12.2 and 12.3 taken from his book: *Crushing and Grinding*, Butterworth (1974).

The Editor *ASHRAE Journal* for Fig. 9.7 taken from *ASHRAE Jnl.*, **14**(11), (1972), 39-45.

George Glew for data in Table 2.1 taken from: *An Experiment in Hospital Catering Using the Cook/Freeze System*, University of Leeds and United Leeds Hospitals, (1970) and Figs. 5.9, 5.10 and 5.13 from *HCIMA Review*, **1** (1), (1974), 39-50.

Küppersbusch Aktiengesellschaft for Figs. 5.20, 5.30, 5.34 and 9.3.

Director, National Engineering Laboratory for Figs. B.1, B.2, B.3, B.4, B.5 and B.6 taken from Dalgleish, N. and Ede, A. J., *NEL Report No. 192*, July 1965, © Controller, Her Majesty's Stationery Office.

Forster-Verlag AG for Fig. 4.15 taken from *Lebensmittel Wissenshaft und Technologie*, 7, (1974), 202-207.

The Metals Society for Fig. 3.15 taken from *Jnl. Iron and Steel Institute*, July 1960, 268-292.

Editor, *Refrigeration and Air Conditioning*, Maclaran Publishers, for Fig. 8.5 taken from Modern Refrigeration, **52**, (1949), 52–55.

International Microwave Power Institute for data in Tables G.6 and H.8 and Fig. G.1 taken from *J. Microwave Power*, **6**(2), (1971), 107-123 and **10**(1), (1975), 93-108.

Womens' Gas Federation & Young Homemakers for Figs. 11.20, 11.26, 11.27 and 11.28 taken from Plant, E. M., *Gas and its Domestic Uses*, 2nd ed., (1971).

L. Keniston, Property Services Agency and the Electricity Council for Fig. 5.24 taken from a paper by L. Keniston to the Caterer in Control Conference, Electricity Council, April 1976.

Editor, *Journal of Food Technology* for Figs. 5.12 and 9.8 taken from *J. Fd. Technol.*, **4**, (1969), 307–318 and **10**, (1975), 375–385 respectively.

Institute of Food Technologists for Fig. 4.14 taken from *Food Technology*, **22** (1968), 37–52 and for Figs. 9.9 and 9.10 taken from *Journal of Food Science*, **41**, (1976), 1047-1053, © by Institute of Food Technologists.

McGraw-Hill Book Company for data in Appendix E taken from Treybal, R. E., *Mass Transfer Operations*, 2nd ed., (1968).

John Wiley & Sons for Table A.3, Figs. A.2 and A.3 and data in Appendix C taken from Welty, J. R. *et al.*, *Fundamentals of Momentum, Heat and Mass Transfer*, (1969).

D. G. Taylor for data in Table 2.2 taken from Energy Measurement Programme, Marketing Department, Electricity Council, May 1977.

Pergamon Press for Figs. 5.3, 5.8 and 6.5 and data in Appendices C and I (Sections 9, 10 and 11) taken from Ede, A. J., An Introduction to Heat Transfer Principles and Calculations, 1967 for Figs. 6.6, D.3.1, D.3.2 and D.3.3 taken from Gray, W. A., and Müller, R., *Engineering Calculations in Radiation Heat Transfer*, 1974.

British National Committee, World Energy Conference for Fig. A.1 and Tables 11.1, A.1 and A.2 and data in Appendices F, I (Sections 1 to 8) and J taken from Rose, J. W. and Cooper, J. R. (eds.), *Technical Data on Fuel*, 7th ed., 1977.

CHAPTER 1

# Context of Equipment Studies

## 1.1 INTRODUCTION

This book is concerned with the design and performance of catering equipment. Much that is written about catering equipment involves what may be termed a 'black box' approach, that is to say unprepared food is placed in the equipment and after a certain period of time emerges as prepared food of certain characteristics. Rarely is the question asked: "What happens inside the equipment?". We are concerned here with what may be termed a 'glass box' approach, that is, we are going to 'look inside' the equipment in order to understand the processes taking place. We shall be looking at ways in which heat is transferred from a gas flame or an electric element to the surface of the food, the way heat is transferred from the surface of the food to its inside, how liquid ingredients are mixed together, what happens inside a mincing machine and so on.

The knowledge already available in the fields of process engineering, chemical engineering and food engineering will be applied to the design of catering equipment. This involves both a conceptual and a numerical approach. We do this because it is possible to understand a good deal about equipment by doing fairly straight-forward arithmetical computations. However, the emphasis is not on the arithmetical computations as such but on what the arithmetic enables us to deduce about what is going on both inside the equipment and the food (conceptual approach).

The reader will find it helpful to keep the overall plan of the book in mind. Chapter 2 is concerned mainly with the 'black box' approach whilst subsequent chapters concentrate on the 'glass box' approach. The flow of fluids (gases and liquids) is considered in Chapter 3, partly because this subject is in itself of importance to caterers but also because it enables us to understand convection heating in a later chapter. Chapters 4 to 8 each deal with a single mechanism of heat transfer and with equipment which relies principally on a single mode of heat transfer for its operation. In Chapter 9 we consider equipment utilising several modes of heat transfer simultaneously. Hence Chapter 9 is a synthesis of Chapters 4 to 8; at this stage the arithmetic is usually sufficiently onerous for use to be made of the computer. A chapter on mass-transfer processes follows. These are sometimes associated with heating or cooling processes or may be carried out for a specific purpose. The methods for generating heat energy and controlling it are surveyed in Chapter 11. The final four chapters (12 to 15) deal with non-heating/cooling equipment, where the theoretical approach is less well-developed than for the topics dealt with in earlier chapters.

## 1.2 RELATIONSHIP TO OTHER SUBJECTS

The study of catering equipment cannot be considered in isolation. It is necessary to relate it to other subjects and, in particular, to the following.

### 1.2.1 Food studies

Equipment design and performance depend on the nature and properties of the food raw materials to be processed. The food scientist and food formulator define the processes the food needs to undergo in terms of: quantities of components required, degree of mixing, size of ingredients and products, types of heat transfer, moisture loss during heating, fat absorption, surface colour required, final texture to be produced, the final temperature-distribution in the food and so on. Once these factors have been specified, the designer can then start to design or select the appropriate equipment or combination of equipment to meet the processing specification. Such specifications of products are not always available to the designer as they have not in general been published. This is a great obstacle in the way of rational equipment design.

### 1.2.2 Financial aspects

The costs of buying and operating catering equipment are important elements in the final cost of a dish of food. One aim should be to balance these out so as to achieve the minimum total cost. The nature of the industry imposes three important constraints, from a financial viewpoint, on equipment design. First, the catering industry, unlike many other food and process engineering plants, usually operates for relatively short periods in the day (a few very large operations are, however, similar to food processing plants). Therefore the throughput of the equipment is small and the capital cost of equipment must be written off on short production runs. This usually means that capital costs must be minimised, otherwise the capital charges per unit of output become intolerably high.

Then the industry has a wide variety of products. This means that usually catering equipment needs to be versatile to cope with this range. If specialised equipment is used, its utilisation efficiency (that is, proportion of time it would actually be in use) would be very low and hence capital charges per unit of output would be very high. Exceptions to this are specialised operations where a very limited range of products is produced in a speciality operation, such as steak, hamburger and pizza restaurants.

Finally it is normal practice to produce food in a non-continuous way as opposed to continuous production. This is essential because of the above factors and initially means that food has to be stored, albeit for short periods only, before service. Therefore proper hot and cold storage equipment has to be designed. Continuous production runs are possible in cook-freeze systems and in very large installations, but caterers tend to be reluctant to use it (except in cook-freeze) for fear of equipment breakdowns.

### 1.2.3 Human aspects

Catering is usually labour intensive. We must distinguish therefore between the manual skills used in a kitchen and the study of catering equipment, which is concerned with the processes taking place inside equipment. The difference between the two can be illustrated by reference to actual processes taking place in a kitchen, for example:

| Process | Manual skills | Catering equipment |
|---|---|---|
| Cleaning potatoes | loading potatoes into rumbler<br>unloading potatoes from rumbler | mechanism for removing outer surface of the potatoes |
| Chipping potatoes | loading chipper | mechanism for cutting potatoes into chips |
| Frying chips | carrying chips to fryer<br>loading fryer<br>unloading fryer | heat transfer from flame to oil<br>heat transfer from oil to chips<br>mechanism for heating up chips |

Operatives with manual skills and equipment are sometimes interchangeable, for example, the loading and unloading of equipment may be mechanised; also human muscle sometimes takes the place of equipment operations, for example, manual as against mechanical chipping.

The human operator is also an important factor of equipment design with regard to the siting of controls, the layout of equipment and safety considerations. Moreover, psychological factors are often crucial in the use of equipment too. If new equipment is introduced then people must be trained to use it properly. There is also much mystique about catering equipment. One has to appreciate this to understand why people favour certain types of equipment.

It is often not that a new piece of equipment has been shown to be better but that caterers believe that it is. People need to be persuaded to make objective judgements about equipment rather than subjective ones. This book should help the reader to make more objective judgements about the selection and use of equipment.

## 1.3 THE AIMS OF EQUIPMENT DESIGN

The aims of equipment design may be summarised, bearing in mind the above discussion, as follows:

(i) To enable the processing of a particular food raw material to be carried out so that the final product is acceptable to the customer.
(ii) To provide as rapid a means as possible of preparing food in order to minimise the amount of equipment and floor space required.
(iii) To provide a safe and acceptable working environment for the operator.
(iv) To make for the hygienic preparation of food with easy cleaning of the equipment.
(v) To minimise total food preparation costs, striking a proper balance between capital cost (including installation) and operating costs (fuel, labour, repairs, maintenance).
(vi) To enable the manufacture of equipment that will be reliable over its planned production life, in relation to the usage it will get.

It must be emphasised that some of these aims can be conflicting and often the equipment designer has to arrive at a compromise.

One means of defining more specifically the aims of equipment design is to establish minimum performance standards for each type of equipment. Such standards are available in several countries and discussions are taking place at international level aimed at achieving agreement on a wider basis. These standards may cover at least the following three aspects of performance:

(a) cooking tests – see Section 1.4,
(b) thermal tests – see Section 2.7,
(c) safety tests – see Section 11.10.

The reader should familiarise himself with the standards, if any, published in his own country.

## 1.4 COOKING TESTS

The ultimate aim of equipment design is to enable food to be prepared to the customer's satisfaction. This is evaluated by the inclusion of cooking tests in the appropriate standards. Several examples of these tests are given below.

(a) **Ovens**: a series of test foods, made up from standard recipes, are cooked at specified thermostat settings in order to illustrate the reaction of the oven to different loadings and temperatures and establish the heat pattern over the whole cooking volume. The performance of the oven with each test food is assessed in terms of cooking time, size, colour and texture of the product.

(b) **Fryers**: the eveness and thoroughness of frying chips are assessed.

(c) **Grillers and toasters**: bread is toasted and the eveness of browning is assessed.

(d) **Steamers**: potatoes and steamed puddings are used. Performance with potatoes is assessed in terms of ease of cutting, shape retention and non-break-up; with puddings, the criteria are size and texture of the product.

Cooking tests must be carried out as part of a performance standard because they relate to the use of the equipment, that is, to produce food of a satisfactory standard for a customer. A further reason is our inability to define a satisfactory product in terms of parameters which can be related directly to equipment-design factors, as noted in Section 1.2.1.

Cooking tests are carried out under strictly defined conditions. Although these may be representative of industrial practice, they do not reproduce the full range of practical operating conditions. Wide changes in product type and equipment loading may give rise to sub-standard products, even if the equipment meets the agreed performance criteria. The standards are based on wide experience however, and should prove applicable in most situations. Occasionally, equipment is manufactured for which there is no agreed standard.

In addition to equipment standards controlled by government bodies, there are also agreed standards, or codes of practice, established between manufacturers and users, or by bodies such as Underwriters' Laboratories and the American Gas Association in the USA and the British Gas Corporation and the Electricity Council in the United Kingdom. These standards are often concerned with thermal performance (Section 2.7) and safety standards (Section 11.10).

CHAPTER 2

# Mass and Heat Balances

## 2.1 INTRODUCTION

In this chapter we look at mass and heat balances of food preparation (mainly cooking) processes. These balances give an overall view of a process in terms of inputs and outputs. They treat a piece of equipment as a **black box**, as illustrated in Fig. 2.1.

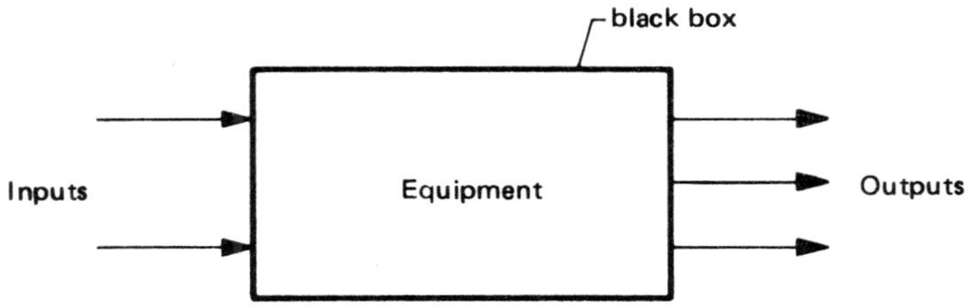


Fig. 2.1 – Equipment as a black box.

In the case of a **mass balance** the inputs are the masses of raw food, air and fuel entering the equipment and the outputs are the masses of cooked food and combustion products leaving the equipment. We are not concerned in a mass balance with what goes on inside the equipment. In the case of a **heat balance** the inputs are usually represented by the heat contained in the gas/electricity entering the equipment and the outputs by the heat contained in the hot food, hot combustion products and heat losses to the kitchen. Again, we are not concerned in a heat balance with what goes on inside the equipment.

This black-box approach (Section 1.1) is useful in that it enables us to define the efficiency of a piece of equipment. It is however of limited value because it does not give any indication of the processes going on inside the equipment and how therefore improved performance may be achieved. Hence most of the remainder of the book is concerned with these processes.

It is important at this stage to distinguish between **batch processing** and **continuous processing** in relation to the black-box approach. In the former, a batch of food is placed in the equipment at one instant in time and is removed at a later instant in time. Subsequently, further single batches may be processed one-at-a-time. In **continuous processing**, food is placed in the equipment on a continuous basis over a period of time with food being removed on a similar continuous basis. Most traditional catering equipment is of the batch type. With recent trends towards the centralisation of food production however, continuous food preparation equipment has been developed and this is referred to, as appropriate, in later chapters.

We also need to distinguish between **steady-state operations** and **unsteady-state operations**. Steady-state operations occur when there is *no* change of physical parameters with time, whilst with unsteady-state operations one or more parameters do change with time. For example, an oven which has just been switched on will for a time be under unsteady-state heating conditions, that is the temperature of the oven structure increases with time. Eventually, the temperature controller (thermostat) will operate to hold the oven temperature at a constant (pre-set) value. When this happens, the temperature of the oven structure will not change with time and the oven has reached steady-state. Most of this book is concerned with steady-state operations but there is one very important exception to this: in Chapter 4 (Sections 4.6ff.) the heating up of food during cooking is considered and this is an unsteady-state process.

## 2.2 MASS (MATERIALS) BALANCES

Mass balances are based on the principle of the conservation of mass which states that mass cannot be created or destroyed. In practice this means that the inputs to the equipment (in terms of mass) are equal to the outputs. Several examples will serve to illustrate this principle.

### 2.2.1 Simple mass (materials) balance on a gas-fired oven

The inputs and outputs are illustrated in Fig. 2.2.

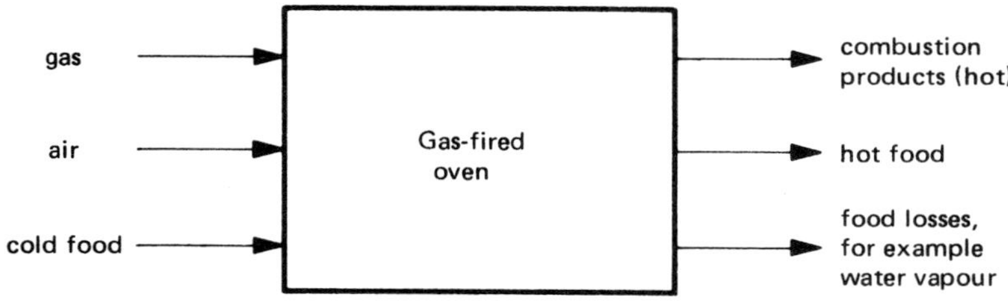


Fig. 2.2 – Inputs and outputs for mass balance on a gas-fired oven.

INPUTS:

mass of gas = $A$ kg
mass of air = $B$ kg
mass of uncooked food = $C$ kg

OUTPUTS:

mass of combustion products = $X$ kg
mass of hot food = $Y$ kg
mass of food losses = $Z$ kg

Mass balance on gas/air mixture and combustion products:

$$\begin{aligned} \text{Inputs} &= \text{Outputs} \\ A + B &= X\,. \end{aligned}$$

Mass balance on food:

$$\begin{aligned} \text{Inputs} &= \text{Outputs} \\ C &= Y + Z\,. \end{aligned}$$

We may define the **yield** of food from an oven by:

$$\text{yield} = \frac{Y}{C} \times 100\%\,.$$

Thus, the yield is the proportion of the uncooked food which is available for use.

The masses involved must be measured over a suitable period of time, for example, over one production period in the kitchen or over the period of time it takes to cook one load of food.

### 2.2.2 Potato rumbler plus de-eyeing operation

The inputs and outputs are illustrated in Fig. 2.3.

Fig. 2.3 – Inputs and outputs for mass balance on potato cleaning.

INPUTS:

mass of raw potatoes = $D$ kg

OUTPUTS:

mass of clean potatoes = $U$ kg
mass of eyes = $V$ kg
mass of peel and dirt = $W$ kg

Mass balance gives:

$$\begin{aligned} \text{Inputs} &= \text{Outputs} \\ D &= U + V + W\,. \end{aligned}$$

$$\text{yield} = \frac{U}{D} \times 100\%\,.$$

The yield is the mass of clean potatoes expressed as a proportion of the mass of raw, uncleaned potatoes.

### 2.2.3 Electric fryer

The inputs and outputs for frying chips are illustrated in Fig. 2.4.

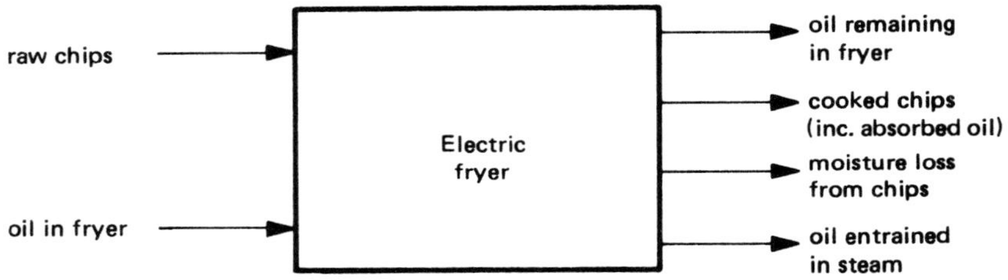


Fig. 2.4 – Inputs and outputs for mass balance on an electric fryer.

Some oil is used up in the fryer, partly by absorbtion and partly by carry-over with the steam evolved during frying. The former loss is included in the mass of cooked chips whilst the latter is shown as a separate output. The moisture loss is also shown separately.

INPUTS:

raw chips $= E$ kg
oil in fryer $= F$ kg

OUTPUTS:

oil remaining in fryer $= Q$ kg
cooked chips (including absorbed oil) $= R$ kg
moisture lost from chips $= S$ kg
oil entrained in steam $= T$ kg

Overall mass balance gives:

$$\begin{aligned} \text{Inputs} &= \text{Outputs} \\ E + F &= Q + R + S + T\,. \end{aligned}$$

The yield may be defined in two ways for this case:

$$\frac{R}{E} \times 100\% \quad \text{or} \quad \frac{R}{E + (F - Q - T)} \times 100\%\,,$$

where $(F - Q - T)$ is the mass of oil absorbed by the chips. The first definition has the merit of simplicity whilst the second definition is more precise as it includes the mass of oil absorbed by the chips. The second definition will give a lower yield than the first one.

### 2.2.4 Example of a mass balance

Tests show that if 4.0 kg potatoes are peeled and eyed a mass loss of 20% results. After chipping, the potatoes were fried and then found to weigh 2.8 kg. Of this final mass of chips, 10% was made up of oil absorbed during frying. Mass balances for the processes involved may be drawn up as follows.

$$\text{Starting mass of potatoes} = 4.0\,\text{kg}\,.$$

Loss is 20%, that is,

$$4.0 \times \frac{20}{100}\,\text{kg} = 0.8\,\text{kg (peel, eyes and dirt)}\,.$$

Therefore,

$$\text{mass of clean potatoes} = (4.0 - 0.8)\,\text{kg} = 3.2\,\text{kg}\,.$$

After frying,

$$\text{mass of chips} = 2.8\,\text{kg}\,.$$

$$\text{loss in mass brought about by frying} = (3.2 - 2.8)\,\text{kg} = 0.4\,\text{kg}\,.$$

This loss = mass gain from the oil absorbed − mass loss caused by the evaporation of water.

$$\text{Mass of oil absorbed} = 2.8 \times \frac{10}{100} = 0.28\,\text{kg}\,.$$

Therefore,

$$\begin{aligned}\text{cooked mass of potatoes} &= \text{mass of fried chips} - \text{mass of oil absorbed}\\ &= (2.8 - 0.28)\,\text{kg}\\ &= 2.52\,\text{kg}\,.\end{aligned}$$

$$\begin{aligned}\text{Mass of evaporated water} &= \begin{bmatrix}\text{mass of}\\ \text{uncooked potatoes}\end{bmatrix} - \begin{bmatrix}\text{cooked mass}\\ \text{of potatoes}\end{bmatrix}\\ &= (3.2 - 2.52)\,\text{kg}\\ &= 0.68\,\text{kg}\,.\end{aligned}$$

These changes are illustrated in Fig. 2.5.

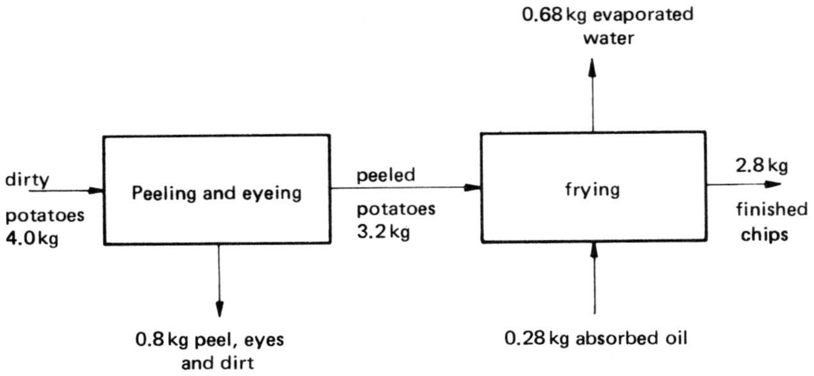


Fig. 2.5 – Mass balance: cleaning and frying potatoes.

$$\text{Yield after peeling and eyeing operation (see Section 2.2.2)} = \frac{3.2}{4.0} \times 100\% = 80\% .$$

$$\text{Yield on frying (see Section 2.2.3)} = \frac{2.8}{3.2} \times 100\% = 87.5\% ,$$

$$\text{or} = \frac{2.8}{3.2 + 0.28} \times 100\% = 80.5\% .$$

The yields, derived from mass balances, can be used to control food wastage caused by either poor operator performance or changing food raw materials quality. It should be possible in most operations to draw up control charts for the preparation of each food raw material. In the case of some foods, such as potatoes, seasonal charts would be required. Control charts will show the mean (mass) yield ($\bar{x}$) and the 95% confidence limits ($\bar{x} \pm 2\sigma$, where $\sigma$ = standard deviation of $\bar{x}$). If in practice yields outside these limits occur, then an investigation should be made to discover the reasons for this. These reasons may be poor operator performance (training could rectify the problem) or abnormally poor/good quality of raw material. The yield is required to calculate back from portion size and number of customers to the mass of raw material to be purchased. Similarly, yield should be taken into account when calculating the cost of food on the plate from the raw materials cost.

Yield is also an important factor in comparing different types of equipment for cooking/preparing the same food material as some methods of cooking give higher yields than other methods, or less 'shrinkage' if one considers cooking losses. It should be used alongside other factors, such as sensory quality of the product, when deciding on the purchase of equipment.

## 2.3 HEAT (ENERGY) BALANCES

Heat balances are based on the principle of conservation of energy, that is, that energy cannot be created or destroyed but only changed from one form to another. In practice this means that the inputs to the equipment (in terms of energy or heat) are equal to the outputs. Several examples will serve to illustrate this principle.

### 2.3.1 Simple heat balance on a gas-fired oven (for oven already raised to its cooking temperature)

The inputs and outputs are illustrated in Fig. 2.6.

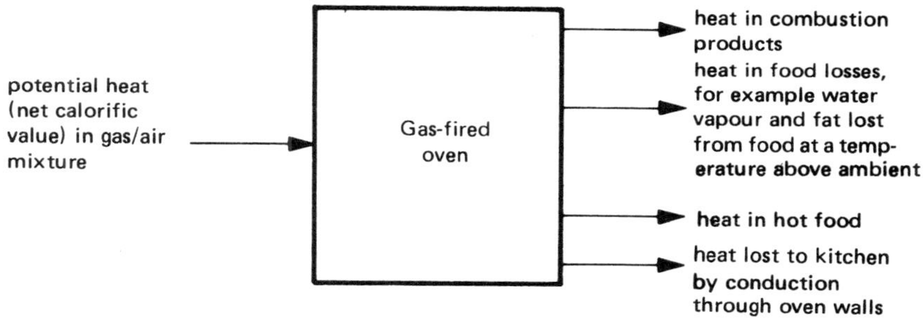


Fig. 2.6 – Inputs and outputs for heat balance on a gas-fired oven.

INPUTS:

potential heat = $G$ J

OUTPUTS:

heat in combustion products = $M$ J
heat in food losses = $N$ J
heat in hot food = $O$ J
heat lost to kitchen by conduction through oven wall = $P$ J

$$\text{Inputs} = \text{Outputs}$$
$$G = M + N + O + P\,.$$

$$\text{Efficiency} = \frac{O}{G} \times 100\% \quad \text{or} \quad \frac{N+O}{G} \times 100\%\,.$$

For some cooking processes $N$ is an essential part of the process and therefore should be added to $O$ in the definition of efficiency, for example the moisture loss when baking cakes, pastries and biscuits. An additional heat input is that required to heat the equipment to the appropriate temperature at the start of production. This heat is eventually lost into the kitchen when the equipment cools down at the end of production. If only one batch of food is being cooked this must be added to *G*. If there is more than one batch than this heat can be apportioned to each batch. This is one reason for using multiple small batches for cooking rather than a single large batch.

### 2.3.2 Electric fryer

The inputs and outputs for frying chips are illustrated in Fig. 2.7.

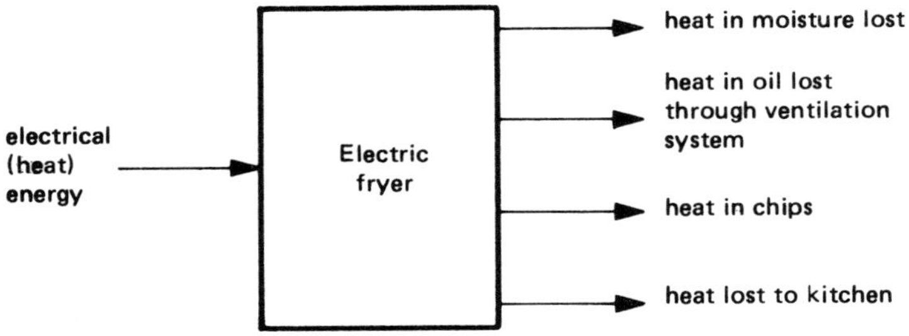


Fig. 2.7 – Inputs and outputs for heat balance on an electric chip fryer.

INPUTS:
electrical = $H$ J

OUTPUTS:
heat in moisture lost = $I$ J
heat in oil lost = $J$ J
heat in chips = $K$ J
heat lost to kitchen = $L$ J

$$\text{Inputs} = \text{Outputs}$$
$$H = I + J + K + L\,.$$

$$\text{Efficiency} = \frac{K}{H} \times 100\%\,.$$

Again, the heat required to heat up the equipment has not been included in this definition of efficiency. An additional heat requirement in the case of a fryer is the heat needed to heat up the cooking oil to the frying temperature. This heat is also lost to the kitchen at the end of the production period. Both these heat input terms should therefore be added to the heat input, by averaging them over the produciton runs made. In some cases the 'heat to hot food' term in the above example will include a latent heat term to cover essential changes of state, as when heating frozen food in an oven. Here again the change of phase (from frozen to liquid) is an essential part of the cooking process.

The heat inputs and heat outputs are usually calculated by one of the following equations.

$$\begin{bmatrix}\text{Thermal heat input (J)}\\ \text{when using gaseous fuel}\end{bmatrix} = \underset{\left(\frac{\text{J}}{\text{m}^3}\right)}{\begin{bmatrix}\text{net calorific}\\ \text{value of gas}\end{bmatrix}} \times \underset{(\text{m}^3)}{\begin{bmatrix}\text{volume of}\\ \text{gas burnt}\end{bmatrix}} .$$

$$\text{Electrical heat input (J)} = \underset{\text{(volts)}}{\text{potential difference}} \times \underset{\text{(amps)}}{\text{current}} \times \underset{\text{(seconds)}}{\text{time}} ,$$

$$= \underset{\text{(watts)}}{\text{power}} \times \underset{\text{(seconds)}}{\text{time}}$$

(1 unit of electricity = 1 kWh = 3 600 000 J) .

$$\text{Sensible heat (J)} = \underset{\text{(kg)}}{\text{mass}} \times \underset{\left(\frac{\text{J}}{\text{kg.K}}\right)}{\text{mass specific heat}} \times \underset{\text{(K)}}{\text{temperature rise}} ,$$

$$\text{or} = \underset{(\text{m}^3)}{\text{volume}} \times \underset{\left(\frac{\text{J}}{\text{m}^3\text{.K}}\right)}{\text{volume specific heat}} \times \underset{\text{(K)}}{\text{temperature rise}} .$$

$$\text{Latent heat of water vapour (J)} = \underset{\text{(kg)}}{\text{mass}} \times \underset{\left(\frac{\text{J}}{\text{kg}}\right)}{\text{latent heat}} .$$

$$\begin{bmatrix}\text{Latent heat of fusion (melting}\\ \text{of ice in frozen food) (J)}\end{bmatrix} = \underset{\text{(kg)}}{\text{mass}} \times \underset{\left(\frac{\text{J}}{\text{kg}}\right)}{\text{latent heat}} .$$

### 2.3.3 Example of an energy balance

A test with a microwave oven showed that when 1 kg of water was heated in the oven for 175 seconds its (the water's) temperature was found to increase from 24°C to 66°C and, at the end of the heating process, it was found that 10 g of water had evaporated. During stand-by, the oven consumed 0.2 kW and during heating the consumption rose to 2.1 kW. When calibrated by its manufacturer the magnetron was found to have a conversion efficiency of 70%. An energy balance and the thermal efficiency of the oven may be calculated under these particular conditions of operation as follows.

Ignoring the effects of any heat absorbed by the container holding the water and also ignoring radiation and convection of heat from the water and its container to the surrounding air and oven compartment, the energy balance for the duration of the test may be represented diagrammatically by Fig. 2.8.

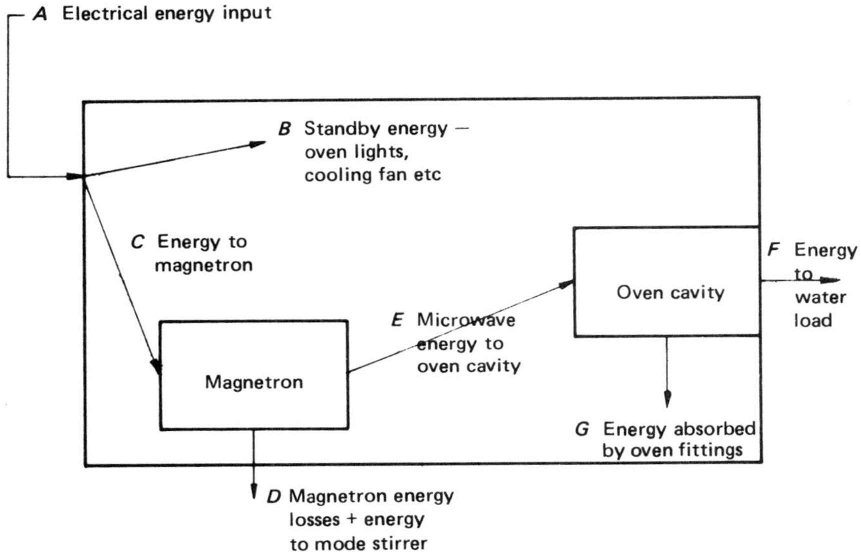


Fig. 2.8 – Energy balance of a microwave oven.

*A* Electrical energy input $= 2.1 \times 10^3 \mathrm{W} \times 175 \mathrm{s} = 367\,500 \mathrm{J}$.

*B* Standby energy consumption (assuming that all components consuming electricity during standby will also do so during the heating of the water load) $= 0.2 \times 10^3 \mathrm{W} \times 175 \mathrm{s} = 35\,000 \mathrm{J}$.

*C* Energy to magnetron $= A - B$

$$= (367\,500 - 35\,000) \mathrm{J}$$
$$= 332\,500 \mathrm{J}.$$

(This includes electrical consumption by transformer, mode stirrer and so on which is small in comparison with consumption by the magnetron.)

*D* Magnetron energy loss $= 332\,500 \times 30/100 = 99\,750 \mathrm{J}$ (using manufacturers efficiency of 70%).

*E* Energy to oven cavity $= C - D$

$$= (332\,500 - 99\,750) \mathrm{J}$$
$$= 232\,750 \mathrm{J}.$$

*F* Energy absorbed by water load = sensible heat gain by water + latent heat loss in water vapour = $[1 \times 4181 \times (66 - 24)] + [0.010 \times 2\,257\,000]$ J

$= (175\,602 + 22\,570)$ J

$= 198\,172$ J

(specific and latent heats of water are given in Appendix F).

*G* Energy absorbed by oven fittings $= E - F$

$= (232\,750 - 198\,172)$ J

$= 34\,578$ J.

These calculations may be summarised in the form of an energy balance:

| | kJ | % | % |
|---|---|---|---|
| *A* Electrical energy input | 367.5 | | 100 |
| *B* Standby energy consumption | 35.0 | 9.5 | |
| *C* Energy to magnetron | 332.5 | | 90.5 |
| *D* Magnetron energy loss | 99.7 | 27.2 | |
| *E* Energy to oven cavity | 232.8 | | 63.3 |
| *G* Energy absorbed by oven fittings | 34.6 | 9.4 | |
| *F* Energy to water load | 198.2 | | 53.9 |

$$\text{Efficiency} = \frac{F}{A} \times 100\% = 53.9\%\,.$$

The thermal efficiency of equipment is a measure of the proportion of the heat input which is actually used to heat up the food. All the remaining heat is lost. In general, the thermal efficiency of catering equipment is low when judged by the standards of some other industries. It is often in the range 40–50% and may be as low as 20% (see Section 2.8). The improvement of equipment efficiency is desirable from two points of view. First, heat losses must be paid for in the same way as useful heat and therefore they represent a waste of money. Also, heat lost to the kitchen increases its temperature, worsens working conditions and lowers productivity, thereby increasing labour costs. The various heat losses from equipment and their reduction are discussed in Section 2.8.

## 2.4 COMBUSTION CALCULATIONS

The combustion of the fuel gas has an important influence on the efficiency of gas-fired equipment. Combustion is also a good example of a mass balance. The chemical equation which expresses the combustion of methane (the principal constituent of natural gas) is an expression of a mass balance because the inputs (left-hand side of the equation) equal the outputs (right-hand side of the equation):

$$CH_4 + 2O_2 = CO_2 + 2H_2O\ ,$$

$$16\,kg + 64\,kg = 44\,kg + 36\,kg\ .$$

In practice methane is burnt with air, and it is easier to work with volumes:

$$CH_4 + 2O_2 + 7.52N_2 = CO_2 + 2H_2O + 7.52N_2\ \cdot$$

$$1 m^3 \quad \underbrace{2\,m^3 \quad 7.52\,m^3}_{\text{theoretical air}} \quad \underbrace{1 m^3 \quad 2\,m^3 \quad 7.52\,m^3}_{\text{theoretical combustion products}}\ .$$

The terms 'theoretical air' and 'theoretical combustion products' refer to the use of just sufficient oxygen (air) to convert all the carbon (C) in the methane to carbon dioxide ($CO_2$) and all the hydrogen (H) to water ($H_2O$), without any excess oxygen being present.

The total volume on the left-hand side of the equation equals the total volume on the right-hand side only by chance. The masses are equal but not necessarily the volumes.

The ratio of nitrogen:oxygen (by volume) in air is 3.76:1 and therefore $2\,m^3$ of oxygen are associated with $2 \times 3.76 = 7.52\,m^3$ of nitrogen.

We can deduce a number of facts from the above equation:

(a) $1 m^3$ of methane requires $2\,m^3 + 7.52\,m^3 = 9.52\,m^3$ of theoretical air for complete combustion.

(b) $1 m^3$ of methane produces $1 m^3 + 2\,m^3 + 7.52\,m^3 = 10.52\,m^3$ of theor-
100.0% .

(c) The composition of the theoretical combustion products (by volume) is:

| | | |
|---|---|---|
| $CO_2$ | $1.00\,m^3$ | 9.5% |
| $H_2O$ | $2.00\,m^3$ | 19.0% |
| $N_2$ | $7.52\,m^3$ | 71.5% |
| | $10.52\,m^3$ | 100.0% |

If the combustion products are dry (as by gas analysis) the composition is:

| | | |
|---|---|---|
| $CO_2$ | $1.00\,m^3$ | 11.7% |
| $N_2$ | $7.52\,m^3$ | 88.3% |
| | $8.52\,m^3$ | 100.0% |

In practice it is not possible to burn methane (or any other fuel) with the theoretical quantity of air and at the same time achieve **complete combustion** (that is, all $C \rightarrow CO_2$ and all $H \rightarrow H_2O$). It is always necessary to use **excess air**, that is, air in excess of the theoretical requirement. The effect of using excess air is to modify the combustion equation. Suppose 25% of excess air is used, then the theoretical combustion equation modifies to:

$$\underset{1\,m^3}{CH_4} + \underbrace{\underset{2\,m^3}{2O_2} + \underset{7.52\,m^3}{7.52N_2}}_{\text{theoretical air}} + \underbrace{\underset{0.5\,m^3}{0.5O_2} + \underset{1.88\,m^3}{1.88N_2}}_{\text{excess air}} =$$

$$\underbrace{\underset{1\,m^3}{CO_2} + \underset{2\,m^3}{2H_2O} + \underset{7.52\,m^3}{7.52N_2}}_{\text{theoretical combustion products}} + \underbrace{\underset{0.5\,m^3}{0.5O_2} + \underset{1.88\,m^3}{1.88N_2}}_{\text{excess air}}$$

The excess air enters on the input side so it must leave on the output side (by the law of conservation of mass) although it takes no part in the chemical reactions.

We can deduce the following from the above equation:

(a) $1\,m^3$ of methane burnt with 25% excess of air uses $9.52\,m^3 + 0.5\,m^3 + 1.88\,m^3 = 11.90\,m^3$ of air.
(b) $1\,m^3$ of methane burnt with 25% excess air produces $10.52\,m^3 + 0.5\,m^3 + 1.88\,m^3 = 12.90\,m^3$ of combustion products.
(c) The composition of the products of combustion when methane is burnt with 25% excess air is:

| | | | | |
|---|---|---|---|---|
| $CO_2$ | 1.00 | = | $1.0\,m^3$ | 7.8% |
| $H_2O$ | 2.00 | = | $2.0\,m^3$ | 15.5% |
| $N_2$ | 7.52 + 1.88 | = | $9.4\,m^3$ | 72.9% |
| $O_2$ | 0.50 | = | $0.5\,m^3$ | 3.9% |
| | | | $12.9\,m^3$ | 100.0% . |

The composition of the dry products is:

| | | |
|---|---|---|
| $CO_2$ | $1.0\,m^3$ | 9.2% |
| $N_2$ | $9.4\,m^3$ | 86.2% |
| $O_2$ | $0.5\,m^3$ | 4.6% |
| | $10.9\,m^3$ | 100.0% |

The effect of using excess air may be summarised as follows:

(a) More air needs to be supplied for the combustion of each $1m^3$ of gas.
(b) More combustion products are formed from the combustion of each $1m^3$ of gas.
(c) The percentages of $CO_2$ and $H_2O$ in the combustion products decrease.
(d) Oxygen appears in the combustion products.
(e) The percentage of $N_2$ in the combustion products changes only slightly.

These conclusions are considered further in Section 2.6.

## 2.5 HEAT LOST IN AND VOLUME OF COMBUSTION PRODUCTS

The combustion products leaving the flame in gas-heated equipment are used to heat the food, cooking oil, water and so on. The combustion products can only transfer heat if the surfaces which are to receive the heat are at a lower temperature than the combustion products. In any piece of gas-heated equipment therefore the combustion products must leave the equipment at some temperature above that of the surface being heated. This heat is lost to the atmosphere. If we assume that the combustion products leave the oven at 200°C, then we can calculate the quantity of heat they contain as follows for the theoretical combustion of $1m^3$ of methane.

Heat lost in combustion products =

$$\begin{aligned}&\text{volume of } CO_2/m^3 \text{ of methane} \times \text{volume specific heat of } CO_2 \times (T_{CP}-20)\\&+\text{volume of } H_2O/m^3 \text{ of methane} \times \text{volume specific heat of } H_2O \times (T_{CP}-20)\\&+\text{volume of } N_2/m^3 \text{ of methane} \times \text{volume specific heat of } N_2 \times (T_{CP}-20)\end{aligned}$$

(where $T_{CP}$ = temperature at which combustion products leave equipment and 20°C = room temperature)

$$= (1 \times 1.80 \times 10^3 \times 180) + (2 \times 1.52 \times 10^3 \times 180) + (7.52 \times 1.31 \times 10^3 \times 180)\ \text{J}$$

$$= 2.64\ \text{MJ}.$$

(The volume specific heats are taken from Appendix Table I.5.)

Expressed as a proportion of the net calorific value, this heat loss in the combustion products is 2.64/35.77 × 100% = 7.4%.

If the above calculation is repeated for the combustion of methane with 25% excess air, the heat loss increases to 3.21 $MJ/m^3$ $CH_4$ or 8.9% of the net calorific value. Alternatively, if 25% excess air is used and the combustion products leave at 300°C (instead of 200°C) the heat loss increases to 5.03 $MJ/m^3$ $CH_4$ or 14.1% of the net calorific value.

The volume of the combustion products leaving the oven depends on their temperature. In combustion calculations the volumes used all assume room temperature (say 20°C). Therefore if the combustion products leave the equipment at 200°C the volume is higher than those calculated in Section 2.4. For example, with theoretical air, the volume of combustion products at room temperature was shown to be $10.52\,m^3/m^3$ $CH_4$. At 200°C this volume becomes $10.52 \times (273 + 200)/(273 + 20) = 16.98\,m^3/m^3$ $CH_4$. The corresponding volume when 25% excess air is used is $20.82\,m^3/m^3$ $CH_4$.

These conclusions are considered further in the next section.

## 2.6 SIGNIFICANCE OF COMBUSTION CALCULATIONS

In Section 2.4 it was stated that it is always impossible, in practice, to achieve complete combustion (that is, all $C \rightarrow CO_2$ and $H \rightarrow H_2O$) by just using the theoretical amount of air for combustion. If we do not achieve complete combustion, then some of the carbon in the methane will appear either as soot (solid carbon) or carbon monoxide and some of the hydrogen will appear as gaseous hydrogen. Incomplete combustion is undesirable for several reasons. First, soot will deposit itself into the appliance, will need to be cleared away periodically and may contaminate the food. Then, if carbon monoxide is formed and discharged into the kitchen, a health hazard arises because it is poisonous. Finally, unburnt carbon, carbon monoxide and hydrogen represent unrealised energy because if they burn they will release heat.

Therefore we achieve complete combustion by using excess air. However, the percentage of excess air used must be minimised because this also presents problems. A larger volume of air is required and must be supplied to the burner by some means, a larger volume of combustion products is formed and provision must be made to extract this from the equipment, including enlarging the kitchen ventilation system. The proportion of the net calorific value lost in the combustion products is increased; this heat is then irrecoverable. For gas-heated equipment, the loss of heat in the combustion products can be as much as 20% of the net calorific value of the fuel at 100% excess air and exit temperatures of 250°C. In this case, the efficiency of the equipment can be no more than 80%, assuming that there are no other heat losses (which there always are). Thus electrically-heated equipment, which produces no combustion products, should always be more efficient than the equivalent gas-heated equipment. In the case of electricity this combustion-products loss takes place at the power station and is reflected in the higher cost of electrical energy to the consumer (although this is not the sole reason for the difference in price between the two forms of energy).

In Section 2.4, attention was drawn to the decrease in the proportion of carbon dioxide and the appearance of oxygen in the combustion products when excess air was used to burn the methane. Graphs of % $CO_2$ and % $O_2$ versus excess air for any fuel can be drawn (Fig. 2.9).

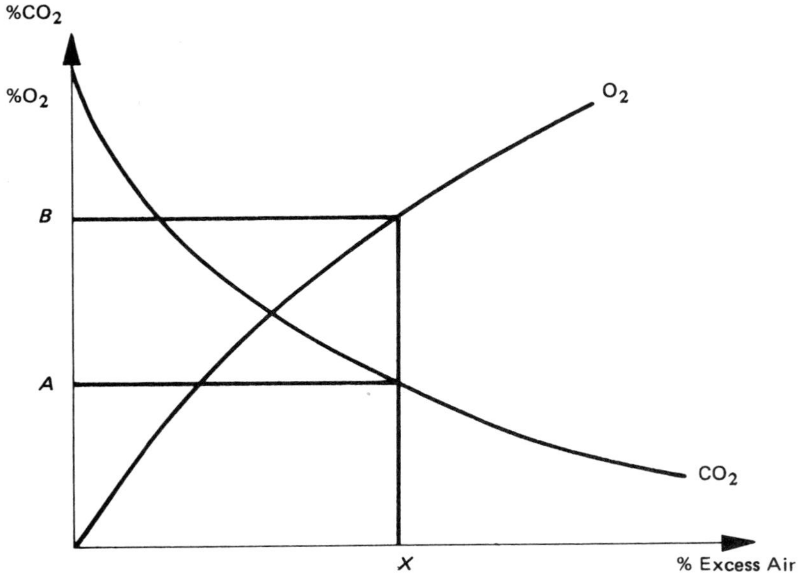


Fig. 2.9 – Variation in % $CO_2$ and % $O_2$ in combustion products of methane.

These graphs can be used to determine the proportion of excess air being used in an item of equipment. If a sample of combustion products leaving the oven is analysed, this enables the % $CO_2$ and % $O_2$ in the sample to be determined. These percentages (*A* and *B* respectively in Fig. 2.9) will give the percentage of excess air (*X*).

The gas analyser normally gives an analysis of the dry gas and hence the reason for calculating this in Section 2.4. Each fuel gas strictly requires its own graph and complete combustion is assumed in this procedure. Small departures in composition of the fuel gas and slight incomplete combustion will not invalidate the use of such curves however.

## 2.7 THERMAL TESTS ON EQUIPMENT

As mentioned in Section 1.3, standardised thermal tests are often specified for measuring the performance of equipment. For gas-fired equipment, the completeness of combustion required may be specified by requiring that the $CO/CO_2$ ratio in the combustion products does not excced, say, 0.02. In this case, if there is 10% of carbon dioxide in the combustion products the implication is that the carbon monoxide content of the combustion products must not exceed 0.2%. Some limiting ratio should apply over the whole range of gas inputs to equipment because complete combustion at high input rates does not necessarily imply complete combustion at low input rates.

For most items of equipment, a maximum heating-up time is specified. It is desirable from an operational point-of-view that this is as short as possible because long heating-up periods will encourage operatives to switch-on equipment too early (see Section 2.8).

The distribution of heat in equipment is usually important. This may be specified, as for the case of a griddle plate, by laying down a maximum temperature variation permitted over its surface. Alternatively, the cooking tests (Section 1.4), by demanding eveness of browning, are an indirect means of specifying eveness of heat-distribution.

The thermal efficiency of equipment is either specified directly as a minimum percentage of the heat input entering the food or indiretly as a maximum quantity of heat (J) required to cook unit mass (kg) of product. Direct specification requires a knowledge of the specific heat of the food. This parameter for a particular food is variable (see Appendices G and H), and therefore direct efficiencies usually are specified only when water is being heated, that is for boiling pans and boilers. In all other cases a maximum heat requirement (J/kg) to cook the food is generally laid down.

The comment made in Section 1.4 regarding the applicability of standardised cooking tests to operational conditions is also pertinent to standardised thermal tests.

## 2.8 THE THERMAL EFFICIENCY OF CATERING EQUIPMENT

Traditionally caterers have not bothered themselves about energy consumption during meal preparation and service. This situation is changing because of the rapid increase in energy costs as a result of the mid-1970's oil crisis. Whitehouse (1976) gave a yearly energy cost of £3640 for a reastaurant serving 1000 meals/week, an average cost of between 7p and 11p per diner. This cost presumably includes restaurant heating, hot water and so on. Crafter (1972) stated that, based upon comparative performance tests in commercial kitchens carried out by various organisations in the USA, the American Gas Association was able to show the following usage for overall cooking and baking:

| | |
|---|---|
| gas-heating | 2035 Btu/meal (2.15 MJ/meal) |
| electrical-heating | 1262 Btu/meal (1.33 MJ/meal) |

With gas charged at 10p/therm and electricity at 1.9p/kWh (his costs), this would give energy costs/meal of 0.20p and 0.70p for gas and electricity respectively.

Keniston (1977) has drawn attention to the large savings possible by using energy more efficiently in the catering industry, for example £3.98 million a year for total U.K. public sector use of oven ranges and grillers. Although the energy cost/meal is not high (less than 1p according to Crafter's data), Keniston's calculations show that in absolute terms it is very high for the whole of the industry in the U.K., as well as in terms of the nation's energy economy.

The concept of efficiency, derived from a heat balance (Section 2.3), is an important absolute standard when comparing the energy consumption of different items of equipment. The heat balance itself indicates how heat is used usefully and also identifies the relative magnitudes of the various heat losses from the equipment. In turn, these indicate where energy savings can be made by correct design and operation.

The various heat losses may be identified as follows.

**(1)** The heat loss in the combustion products leaving the equipment is increased by an increase in the proportion of excess air used as considered in Section 2.5. Thus, burner design, the siting of the burner in the equipment and the availability of air for complete combustion are all important design considerations. The overall aim is to achieve complete combustion with the minimum of excess air. The practical aspects of this are discussed in Chapter 11. Good burner maintenance will also help to ensure efficient combustion.

**(2)** Heat is lost through the walls of the equipment into the kitchen. This loss occurs in three ways:

(a) heat lost during heating-up, made up of two components:
  (i) heat stored in the structure of the equipment, which is lost at the end of the production period when the equipment cools down;
  (ii) heat lost by conduction through the walls during the heating-up period;
(b) heat lost during food production by conduction through the walls of the equipment into the kitchen;
(c) heat lost during idling periods when the equipment is kept hot but is not producing food – same loss as in (b).

Losses (a)(ii), (b) and (c) may all be improved by better insulation using better quality insulating materials and/or greater thickness of insulation. This is considered in more detail in Chapter 4.

Loss (a)(i) will depend on the structure of the oven and can be improved by using materials of low heat-storage capacity. These materials usually have good insulating properties and hence these two factors work in unison. In deep fat fryers, the problem is magnified by the presence of a large volume of oil which needs to be pre-heated and is then allowed to cool at the end of the production period. The same is true of storage water heaters/boilers. Loss (a) is accentuated if equipment is only used for a short period of time such that the heat stored is relatively large compared with total consumption. As a general principle therefore, food should be cooked in as many small batches as possible, using equipment of smaller capacity, rather than in one or two large batches in a large piece of equipment. This also gives the advantage of the food being held for shorter periods before service to the customer with increased customer satisfaction. Fischer (1967) considered that the logical outcome of this principle is the development of **continuous** equipment for situations where there are large numbers of customers.

Heat is not stored in the structure of microwave ovens and these offer potential advantages in this direction (see Chapter 7). Convection ovens also heat up more quickly than conventional ovens and therefore loss (a)(ii) is reduced in this case. Modern types of grills heat up more quickly than older types as well.

Loss (c) is also very significant in many catering situations (see for example, Keniston (1977)), as when equipment is left on over night or turned on/off when the chef comes on/goes off duty, regardless of when the equipment is required for cooking. The University of Leeds and the United Leeds Hospitals (1970) reported some data which also support this view (Table 2.1). The data were deduced from observations which the University of Leeds made on each piece of catering equipment in Leeds Hospital for Women at 15-minute intervals throughout six working days.

**Table 2.1 – Equipment utilisation in a hospital.**

| Item of equipment | Time in use (%) | Time on but not in use (%) | Time off (%) |
|---|---|---|---|
| 4 general purpose ovens | 28% | 32% | 40% |
| Baker's oven | 32% | 45% | 23% |
| 3 boiling pans | 59% | 7% | 34% |
| 2 steaming ovens | 39% | 12% | 49% |
| 2 fish fryers | 4% | 16% | 80% |
| 4 boiling tables | 39% | 4% | 57% |
| Grill | 11% | 43% | 46% |
| Mixer | 14% | – | 86% |

These data show that idling time often exceeds the time in use. In practice, equipment should only be turned on in time for the start of cooking and should be turned off immediately cooking has finished (or before the end of cooking in cases where the equipment has high heat-retention characteristics). Certain design features aid this, for example automatic ignition for gas burners, automatic timers and pan sensors on boiling rings (see Chapter 11). In addition to these devices, scheduling of the use of equipment should be practiced, including times to switch on/off individual items of equipment. This will have the additional advantage of eliminating the tendency of the chef to prepare food before it is required, which causes a loss of quality on holding before the customer is served. Considerable operative training may be required to achieve this. Equipment which is not fully loaded may be regarded as partially idling. Therefore the size of the equipment should be selected so that its capacity is fully utilised and this point should be observed during use.

**(3)** The loss of hot combustion products or hot air to the kitchen impairs efficiency in the same way as the heat lost in the combustion products leaving by the ventilation system. Conversely, the infiltration of cold air into the equipment will act effectively as additional excess air. Effective door seals and general 'gas tightness' are required therefore to prevent these losses. Maintenance is needed if door seals are to stay effective.

**(4)** Doors should never be left open unless food is being transferred into/out of the equipment. The losses of (3) may take place whilst the door is open but, in addition to this, there will be a radiation loss from the hot walls of the oven (radiation is considered in Chapter 6). Operative training is required to ensure that this is done.

**(5)** Cold equipment, such as refrigerators or ice makers, should not be placed next to hot equipment, for example cookers or steamers, otherwise the cold equipment will receive heat from the hot equipment and this heat must be removed from the cold equipment.

**(6)** Water heaters should be placed near to the point where the water will be used so as to minimise heat losses from the pipes. Hot water pipes should be insulated (see Chapter 4). Hot water taps should not be left running otherwise the heat energy of the water literally runs down the drain.

A further use of thermal efficiency (combined with mass efficiency and sensory evaluation) is the comparison of different types of equipment used for preparing the same foodstuff. Such complete comparisons do not appear in the literature for large-scale catering equipment, although Drew and Rhee (1978) reported such a study for domestic-type equipment. Taylor (1977) has introduced the concept of an **energy consumption factor** as a measure of the efficiency of equipment. This is the energy (kWh) required to cook one kilogram of food. The preheat energy is not included. Values of energy consumption factor for different types of equipment as reported by him are summarised in Table 2.2. The values recorded for each item of equipment vary depending on food raw materials used and the loading of the equipment. Although there are some large variations in the energy consumption factor for each particular item of equipment, the results for the brat pan, convection oven, pressure steamer and fryer do suggest that, from an energy-use viewpoint, these are the most efficient types of equipment.

**Table 2.2 – Measured energy consumption factors.**

| Equipment | Energy consumption factors (kWh/kg)† | | | Mean energy consumption factor (kWh/kg) |
|---|---|---|---|---|
| Brat pan | 0.36 | 0.27 | 0.11 | 0.29 |
| | 0.13 | 0.29 | 0.60 | |
| Convection oven | 0.30 | 0.30 | | 0.30 |
| Pressure steamer | 0.36 | 0.44 | 0.36 | 0.43 |
| | 0.67 | 0.33 | 0.44 | |
| Boiling table/range top | 0.42 | 0.26 | 0.94 | 0.52 |
| | 0.36 | 0.63 | | |
| Fryer | 0.49 | 0.58 | | 0.54 |
| Range (oven) | 0.60 | 0.60 | 0.41 | 0.61 |
| | 0.58 | 0.84 | | |
| Atmospheric steamer | 0.60 | 0.62 | | 0.61 |
| Microwave/forced air | 0.92 | 0.85 | | 0.88 |

† Results of individual tests – a range of food raw materials and equipment loadings were used for each item of equipment.

Collison and Wilson (1977) have studied the energy consumption when baking potatoes to an end temperature of 95°C in two electric forced-convection ovens, one gas forced-convection oven and a conventional electric oven. They also studied frying potatoes in a deep-fat fryer and boiling potatoes in water. Different temperature settings and loads were examined. They included preheat energy in the energy consumption, which is not an entirely satisfactory basis of comparison because for one load of potatoes this is not only unrealistic but is a significant proportion of the total energy consumption. The heat balances (excluding preheat energy) reported give efficiencies of 33–45% for the two electric forced-convection ovens, 19–21% for the gas forced-convection oven and 50–71% (two loads) for deep-fat frying. No clear relationship existed between temperature setting and efficiency although a more highly loaded oven may be slightly more efficient than one under-loaded. In fact, there is probably an optimum loading for maximum efficiency of any oven, as illustrated in Fig. 2.10.

If the preheat energy is included, then energy consumption is lower at lower temperature settings because of the lower preheat energy required at these temperatures.

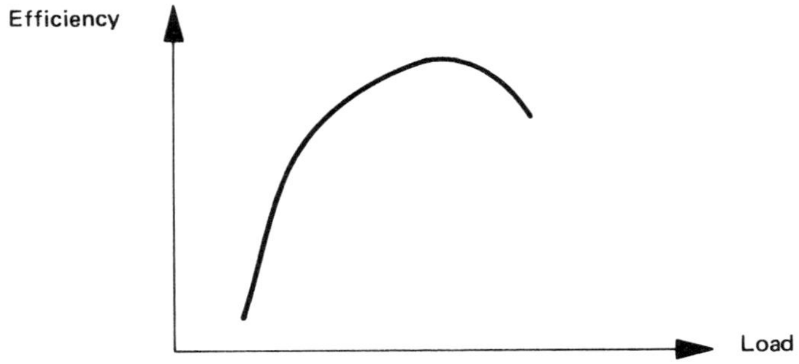


Fig. 2.10 – Effect of loading on equipment thermal efficiency.

The authors also estimated the energy loss for the two electric forced-convection ovens caused by the door being opened for loading purposes. Their data give values in the range 2 to 20% (excluding preheat energy). This loss appears to be dependent on oven setting (higher losses at higher temperatures) as one would expect.

## 2.9 PROBLEMS

**(1)** In Sections 2.4 and 2.5 the following parameters were calculated for the combustion of methane with 0% (theoretical) and 25% excess air:

(i) volume of combustion air/$m^3$ gas burnt,
(ii) volume of combustion products/$m^3$ gas burnt,
(iii) % $CO_2$ and % $O_2$ in dry combustion products,
(iv) % $H_2O$ in wet combustion products,
(v) heat contained in combustion products at 200°C, MJ/$m^3$ gas burnt,
(vi) heat in combustion products as % of net calorific value,
(vii) volume of combustion products at 200°C/$m^3$ gas burnt.

(a) Calculate all the above parameters for 50% and 100% excess air.
(b) Plot all the results versus percentage of excess air.
(c) Comment on the significance of these calculations.

**(2)** A high-pressure steamer, with an external steam generator, was used to cook batches of carrots of varying size. Measurements were made of the energy comsumed under several conditions with the following results:

Steam generator: energy used in boiler to raise steam 0.85 kWh.
Cooking compartment: batch of 0.5 kg energy consumption 0.2 kWh
batch of 1.0 kg energy consumption 0.3 kWh
batch of 2.0 kg energy consumption 0.4 kWh.

Cooking time was 11 minutes in all cases; weight loss during cooking was 3.6% in all cases. Assume the carrots were at 0°C (but not frozen) at the start of each test.

(a) What is the theoretical energy requirement for cooking 1.0 kg of carrots?
(b) Compare the energy consumption/kg for the following cooking processes, assuming that the data given above are additive:

(i) one batch of 0.5 kg
(ii) four successive batches each of 0.5 kg
(iii) one batch of 1.0 kg
(iv) two successive batches each of 1.0 kg
(v) one batch of 2.0 kg
(vi) four successive batches each of 2.0 kg
(vii) eight successive batches each of 2.0 kg.

Mass specific heat of carrot = 3.72 kJ/kg.K.

Note that the total energy used to cook the carrots = energy used in the steam generator + energy used in the cooking compartment.

(c) Comment on the results of the calculations.

## 2.10 REFERENCES

Collison, R. and Wilson, G. (1977), paper in Glew, G. (ed.), *Catering equipment and systems design*. Applied Science Publishers.
Crafter, R. J. (1972), paper in proceedings of the Public Works Congress.
Drew, F. and Rhee, K. S. (1978), *Jnl. American Dietetic Assn.*, **72**, 37–44.
Fischer, S. (1977), paper in Glew, G. (ed.), loc. cit.
Keniston, L. (1977), *HCIMA Journal*, December, 11, 13-15, 17.
Taylor, D. G. (1977), paper in Glew, G. (ed.), loc. cit.
Taylor, D. G. (1977), *Energy measurement programme*, Project Planning Unit, Marketing Department, Electricity Council, London.
University of Leeds and the United Leeds Hospitals (1970), *An experiment in hospital catering using the cook/freeze system.*
Whitehouse, E. (1977), *Caterer & Hotelkeeper*, October 7th, 58–59.

CHAPTER 3

# Fluid Flow

## 3.1 INTRODUCTION

In this Chapter we consider the flow of fluids (gases and liquids). First, a distinction is made between Newtonian fluids, such as air and water, and non-Newtonian fluids. Many liquid foodstuffs are non-Newtonian in character. Fluids are then considered using the principle of conservation of energy. The concept of a boundary layer is next discussed. Understanding this concept is a necessary pre-requisite to understanding the heating/cooling of fluids discussed in a later chapter. A method of characterising fluid flows is introduced before the flow of fluids in pipework systems and the like is considered. There are a number of important applications of fluid flow in catering situations, for example the supply of water to dishwashers, air extraction (ventilation) systems for kitchens and the disposal of waste liquids through drains. The flow of gases in catering equipment, for example ovens, is also discussed. The Chapter concludes with a consideration of pumps, fans and mechanical separations.

## 3.2 NEWTONIAN FLUIDS

In order to understand fluid flow we need to understand the concept of viscosity. Viscosity arises because, in a moving fluid, layers of fluid are moving over other layers of fluid and there is friction between the layers in contact with each other just as there is when a solid surface moves over another solid surface (Fig. 3.1).

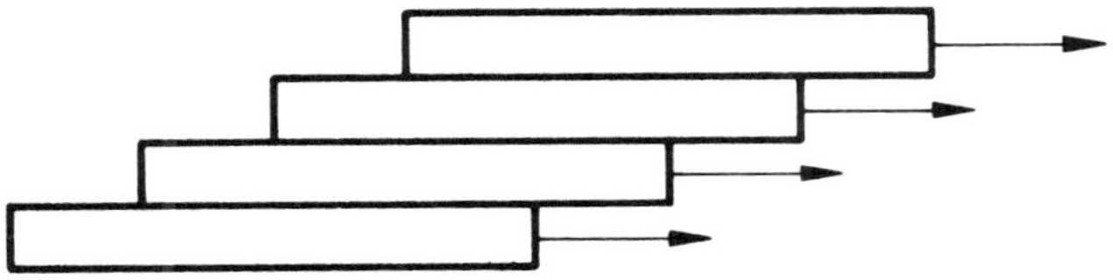

Fig. 3.1 – Fluid layers moving relative to each other.

This situation can be represented by considering fluid held between two parallel plates, one of which is stationary whilst the other moves (Fig. 3.2). A shearing force ($F$) needs to be exerted on the moving plate so that it overcomes the friction between the layers of fluid. This causes the layer of fluid in contact with it to move at a certain velocity (d$u$).

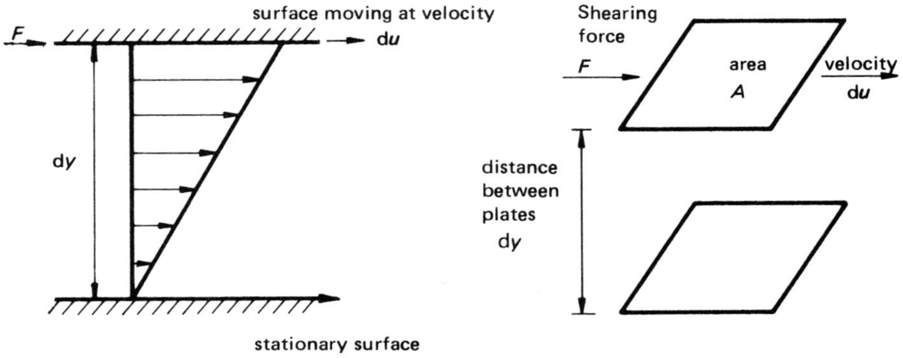


Fig. 3.2 – Fluid flow between parallel plates.

For Newtonian fluids:

shear stress ∝ rate of strain (∝ means 'is proportional to'),

therefore,

$$\tau \propto \left(\frac{\mathrm{d}u}{\mathrm{d}y}\right)$$

or

$$\tau = \mu\left(\frac{\mathrm{d}u}{\mathrm{d}y}\right)$$

where $\mu$ = dynamic/absolute coefficient of viscosity (units: kg/m.s). We also define kinematic viscosity, $\eta = \mu/\rho$ (units: $m^2/s$), where $\rho$ = density of fluid ($kg/m^3$).

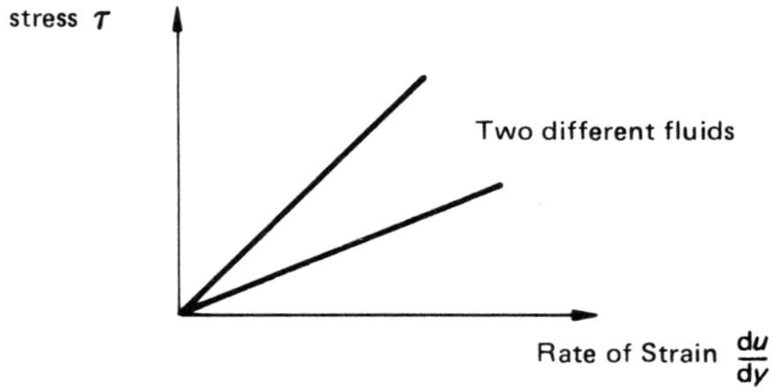


Fig. 3.3 – Newtonian fluids.

### 3.3 NON-NEWTONIAN FLUIDS

Many fluids are non-Newtonian (that is, for them shear stress is not proportional to rate of strain). Many foodstuffs are non-Newtonian fluids. Three types of non-Newtonian fluid may be described as shown in Fig. 3.4.

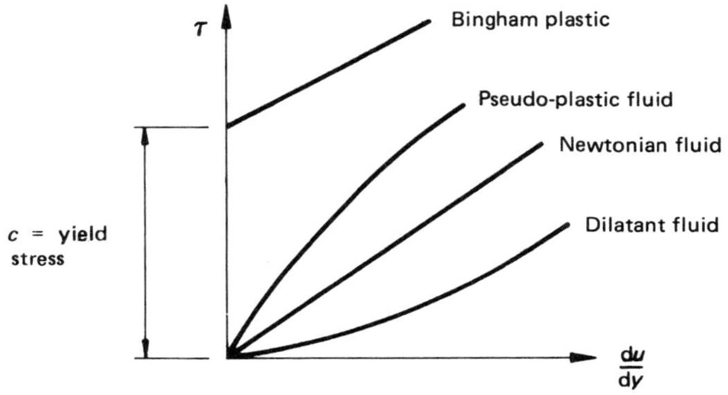


Fig. 3.4 – Non-Newtonian fluids.

The relationships between the shear stress, $\tau$, and the strain for these three types of non-Newtonian fluid are:

| | | |
|---|---|---|
| Bingham plastic | $\tau = b\left(\frac{\mathrm{d}u}{\mathrm{d}y}\right) + c$ | $b$ = proportionality factor<br>$c$ = yield stress. |
| Pseudo-plastic fluid | $\tau = b\left(\frac{\mathrm{d}u}{\mathrm{d}y}\right)^s$ | $0 < s < 1$. |
| Dilatant fluid | $\tau = b\left(\frac{\mathrm{d}u}{\mathrm{d}y}\right)^s$ | $1 < s < \infty$. |

The term $b$ is called a proportionality factor and has the same units as $\mu$, while $s$ is called the power-law-fluid constant. The yield stress ($c$) must be applied to a Bingham plastic before it will start to strain. Since the equations for pseudo-plastic and dilatant fluids are the same, they are often grouped together under the heading **power-law fluids**.

All four types of fluid described here (including Newtonian fluids) can be expressed generally by the following equation:

$$\tau = b\left(\frac{\mathrm{d}u}{\mathrm{d}y}\right)^s + c\ .$$

In the case of Newtonian fluids $b = \mu, s = 1, c = 0$. Most food materials are described by this last equation. Examples include:

| | |
|---|---|
| Newtonian fluid | water |
| Bingham plastic | suspensions |
| pseudo-plastic fluid | mayonnaise |
| dilatant fluid | uncooked corn starch paste. |

These equations are called **models** of the fluids' behaviour. They are not the only models used but others are more complex. From our point of view however, these simple models will be adequate. The accuracy of any predictions which result from their application will be determined by the accuracy of the particular model used. Care must be taken in using these models, since extrapolation of the model to conditions outside those for which it has been determined experimentally (that is, to other values of shear stress and shear strain) may be unreliable. Under these new conditions the model may no longer apply. It may also be inadvisable to infer the model of one non-Newtonian fluid from that of another non-Newtonian fluid, even though they appear to be approximately the same fluids.

### 3.4 BERNOULLI'S EQUATION

A fluid (gas or liquid) may be looked at in several ways. In this section, it is looked at from the point of view of the conservation of energy. This approach is similar to the heat balances of Section 2.3 but in this case thermal energy is not involved. In some situations it is necessary to allow for changes in the height of a fluid and changes in its pressure and velocity. This may be done by using Bernoulli's equation:

$$\underset{\text{kinetic energy}}{\frac{\rho u^2}{2}} + \underset{\text{pressure energy}}{p} + \underset{\text{potential energy}}{z\rho g} = \text{constant},$$

where $\rho$ = density of fluid (kg/m$^3$), $u$ = velocity of fluid (m/s), $p$ = pressure of fluid (N/m$^2$), $z$ = height of fluid above a datum level (m). The kinetic energy term allows for changes in velocity, the pressure energy term for changes in pressure and the potential energy term for changes in height.

This equation states that the total energy in a fluid system is constant but that it may change from one form to another. Each term has the units of pressure or force per unit area, that is N/m$^2$. Each term also has the units of energy per unit volume, that is J/m$^3$.

If we consider any two points in a connected flow system then Bernoulli's equation may be written as:

$$\frac{\rho u_1^2}{2} + p_1 + z_1 \rho g = \frac{\rho u_2^2}{2} + p_2 + z_2 \rho g$$

where the subscript 1 refers to point 1 in the system and subscript 2 refers to point 2 in the system. Bernoulli's equation involves three important assumptions:

(a) the fluid is **inviscid**, that is, it has zero viscosity;
(b) the flow is **steady**, that is, it is not changing with time;
(c) the flow is **incompressible**, that is, it does not apply to gases at high pressures.

Since one assumption is that the fluid in inviscid, it only applies to situations where fluid friction (viscosity) has negligible effect.

An **ideal fluid** has zero viscosity and does not produce fluid friction, that is, layers of fluid do not slide over each other as discussed in Section 3.2. A **real fluid**, on the other hand, has a measurable viscosity. Bernoulli's equation, as given above, applies only to ideal fluids, that is, situations where the effects of fluid friction are negligible. It can however, be modified to take into account viscosity effects (real fluids) by adding two additional terms:

$$\frac{\rho u_1^2}{2} + p_1 + z_1 \rho g + \Delta p_p = \frac{\rho u_2^2}{2} + p_2 + z_2 \rho g + \Delta p_f ,$$

where $\Delta p_p$ = pressure difference (increase) produced by a pump or fan (Sections 3.10 and 3.11), $\Delta p_f$ = pressure loss due to friction (obtained from Section 3.7).

As an example of the use of Bernoulli's equation, it may be used to calculate the time taken to empty a sink. The situation is illustrated in Fig. 3.5.

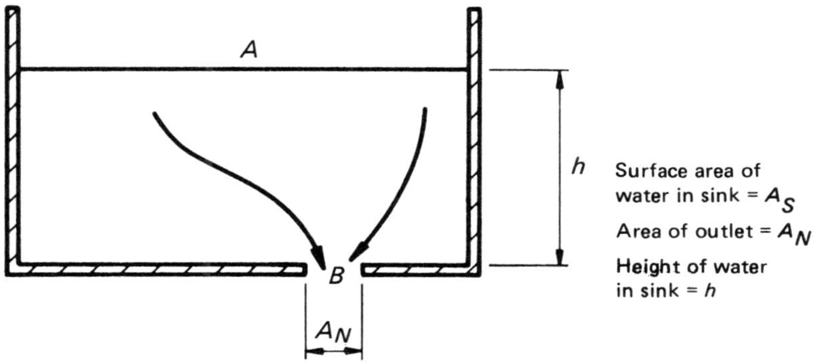


Fig. 3.5 – Emptying of a sink.

At $A$: pressure, $p_1 = 0$; velocity, $u_1 = 0$; $z_1 = h$ .

At $B$: pressure, $p_2 = 0$; velocity, $u_2 = u_B$; $z_2 = 0$ .

Applying Bernoulli's equation gives:

$$\frac{\rho u_1^2}{2} + p_1 + z_1 \rho g = \frac{\rho u_2^2}{2} + p_2 + z_2 \rho g .$$

Therefore,

$$h\rho g = \frac{\rho u_B^2}{2},$$

so,

$$h = \frac{u_B^2}{2g} \quad \text{or} \quad u_B = \sqrt{2gh} .$$

The rate of flow through the nozzle,

$$Q = C_d \times A_N \times u_B$$

where $C_d$ = coefficient of discharge of outlet, which allows for entry losses due to friction as the water passes through the nozzle, that is, it allows for the fluid to be real rather than ideal. The rate of flow through the nozzle is equal to the rate at which the sink empties.

The rate of emptying the sink,

$$Q = A_s\left(-\frac{\mathrm{d}h}{\mathrm{d}t}\right).$$

Therefore,

$$-A_s \frac{\mathrm{d}h}{\mathrm{d}t} = C_d A_N \sqrt{2gh} ,$$

so,

$$\frac{-A_s}{C_d A_N \sqrt{2g}} \frac{\mathrm{d}h}{\sqrt{h}} = \mathrm{d}t .$$

Integrating:

$$\frac{A_s}{C_d A_N \sqrt{2g}} \left[2\sqrt{h}\right]_{h_1}^{h_2} = t$$

where $t$ represents the time taken for the layer of water to drop from $h_2$ to $h_1$ .

That is,

$$t = \frac{A_s . 2}{C_d A_N \sqrt{2g}} (\sqrt{h_2} - \sqrt{h_1}) .$$

This equation may be used to calculate the time it takes to empty, or partially empty, a sink or tank.

## 3.5 BOUNDARY LAYERS

In Section 3.2, the idea of viscosity was considered in terms of layers of fluid sliding over each other. At a stationary surface, the fluid layer adjacent to the surface has zero velocity (Fig. 3.2). Further away from the surface, fluid layers move over each other. In Section 3.4, we distinguished between ideal fluids (zero viscosity, no sliding layers) and real fluids (exhibiting viscosity). Let us now consider the situation arising when a wide stream of Newtonian fluid of uniform velocity enters a pipe (Fig. 3.6). Some of the fluid enters the pipe and it is this fluid which is of interest. At the entrance the fluid still has a uniform velocity. The velocity is indicated by the lengths of the lines with arrows in the Figure. It can be seen that at the entrance all of these have the same length. As soon as the fluid comes into contact with the pipe wall, beyond the entrance, the fluid near the wall is slowed down, that is, its velocity is reduced. This is indicated in the Figure by the relative shortening of the arrowed lines. The velocity away from the wall still shows a flat profile (although the velocity is increased to compensate for the decrease in velocity near the wall). The layer of fluid affected by the wall is termed a **boundary layer**. Within the boundary layer, it is the effect of viscosity which causes the fluid to slow down, that is, within the boundary layer the fluid is real. Outside the boundary layer, the effect of viscosity is negligible and the fluid acts as an ideal fluid. The larger the diameter of the pipe, the smaller becomes the proportion of fluid affected by the wall, that is, the larger the proportion of ideal fluid and the smaller the proportion of real fluid.

The thickness of the boundary layer is zero at the entrance of the pipe and increases in thickness with distance from the entrance until is occupies the full cross-section. At this stage the flow is said to be fully-developed. The section of the pipe in which the boundary grows is termed the entrance length (see Section 3.6).

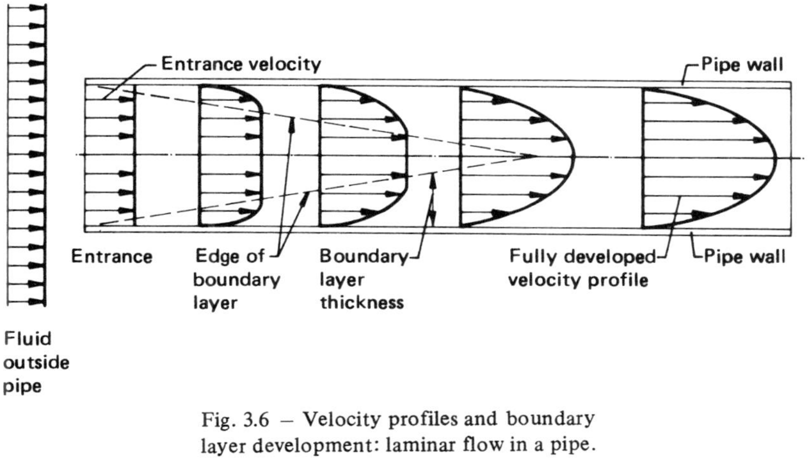


Fig. 3.6 – Velocity profiles and boundary layer development: laminar flow in a pipe.

## 3.6 FLOW CHARACTERISATION

A fluid flowing in a pipe at very low velocities may be imagined to be made up of little pockets of fluid each flowing in a straight line parallel to the walls of the tube. This type of flow is called a laminar (or streamline) flow (Fig. 3.7).

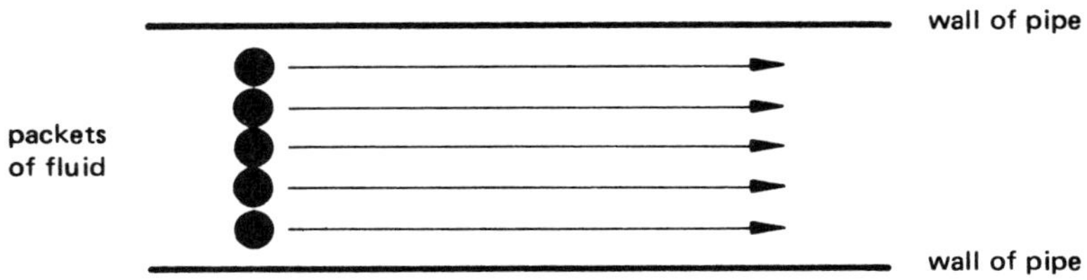


Fig. 3.7 – Laminar flow of a fluid in a pipe.

If the velocity of the fluid in the pipe in increased, a stage is reached when the flow is no longer laminar. There is mixing between the pockets of fluid across the diameter of the pipe. In this case the concept of viscosity so far described breaks down. This situation, where pockets of fluid mix across the pipe, is called **turbulence** and the flow is said to be turbulent.

There is an intermediate stage between laminar and turbulent conditions in which the flow is alternately one or the other. This is referred to as a **transitional flow**. At the laminar end of the transitional-flow region, the flow is laminar for a large proportion of the time. At the turbulent end of the region the converse is true. Laminar flows and turbulent flows may be distinguished by measuring how the velocity at any point in the fluid varies with time (Fig. 3.8). The flow is constant with time for a laminar flow but fluctuates rapidly for a turbulent flow.

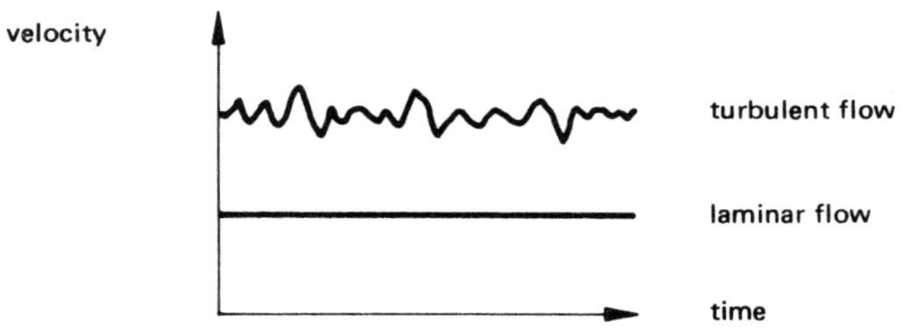


Fig. 3.8 – Velocity changes with time: laminar and turbulent flows.

In practice, it is not only velocity which distinguishes between a laminar and a turbulent flow in a pipe. Four parameters in all are involved:

fluid velocity, $u$ (m/s),
diameter of pipe, $d$ (m),
density of fluid, $\rho$ (kg/m$^3$),
dynamic viscosity of fluid, $\mu$ (kg/m.s).

These may be combined into a single parameter, the Reynolds number ($Re$), which may then be used to characterise the flow situation:

$$Re = \frac{du\rho}{\mu}\,.$$

The Reynolds number will be considered in more detail in the next section. For the flow of Newtonian fluids in pipes:

| | |
|---|---|
| if $Re < 2000$ | laminar flow |
| if $Re = 2000$ to 4000 | transitional flow |
| if $Re > 4000$ | turbulent flow. |

The velocity profile in a turbulent flow (Fig. 3.9) is different to that in laminar flow (Fig. 3.6). It is much steeper near the tube wall and flatter in the centre. centre.

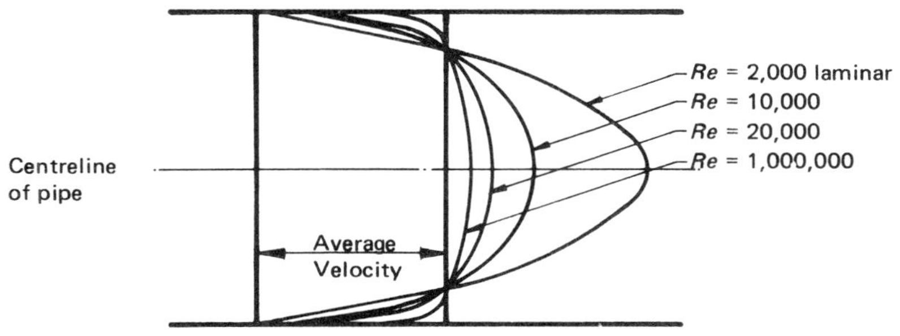


Fig. 3.9 – Velocity profiles in a pipe: Newtonian fluid in turbulent flow.

The velocity is zero at the wall and there is a 'laminar sub-layer' in the region of the wall where the flow velocity is small and approximates to a laminar flow. In the centre of the tube the flow is turbulent giving rise to a 'turbulent core'. Between the turbulent core and the laminar sub-layer is a 'buffer zone' which represents the transition between the two flows. The three zones are not absolutely distinct but tend to merge into each other. We will meet this situation again in Section 3.8 when considering the flow of a fluid over a flat plate.

In Section 3.5, the term entrance length was defined. The entrance length ($l_e$) for a fully developed velocity profile to form in laminar flow is given by:

$$\frac{l_e}{d} = 0.0575\,Re\,.$$

In turbulent flows, an additional factor affecting entrance length is the nature of the entrance itself. In general it appears that, in turbulent flows, entrance lengths of at least 50 pipe diameters are required before the flow becomes fully developed. For non-Newtonian fluids the entrance length ($l_e$) is smaller for pseudo-plastic fluids and greater for dilatant fluids compared with those for Newtonian fluids under the same conditions of pipe diameter and Reynolds number. However, insufficient data are available to give a numerical value for such differences.

## 3.7 FRICTION LOSSES IN PIPES AND FITTINGS

In Section 3.4, it was shown that Bernoulli's equation could be modified to take account of pressure losses (viscosity effects). We now consider how these pressure losses may be estimated. The appropriate equations for laminar flows in straight pipes can be derived, for both Newtonian and non-Newtonian fluids, from first principles. These equations may be summarised as follows.

Newtonian fluids:

$$Q = \frac{\Delta_p \pi r_w^4}{8\mu l} \text{ (Hagen-Poiseuille equation)}$$

where $Q$ = volume flow rate (m$^3$/s), $\Delta_p$ = pressure drop along a pipe (N/m$^2$), $r$ = radius of pipe (m), $l$ = length of pipe (m), $\mu$ = dynamic viscosity of fluid (kg/m.s).

Bingham plastics:

$$Q = \frac{\pi r_w{}^4\,\Delta_p}{8lb}\left[1 - \frac{4}{3}\left(\frac{2lc}{r_w\Delta_p}\right) + \frac{1}{3}\left(\frac{2lc}{r_w\Delta_p}\right)^4\right]$$

where $b$ and $c$ are the constants defined in Section 3.3. For a Newtonian fluid, $c = 0$ and this equation reduces to the Poiseuille equation.

Power law fluids (pseudo-plastics and dilatants):

$$Q = \frac{s\pi r_w{}^3}{3s+1}\left(\frac{r_w \Delta_p}{2lb}\right)^{\frac{1}{s}}$$

where $b$ and $s$ are the constants defined in Section 3.3. For a Newtonian fluid $s = 1$ and $b = \mu$ and this equation reduces to the Hagen-Poiseuille equation.

For transitional and turbulent flows, the mathematical models describing them are much more complex than those for laminar flows (see Section 3.3 where the models given are for laminar flows). Although it is possible to solve turbulent flow problems numerically, the procedures are complex and require powerful computers. A method of approach which is commonly used for such problems involves the technique of **dimensional analysis**. This technique is empirical and not theoretical. It entails selecting the variables we think will be involved in a particular situation and manipulating them is such a way as to reduce them to a smaller number of dimensionless groups (Reynolds number is a dimensionless group – see Section 3.6). The technique is best described by reference to a specific example. In the following we consider a Newtonian fluid flowing in a straight, horizontal, circular pipe of constant cross-section. From experience of this situation we would expect the following seven variables to affect the flow in this particular case. It is important to realise that if we have missed out one variable at this stage, the analysis can be ruined. Each variable is given a symbol and its dimensions are expressed in terms of mass ($M$), length ($L$) and time ($T$).

| Variable | Symbol | Dimensions |
|---|---|---|
| pressure drop | $\Delta_p$ | $[ML^{-1}T^{-2}]$ |
| fluid density | $\rho$ | $[ML^{-3}]$ |
| velocity of fluid | $u$ | $[LT^{-1}]$ |
| pipe diameter | $d$ | $[L]$ |
| pipe length | $l$ | $[L]$ |
| pipe roughness (height of projections) | $e$ | $[L]$ |
| fluid viscosity (absolute) | $\mu$ | $[ML^{-1}T^{-1}]$ |

Because there are seven variables and three primary quantities ($M$, $L$, $T$), there will be $7 - 3 = 4$ dimensionless groups. More generally, this may be expressed as the Buckingham Pi theorem: if there are $p$ primary quantities in a physical phenomenon which involves $n$ variables, then there will be $(n - p)$ dimensionless groups such that $\phi(\pi_1, \pi_2, \pi_3, \ldots \pi_{(n-p)}) = 0$ where each $\pi$ is an independent dimensionless group made up of some of the variables.

We may now write:

$$[\Delta_p]^1 [\rho]^\alpha [u]^\beta [d]^\gamma [l]^\delta [e]^\epsilon [\mu]^\theta = [1]^0$$

where [ ] means 'the dimensions of'. Each variable is raised to some power ($\alpha$, $\beta$, $\gamma$, and so on) with one variable raised to the power 1. If we now insert the dimension of each variable we get:

$$[ML^{-1}T^{-2}]^1 [ML^{-3}]^\alpha [LT^{-1}]^\beta [L]^\gamma [L]^\delta [L]^\epsilon [ML^{-1}T^{-1}]^\theta = [1]^0 \ .$$

Equating the indices of the primary quantities gives:

$$\begin{aligned} \text{Condition on } M\text{:} \quad & 1 + \alpha + \theta = 0 \\ L\text{:} \quad & -1 - 3\alpha + \beta + \gamma + \delta + \epsilon - \theta = 0 \\ T\text{:} \quad & -2 - \beta - \theta = 0 \ . \end{aligned}$$

There are three equations and six unknowns and therefore the equations can be solved in terms of three unknowns. Solving in terms of $\theta$, $\delta$ and $\epsilon$:

$$\begin{aligned} \text{From } T\text{:} \quad \beta &= -2 - \theta \\ M\text{:} \quad \alpha &= -1 - \theta \\ L\text{:} \quad \gamma &= 1 + 3\alpha - \beta - \delta - \epsilon + \theta \\ &= 1 + 3(-1 - \theta) - (-2 - \theta) - \delta - \epsilon + \theta \\ &= 1 - 3 - 3\theta + 2 + \theta - \delta - \epsilon + \theta \\ &= -\theta - \delta - \epsilon \ . \end{aligned}$$

Substituting back into the original equation:

$$[\Delta_p]^1 [\rho]^{-1-\theta} [u]^{-2-\theta} [d]^{-\theta-\delta-\epsilon} [l]^\delta [e]^\epsilon [\mu]^\theta = [1]^0 \ .$$

Collecting up variables raised to the same power, that is 1, $\theta$, $\delta$, and $\epsilon$, gives:

$$\left[\frac{\Delta_p}{\rho u^2}\right]^1 \left[\frac{\mu}{\rho u d}\right]^\theta \left[\frac{l}{d}\right]^\delta \left[\frac{e}{d}\right]^\epsilon = [1]^0.$$

Thus we have collected up all the variables raised to a number power ($\Delta_p$, $\rho$ and $u$) to form the group $[\Delta_p/\rho u^2]$. Similarly, all the variables raised to the power $\theta$ ($\rho$, $u$, $d$ and $\mu$) were collected to form the group $[\mu/\rho u d]$. All the groups of variables in the square brackets are dimensionless and are called **dimensionless groups**. The above equation may be written as:

$$\frac{\Delta_p}{\rho u^2} = \phi\left(\frac{du\rho}{\mu}, \quad \frac{l}{d}, \quad \frac{e}{d}\right)$$

| Name of dimensionless group | Euler number | Reynolds number | length/ diameter ratio | relative roughness |
|---|---|---|---|---|
| Abbreviation | *Eu* | *Re* | | |

where $\phi$ means 'a function of'.

The Reynolds number characterises the flow situation (Section 3.6) and is the ratio of inertial to viscous forces experienced by the fluid. The Euler number contains the pressure drop term. Experimental work shows that Euler number is directly proportional to $l/d$, and therefore the relationship may be written as:

$$\frac{\Delta_p}{\rho u^2} = \frac{l}{d}\phi'\left(\frac{e}{d}, Re\right)$$

where $\phi'$ means 'a function of'.

The function $\phi'$ is called the **friction factor** and is designated by the term $2C_f$ where $C_f = \tau/\rho u^2$. The friction factor is a function of $Re$ and $e/d$. This relationship has been determined experimentally and may be obtained from Fig. A.1 (Appendix A). The relative roughness $\epsilon = b/d$, where $d$ = diameter of the pipe and values of $b$ for several materials are recorded in Table A.1 (Appendix A).

Re-arranging the equation gives:

$$\frac{\Delta_p}{\rho u^2} = \frac{l}{d}.2C_f \quad \text{or} \quad \Delta_p = 4C_f\frac{l}{d}\frac{\rho u^2}{2}.$$

This equation gives the pressure drop for a Newtonian fluid flowing in a straight, jointless, horizontal pipe when the fluid is flowing under fully-developed turbulent conditions. The experimental data for the laminar flow region are represented by a single straight line (Fig. A.1, Appendix A). This line gives the (experimental) relationship $C_f = 32/Re$. If this relationship is substituted in the above pressure-drop equation (derived *via* dimensional analysis), the theoretical Hagen-Poiseuille equation may be derived, thus confirming experimentally a theoretically-derived equation.

The equation given relates to straight lengths of pipe only. Pipework systems, however, often consists of valves, constrictions, bends and other fittings as well as straight pipe. These may be taken into account by incorporating into the flow equation '$q$ factors', standard values which make an allowance for fittings. The modified equation is:

$$\Delta_p = \Big(\underbrace{4C_f \frac{l}{d}}_{\text{straight pipe}} + \underbrace{q}_{\text{fittings}}\Big) \frac{\rho u^2}{2} .$$

Values of $q$ for various situations are given in Table A.2 (Appendix A).

The above treatment assumes that the flow in the tube is fully developed but it does not apply to the entrance length. The friction factor is greater in this region than in the fully-developed flow region of the pipe and may be calculated from the data sources given in Appendix A.

The foregoing discussion applies to Newtonian fluids only. The flow equations for the laminar flow of non-Newtonian fluids were given earlier in this section. It is possible to carry out a dimensional analysis for the turbulent flow of a non-Newtonian fluid but this is unlikely to be of interest in catering situations and, in any case, it is much more complex than the Newtonian-fluid case.

A final note of caution must be sounded with regard to data and equations which have their origins in dimensional analysis. Because they are based on experimental measurements they should not be used outside the range for which they were determined. This will become more apparent in Chapter 5 where the ranges of the dimensionless groups for which data correlations have been derived are quoted.

## 3.8 EXTERNAL FLOWS

In Section 3.6, the flow of fluids in pipes was discussed. When the flow is contained within, say, a pipe or duct it is referred to as an **internal flow**. On the other hand, **external flows** arise when a large (infinite) flow of gas passes over a solid object, such as a flat plate or cylinder or sphere. Consider Fig. 3.10 which shows a parallel laminar flow of air (at constant velocity over its cross-section) coming into contact with a plate. The fluid immediately in contact with the plate is stationary, whereas layers of fluid further removed from the plate will have a greater velocity, this velocity inreasing as the distance from the plate increases until the distance from the plate is such that the plate no longer affects the flow. As the flow passes along the plate the depth of fluid affected gets larger until it eventually becomes constant. This depth of fluid affected by the plate is again called the **boundary layer**. The velocity of the fluid within the boundary layer is less than in the bulk of the (unaffected) fluid.

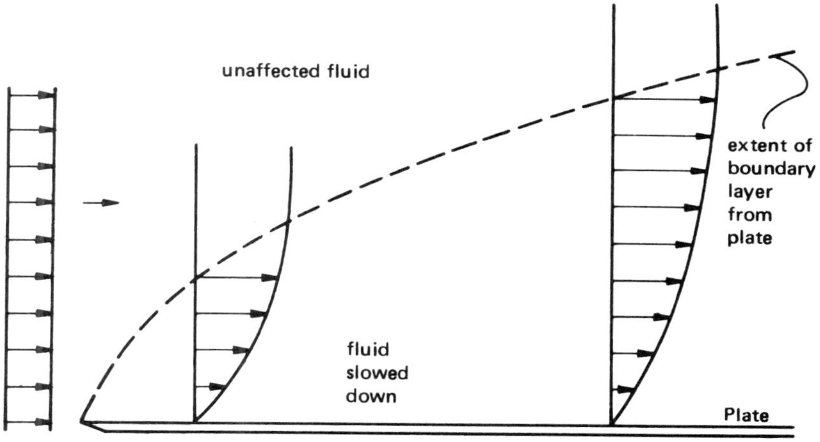


Fig. 3.10 – Development of a boundary layer on a flat plate: laminar flow.

The structure of a turbulent boundary layer on a flat plate is illustrated in Fig. 3.11. Initially, the layers of fluid next to the plate are slowed down so much by contact with the plate that a laminar boundary layer exists. The boundary layer then passes through a transition region before it becomes a fully-developed turbulent boundary layer. This consists of three zones: a laminar sub-layer next to the plate, a buffer zone and an outer turbulent zone. These correspond to the three zones described for a turbulent boundary layer in a pipe in Section 3.6.

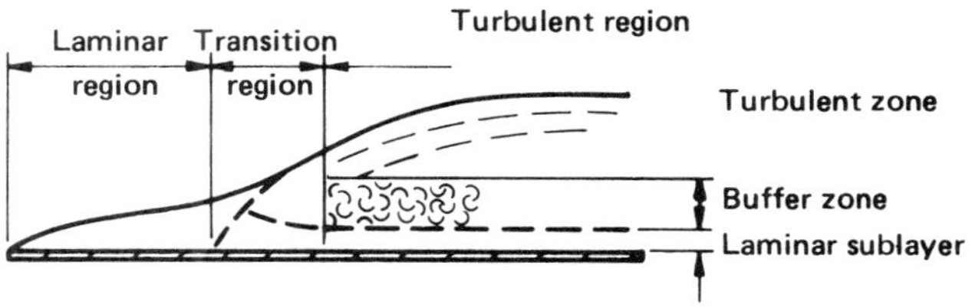


Fig. 3.11 – Structure of a tubulent boundary layer on a flat plate.

All flows over solid objects, whatever the shape of the object, produce boundary layers. These boundary layers, as will be explained in a later chapter, are very important in deciding the rates of convective heat transfer.

## 3.9 FLOWS OF GASES IN CATERING EQUIPMENT

The flow (circulation) of combustion products/air in ovens is caused by means of the flame itself (conventional ovens) or by means of a fan (forced-convection ovens). A single flame, from a fluid-flow point of view, acts as a jet of gas mixing with surrounding air ('free-jet' flow) as shown in Fig. 3.12.

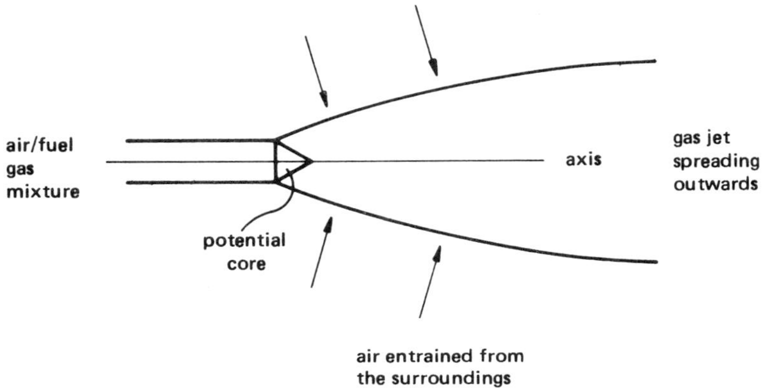


Fig. 3.12 – Flame as a gas jet.

The velocity distribution along the axis of the jet first remains constant in the potential core and then falls (Fig. 3.13).

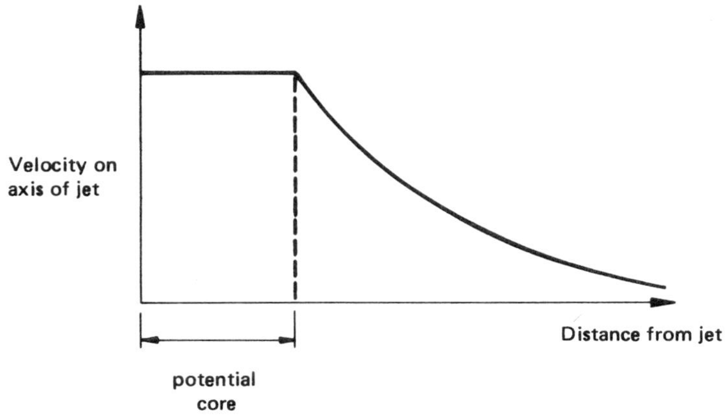


Fig. 3.13 – Velocity along axis of jet of gas.

Across the jet, the velocity has the shape shown in Fig. 3.14, although the shape of this curve is more complicated than this within the potential core.

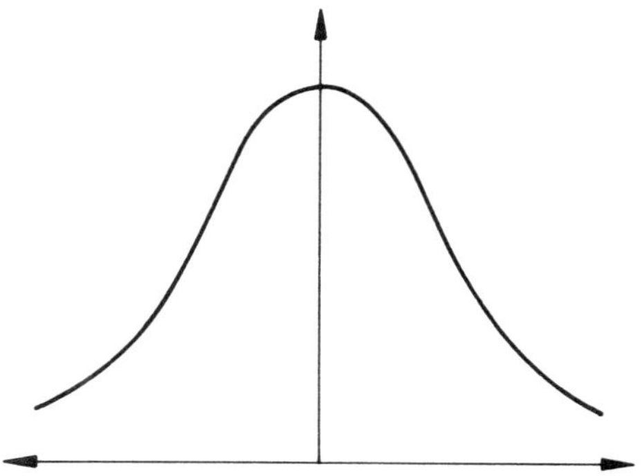


Fig. 3.14 – Velocity distribution across a jet of gas.

A gas burner as used in a conventional oven consists of a row of single jets and these tend to merge with each other, giving rise to a rather complicated flow pattern. This flow pattern is further complicated when it comes into contact with the sides of the oven and is deflected in other directions. The **stirring effect** of this flow is not great however, as the velocity of the gases leaving the burner is low and decays quite rapidly, as in the case for a single jet. This is partly the reason why temperature is not evenly distributed in a conventional oven.

The implication of the last sentence is that, if the gas-flow is speeded up, then there will be a more even distribution of temperature in the oven space. This is the principle of the **forced-convection oven**. Two types of oven are manufactured. In the first, a fan is situated inside the oven space which draws in gases from the combustion space and recirculates them at a higher velocity than is achieved in a normal oven. The characteristics of the fan and the size and shape of the oven govern the flow pattern set up, which is more complicated than pipe flows and flows on flat plates.

In the second type, air enters the oven space through a series of circular holes or slots on one side of the oven and is removed through a similar series of holes/slots in the other side of the oven. The flow pattern set up is of a multitude of small jets all merging with each other and tending to form a parallel flow across the oven. This flow pattern is much simpler than the others mentioned above.

While no published data on gas/air circulation patterns in ovens are available, some work has been reported on the flow patterns in a cube, a sphere, a hemisphere, a cone, an ellipsoid and a cylinder (Bacon, 1960). The only shape relevant to catering is the cube and this was represented as having a burner in the centre of the base. The complex flow pattern observed is reproduced in Fig. 3.15.

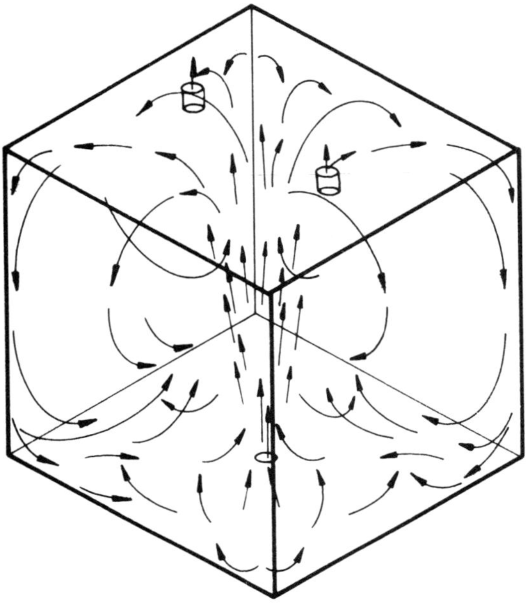

Fig. 3.15 – General flow pattern in a cube.

The basic flow pattern shows a central expanding conical jet, which on reaching the upper face of the cube spreads radially to the four side walls, return flow being initially downwards over all faces of the cube. The main return flow is however, confined to the corners formed by the side faces. On hitting the bottom face of the cube these four streams spread radially and meet on the arms of a cross, the centre of which is at the inlet jet and the arms of which are parallel to the edges of the base of the cube. The flow then turns upwards, much of the fluid being entrained into the inlet jet. The length of the entrainment zone would appear to be about two-thirds the cube length.

In practice, of course, an oven is not an empty cube but a cube-shaped box containing food items. These food items will affect the flow pattern considerably, even in a forced-convection oven. The flow pattern established for one set of loading conditions will not necessarily apply under other loading conditions. These changes in flow pattern will affect the convective heat transfer pattern in the oven, as we shall see in Chapter 5.

Another flow situation which is commonly met with in catering equipment occurs where a free jet impinges on a plate, for example a flame impinging on the base of a pan. In this case the jets of gases leaving the burner are deflected by the plate (Fig. 3.16) as described by Chigier and Milson (1973).

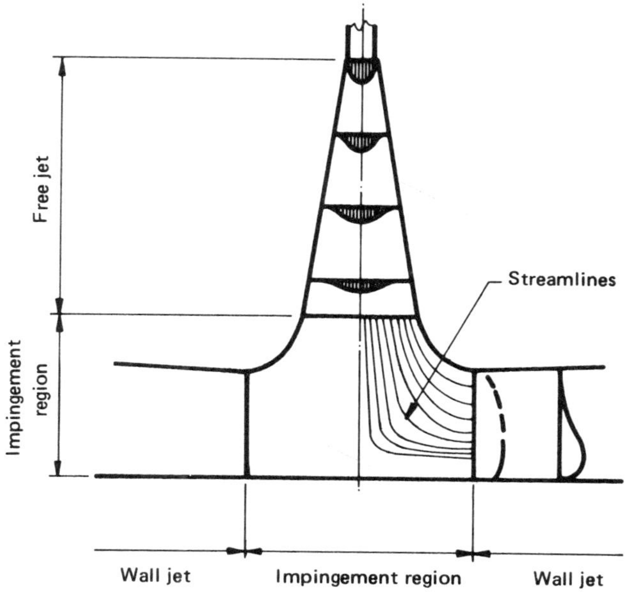

Fig. 3.16 – Flow pattern of an impinging jet.

The free jet leaving the burner is deflected, forming an impingent region near to the plate. As the jet spreads outwards a wall jet is formed. The velocity profile in the wall jet is shown in Fig. 3.17.

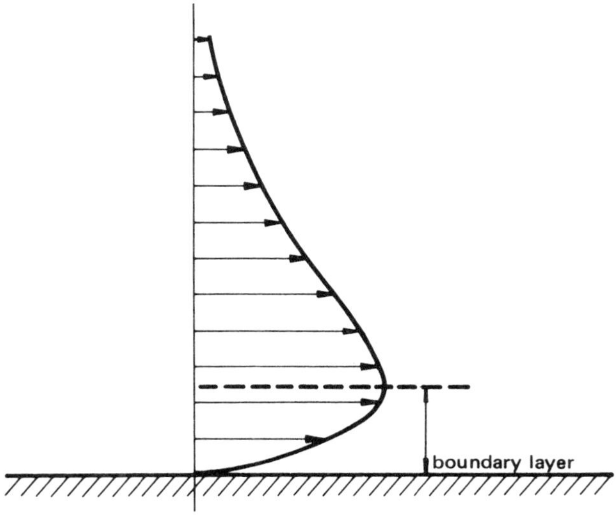

Fig. 3.17 – Velocity profile of a wall jet.

There is a boundary layer near to the wall because the gases next to the wall are slowed down (as when a parallel flow moves over a flat plate). Further away from the wall there is half a free jet. The thickness of the boundary layer increases with distance from the impingent region. The implications of the impingent region for heat-transfer will be discussed in Chapter 5.

## 3.10 PUMPS

Pumps find limited use in catering, primarily being reserved for pumping water in dishwashers, where the type used is a **centrifugal pump**. These are very efficient when operated at high speeds (1200 to 3600 rev/minute). They are directly coupled *via* a shaft to an electric motor. This type of pump is smaller in size than other types of the same capacity and are comparatively low in cost. They are reliable in operation, easy to clean and maintain and can be mounted in any position.

The flow pattern inside a centrifugal pump is shown in Fig. 3.18. The liquid is sucked into the casing of the pump along the impeller axis. It moves into a series of vanes (mounted on an impeller) which are rotating. As the liquid moves from near the pump axis towards the outside of the vanes, its speed and hence kinetic energy increases. It passes into the space between the impeller and the casing. The volute is of gradually increasing cross-section until it reaches the point of discharge from the pump. Thus its velocity falls and kinetic energy is converted into pressure energy. It reaches the point of discharge on the casing and is flung out of the pump at high speed and hence high kinetic energy. A certain amount of energy is lost because turbulence, as kinetic energy, in transformed to pressure energy.

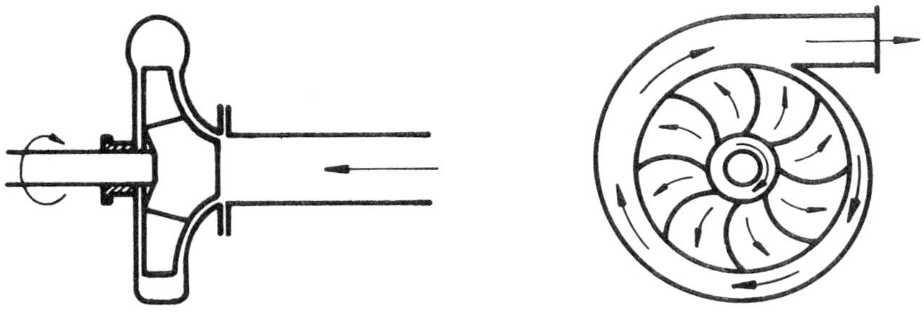

Fig. 3.18 – Centrifugal pump.

The efficiency of a centrifugal pump is greater at higher speeds and therefore the smallest pump capacity, consistent with the requirements, should be selected and this should be operated at high speed. The capital cost of a smaller pump is less and its power consumption is also less, because of its greater efficiency.

The efficiency of a centrifugal pump is highest at its designed capacity and less than this at flow rates above and below the design figure. The characteristic curves of a centrifugal pump are shown in Fig. 3.19.

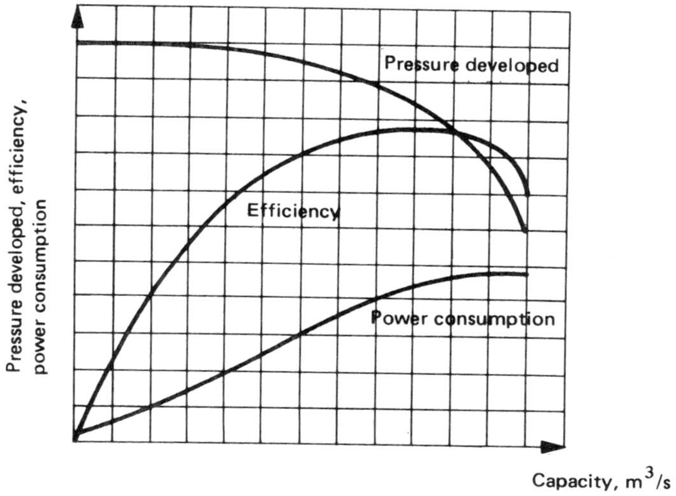


Fig. 3.19 – Characteristic curves of a centrifugal pump.

## 3.11 FANS

Fans are used in convection ovens, where they are used to circulate hot air over the food. They are also used in freezers and chillers to circulate cold air, carbon dioxide or nitrogen (Chapter 8). The simplest fans are of the propeller type, consisting of blades mounted on a hub. They cannot overcome large resistances to flow and are suitable for application in refrigerated stores where the circulation of air need not be too great.

Larger fans are usually of the centrifugal type and operate on the same principle as centrifugal pumps (Section 3.10). The construction of a fan is much lighter than that of a pump, light metal sheet being sufficiently strong for catering applications. It is possible to achieve discharge pressures in the range 30 to 400 $N/m^2$ with this type of fan, all the added energy from the electric motor being converted to kinetic energy and a negligible amount to pressure energy. The blades of the fan may be mounted in three different ways: straight, forward-curved or backward-curved. The forward-curved fan (Fig. 3.20) is relatively quiet in operation but can be overloaded and damaged easily. The backward-curved fan (Fig. 3.20) has blades which are less curved (for the same diameter) and operates at 1.5 to 2 times the speed of the forward-curved type.

The higher speed demands heavier construction and a larger space requirement. It is usually used where large variations in airflow volumes are required. Straight-bladed fans are used for the movement of low volumes against high pressures.

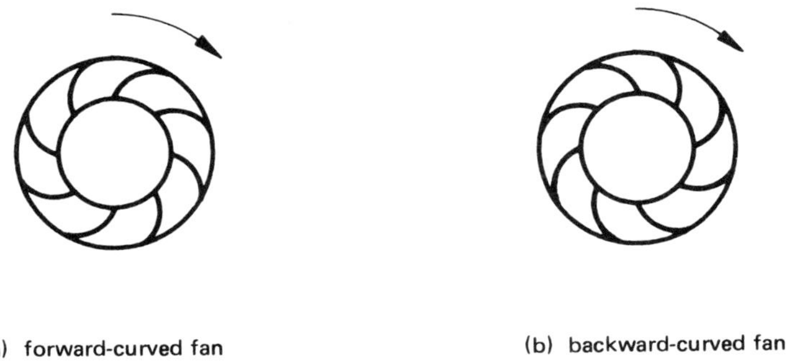


Fig. 3.20 – Centrifugal fans.

## 3.12 MECHANICAL SEPARATIONS

Mechanical separations are used mainly to separate solid particles from a liquid. Simple strainers (thin metal plates with perforations) are used to separate solid particles from water in dishwashing machines and sinks or other liquids. The main aim of these is to prevent blockages in pipes and pumps. The size of the perforations will determine the size of the particles retained on the metal screen. The process of straining is termed **filtration**. It is important to clean a strainer regularly, otherwise the flow of water through it is reduced and eventually the strainer itself may become blocked. The liquid flows through the strainer because of the pressure difference across it: usually the height of water above the strainer is the driving force on one side ($z\rho g$ of Bernoulli's equation), with atmospheric pressure ($p$ of Bernoulli's equation) acting on both sides of it. The potential energy ($z\rho g$) will be dissipated as kinetic energy (velocity of flow through the perforations) and frictional losses through the strainer and solid material retained on it. The situation is analogous to that for emptying a sink (Section 3.4) but in this case there are many outlets (holes) for the water. As solid material builds up on the strainer, the passage for the water to flow through becomes more restricted, effectively reducing the value of $C_d$, the discharge coefficient.

An important application of filtration is the use of filters for removing solid particles from frying oil. Excessive exposure of the oil to atmospheric oxidation must be avoided in such units and so open filtration systems (open gauze and cloth filters and open mesh strainers) are at a disadvantage from this viewpoint. Enclosed filters or mechanical filter units reduce the effect of oxidation as described by Fettes (1977), whose diagrams of this type of equipment are reproduced in Figs. 3.21 and 3.22. Other arrangements of the filter are also available, for example, fitted horizontally with the oil feed through the top. A continuous unit involves the use of a centrifugal pump for circulating the oil between the fryer and the filtration unit.

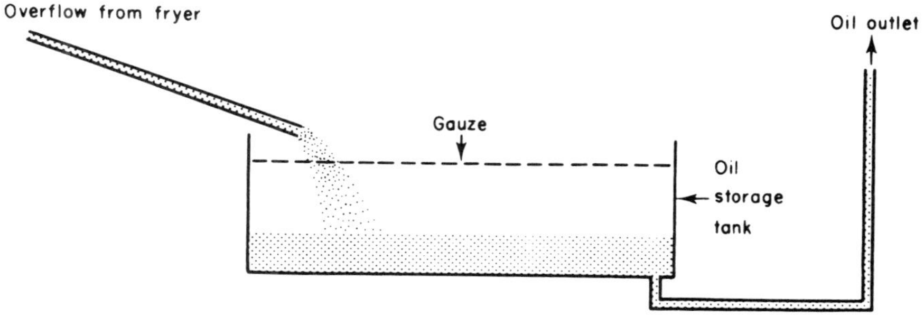


Fig. 3.21 – Open gauze strainer.

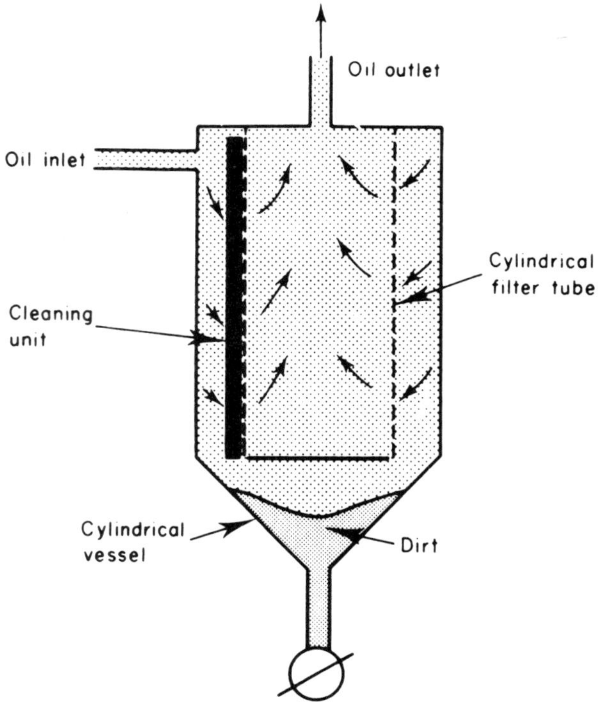


Fig. 3.22 – Mechanical filter unit.

It is often necessary to wash and de-water large quantities of salad materials and vegetables. The water at the end of washing will not drain away naturally and therefore equipment is available to speed up this process (Fig. 3.23). The salad is contained in an inner perforated drum which is rotated by an electric motor. The water is subjected to a centrifugal force and moves from the salad materials onto the inner surface of the inner drum. It passes through the perforations and is collected in an outer drum before if drains away.

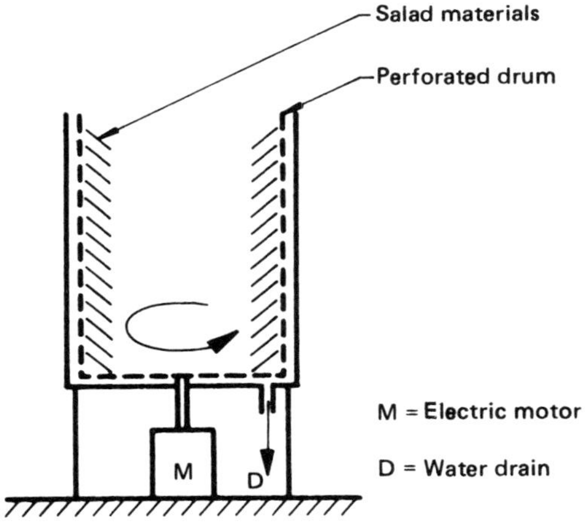


Fig. 3.23 – Vegetable centrifuge.

## 3.13 NOMENCLATURE

| | |
|---|---|
| $b$ | proportionality factor (kg/m.s) |
| $c$ | yield stress (N/m$^2$) |
| $d$ | pipe diameter (m) |
| $dy$ | distance (m) |
| $e$ | pipe roughness (m) |
| $g$ | acceleration due to gravity (m/s$^2$) |
| $h$ | height of water in a sink (m) |
| $l$ | length of pipe (m) |
| $l_e$ | entrance length of a pipe (m) |
| $p$ | pressure of a fluid (N/m$^2$) |
| $\Delta_p$ | pressure drop/loss over a length of pipe (N/m$^2$) |
| $\Delta p_p$ | pressure difference produced by a pump or fan (N/m$^2$) |
| $\Delta p_f$ | pressure loss due of friction (N/m$^2$) |
| $r$ | radius (m) |
| $r_w$ | inner radius of a pipe (m) |
| $s$ | power law constant (–) |

| | |
|---|---|
| $t$ | time (s) |
| $u$ | velocity (m/s) |
| $z$ | height of fluid above a datum level (m) |
| $A$ | area ($m^2$) |
| $A_N$ | area of a sink outlet ($m^2$) |
| $A_s$ | area of water surface in a sink ($m^2$) |
| $Eu$ | Euler number (–) |
| $F$ | shear force (N) |
| $Q$ | volume flow rate ($m^3/s$) |
| $Re$ | Reynolds number (–) |
| $\eta$ | kinematic vicosity ($= \mu/\rho$) ($m^2/s$) |
| $\mu$ | dynamic/absolute viscosity (kg/m.s) |
| $\rho$ | density ($kg/m^3$) |
| $\sigma$ | surface tension ($kg/s^2$) |
| $\tau$ | shear stress ($N/m^2$) |

## 3.14 PROBLEMS

**(1)** Apple sauce is a non-Newtonian fluid with constants, $b = 0.6$ and $s = 0.5$. How much will flow down a tube in laminar flow if the diameter of the tube is 4 cm, its length is 5 m and the pressure drop down the tube is 10 000 $N/m^2$. Express the answer in $m^3/h$. ($\pi = 3.142$).

**(2)** Water is to be supplied from a 2 in diameter water main to a continuous dishwasher. The diameter of the connecting pipe is 1 in (2.5 cm) and its length is 20 m. The pipe starts as an abrupt contraction from the main and contains four 90° bends of radius 2 in. The water is supplied at the rate of 0.7 $m^3/h$. Calculate the pressure drop from the main to the dishwasher. Viscosity of water = 1 g/m.s and its density is 1000 $kg/m^3$. The pipe is made of commercial steel.

**(3)** A sink has the dimensions 2 m X 0.5 m and has a discharge hole of diameter 5 cm. When full the depth of water is 60 cm. Calculate

(i) the time for the first 25% of the water to be emptied;
(ii) the time for the second 25% of the water to be emptied;
(iii) the time for the third 25% of the water to be emptied;
(iv) the time for the fourth 25% of the water to be emptied.

($C_d = 0.8, g = 9.81\,m/s^2$.)

**(4)** Carry out a dimensional analysis for the flow of a fluid assuming that surface tension is important. Use the information given below and solve in terms of $\delta$, $\epsilon$ and $\theta$:

| Variable | Symbol | Dimensions | Exponent |
|---|---|---|---|
| fluid velocity | $u$ | $[Lt^{-1}]$ | $\alpha$ |
| linear dimension | $l$ | $[L]$ | $\beta$ |
| force | $F$ | $[MLt^{-2}]$ | 1 |
| fluid density | $\rho$ | $[ML^{-3}]$ | $\gamma$ |
| fluid viscosity | $\mu$ | $[ML^{-1}t^{-1}]$ | $\delta$ |
| surface tension | $\sigma$ | $[Mt^{-2}]$ | $\epsilon$ |
| acceleration due to gravity | $g$ | $[Lt^{-2}]$ | $\theta$ |

## 3.15 REFERENCES

Bacon, N. P. *et al.* (1960), *Jnl. Iron Steel Institute*, July, 268–292.

Chigier, N. A. and Milson, A. (1973), *La Rivista dei Combustibili*, 27 (4–5), 137–146.

## 3.16 FURTHER READING

Muller, H. G. (1973), *An introduction to food rheology*. Heinemann.

Richardson, J. F. (1971), *Chemical Engineering*, Vol. 3, Ch. 6. Pergamon Press.

Skelland, A. H. P. (1967), *Non-Newtonian flow and heat transfer*. John Wiley & Sons.

Wilkinson, W. L. (1960), *Non-Newtonian fluids*. Pergamon Press.

CHAPTER 4

# Conduction

## 4.1 INTRODUCTION

This chapter is concerned with heat transfer by conduction, that is, it is concerned with heat transfer in solid materials. Two types of conduction heat transfer are considered. First, **steady-state conduction** (Sections 4.2 to 4.5) where the temperature distribution in the material does not change with time. The main applications of steady-state conduction include heat losses through the walls of equipment (into the kitchen) and the converse process in freezer cabinets and refrigerators. A further important example is the use of extended surfaces in fryers to increase heat-transfer rates. In **unsteady-state conduction**, on the other hand (Sections 4.6 to 4.8), the temperature distribution in the material does change with time. This applies to all those occasions when food is heated up during cooking or cooled down after cooking. The processes of freezing and thawing are not considered in this chapter but in Chapter 8.

## 4.2 STEADY-STATE CONDUCTION: SINGLE WALL

The steady-state flow of heat through a single wall may be represented as shown in Fig. 4.1. Since this is steady-state conduction, the wall temperatures $\theta_1$ and $\theta_2$ do not vary with time. The equation which is used to calculate the flow of heat through the wall is:

$$q = \frac{\lambda A(\theta_1 - \theta_2)}{l}.$$

This equation is for the heat flow in one direction only, that is, in the direction of the arrow in Fig. 4.1. It is assumed that there is no heat flow in the other two directions, perpendicular to that arrow, that is, into the paper, or vertically upwards. This assumption is valid if the thickness of the wall is small relative to the width and length of the wall. This is so for most catering applications of steady-state conduction. The above equation will allow us to calculate the heat conducted through the walls of an oven (from the hot inside face to the cooler outside face) provided that it is made of only one material. Usually more than one material is employed and this is dealt with in Section 4.3.

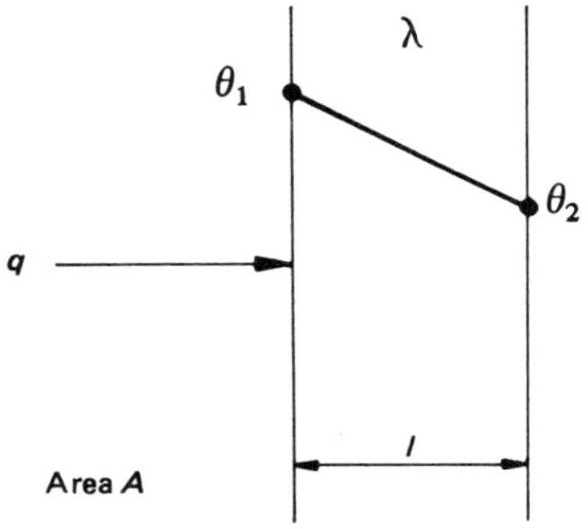


Fig. 4.1 – Steady-state conduction: single wall.

The other assumption made in the above equation is that the thermal conductivity of the material of the wall (λ) is independant of temperature. This is usually not so and therefore the value of thermal conductivity (λ) used in a calculation is the mean value between the two termperatures involved ($\theta_1$ and $\theta_2$).

Although the equation given above is convenient to use, it is advantageous to express it in a slightly different form for reasons which will become more apparent in the next Section and in later Chapters. It may be re-written as:

$$q = \frac{(\theta_1 - \theta_2)}{\left(\dfrac{l}{\lambda A}\right)}$$

or

$$q = \frac{(\theta_1 - \theta_2)}{R}$$

where

$$R = \left(\frac{l}{\lambda A}\right).$$

This equation is of the same form as Ohm's law in electricity (Fig. 4.2).

$$I = \frac{V}{R},$$

where $I$ = current (amperes), $V$ = potential difference (volts), $R$ = resistance (ohms), that is,

$$\text{flow of electricity (current)} = \frac{\text{driving force (potential difference)}}{\text{resistance}} .$$

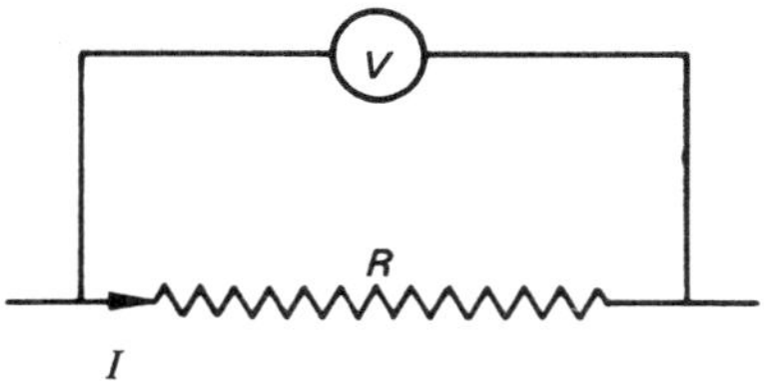


Fig. 4.2 –Electrical analogy of conduction: single wall.

Thus, the heat conduction equation in this new form is made up of a **flow** of heat ($q$), a **driving force** ($\theta_1 - \theta_2$) and a **resistance** to heat flow ($l/\lambda A$). This analogy between electricity and heat will recur several times in later Chapters. The driving force in conduction heat transfer is the temperature difference ($\theta_1 - \theta_2$). The resistance to heat flow in the case of a single wall is ($l/\lambda A$).

The units of heat flow are W, those of driving force are °C or K and the units of the resistance are K/W.

## 4.3 STEADY-STATE CONDUCTION: COMPOSITE WALL

When we consider catering equipment, in practice we find the walls are not made of a single material but of several materials, one behind the other. Such a wall is termed a composite wall. The wall of an oven for example is made up of an inner layer of stainless steel, a middle layer of insulating material and an outer layer of stainless steel. We can adapt the single-wall equation for use with composite walls as follows.

Consider a composite wall made up of three different materials (A, B and C) as shown in Fig. 4.3.

The areas ($A$) of all three walls are the same, as is the heat flow ($q$) through all three, otherwise the interface would receive more heat than it lost, the temperature would rise and unsteady-state conditions would occur.

We can apply the single-wall equation to each material in turn.

$$\text{For material A:} \quad q = \frac{(\theta_1 - \theta_2)}{\left(\dfrac{l_A}{\lambda_A A}\right)} = \frac{(\theta_1 - \theta_2)}{R_A} .$$

For material B: $$q = \frac{(\theta_2 - \theta_3)}{\left(\dfrac{l_B}{\lambda_B A}\right)} = \frac{(\theta_2 - \theta_3)}{R_B}.$$

For material C: $$q = \frac{(\theta_3 - \theta_4)}{\left(\dfrac{l_C}{\lambda_C A}\right)} = \frac{(\theta_3 - \theta_4)}{R_C}.$$

Re-arranging the above equations:

For material A: $\theta_1 - \theta_2 = q R_A$

For material B: $\theta_2 - \theta_3 = q R_B$.

For material C: $\theta_3 - \theta_4 = q R_C$.

Adding: $\theta_1 - \theta_4 = q(R_A + R_B + R_C)$

or

$$q = \frac{(\theta_1 - \theta_4)}{R_A + R_B + R_C}.$$

This equation is very similar to the single-wall equation:

(i) $q$ is the same;
(ii) $\theta_1 - \theta_4$ now represents the temperature drop across all the walls rather than just one;
(iii) the total resistance to heat flow is obtained by adding the individual resistances of the single walls together.

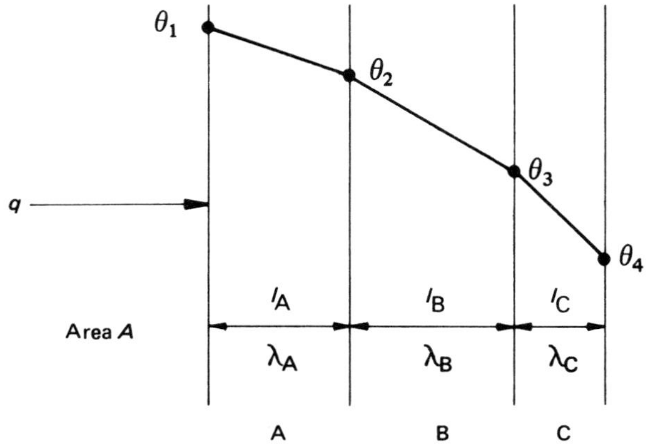


Fig. 4.3 – Steady-state conduction: composite wall.

Point (iii) is exactly what one does to find the overall resistance of a series circuit in electricity (Fig. 4.4).

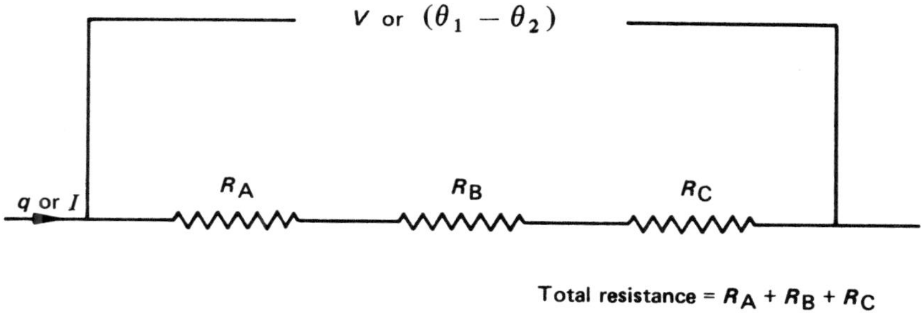


Fig. 4.4 – Electrical analogy of conduction: composite wall.

Thus the more walls we add, the more resistance we add to heat flow and hence for a given temperature drop $(\theta_1 - \theta_4)$, the lower will be the heat-flow $(q)$. This is the purpose of insulation. It increases the resistance of a wall to heat flow and so reduces the heat loss (heat passing through the wall).

The above equations apply to plane (flat) walls. The same idea can be applied to a curved wall, say a steam pipe covered with insulation. In this case the resistance of the pipe or insulation is given by:

$$R = \frac{\log_e(r_2/r_1)}{2\pi l\lambda}$$

where $r_2$ = outside radius of pipe (m), $r_1$ = inside radius of pipe (m), $l$ = length of pipe (m), $\lambda$ = thermal conductivity of pipe material (W/m.K).

If there is more than one material, as sometimes there is with an insulated steam pipe, then the resistance for each material (as given above) may be added up as for the plane-wall case, that is

$$q = \frac{(\theta_i - \theta_o)}{R_A + R_B + R_C + \cdots}$$

where $\theta_i$ = inner wall temperature, $\theta_o$ = outer wall temperature. If two insulating materials are used in this way, then there is an optimal way of arranging them to give the best insulating effect.

## 4.4 INSULATION OF EQUIPMENT

In Section 4.2, it was seen that the heat loss through an oven wall is given by the equation:

$$q = \frac{\lambda A(\theta_1 - \theta_2)}{l}.$$

This equation shows that for the heat loss by conduction through the oven wall ($q$) to be low we require:

(a) a wall material of low thermal conductivity ($\lambda$);
(b) walls of low surface area ($A$);
(c) a small temperature difference ($\theta_1 - \theta_2$);
(d) thick insulation ($l$).

Smaller temperature differences are achieved by cooking/warm holding food at low temperatures ($\theta_1$) and operating with high temperatures on the surface of the equipment ($\theta_2$). Cooking at lower temperatures will give longer cooking times and lower equipment throughput (and possibly a danger of microbiological hazard if the temperatures are low enough). Warm-holding at lower temperatures may give rise to poor customer satisfaction (if food is too cold) and the possibility of microbiological hazards. Many countries have legislation which specifies the minimum temperature at which warm foods may be held; the temperature is intended to avoid any danger from the growth of food poisoning bacteria.

High equipment surface temperatures may be unacceptable because of operator safety (burns and fire risk). This latter point may be covered in equipment specifications by specifying maximum temperatures for parts of the equipment which need to be handled when in use. Moreover, the maximum temperatures permitted for adjacent walls and floors and other specific parts of the appliance could be stated, for example, electrical insulation and leather diaphragms in gas carrying components.

Small temperature differences are also required for cold stores and refrigerators because, in this case, heat entering the cold space has to be removed. The surface temperature of the store is controlled by the temperature of the environment (the cooler the better) whereas the inside temperature is controlled by the requirements of the food being stored. In general the higher the temperature in the store the less effective it is in preventing quality deterioration. The cost of using lower temperatures needs, in this case, to be balanced against food quality. Here again legislation is often used to control food poisoning risks in susceptible foods stored at low temperatures.

Walls of low surface area are obtained by purchasing equipment which is as small as possible consistent with the required throughput. If the thickness of the insulation is doubled then the heat loss will be halved, according to the above equation for a single material. Increasing the thickness of the insulation will only be obtained at extra capital cost for the insulating material plus additional costs. Among these are the cost of extra floor space required (if oven volume remains the same then the outside dimensions of the equipment increases as extra insulation is added) and the greater cost for the outer metal casing of the equipment. To make the investment worthwhile, these costs should be less than the savings made in fuel charges obtained as a result of the extra insulation, or offset some other way such as improved working conditions. An alternative means of reducing the heat loss is to use insulating materials of lower thermal conductivity but this may also lead to a higher capital cost for the insulating material. Thus costs and savings should be balanced over the life of the equipment, looking at the problem from a purely accounting viewpoint. This is not necessarily the only viewpoint; working conditions in the kitchen also need to be considered whilst in other cases Government energy policy may also be a factor.

One example of using material of lower thermal conductivity is the substitution of conventional mineral wool insulation by polyurethane foam in electrically-heated *bain maries* (Keniston, 1977). The measured energy saving was 18% and improved working conditions were claimed from the lower surface temperatures of the equipment. The thermal conductivities of mineral wool and polyurethane are approximately 0.022 and 0.013 W/m.K respectively. These data would indicate a larger saving of energy than that reported above. The reason why this is not achieved in practice will be explained in Chapter 5. Fibreglass (with thermal conductivity approximately that of mineral fibre) is also used for insulation in *bain maries* and hot cupboards.

The *bain marie* and hot cupboard are interesting examples of the use of insulation as a factor in determining capital cost and operating costs. The simplest and cheapest (capital cost) type comprises a single sheet of metal. Next come those comprising two layers of metal separated by an air gap. The air gap acts as an insulating medium but without the expense of a filling material. The type with two metal layers and a gap filled with mineral wool/fibreglass follows; this has better insulating properties than the air gap but at higher capital cost. Finally we have foam insulation, which is injected into the cavity between the plates and holds them rigid without the metal frame-work required for the mineral wool-type filling. Hence, this has advantages both during construction and for insulation purposes.

## 4.5 STEADY-STATE CONDUCTION: EXTENDED SURFACES

In Section 4.2 we considered steady-state conduction in one direction (dimension) only. There is an important application of steady-state conduction in two directions (dimensions) in catering. This occurs in fryers where extended surfaces are used to increase the rate of heat transfer from the hot combustion gases to the frying oil. An extended surface, as the name implies, is a modification to a surface such that its area is increased. This means that there is a larger area to receive/lose heat, compared with the situation when a simple flat plate is used, as illustrated in Fig. 4.5.

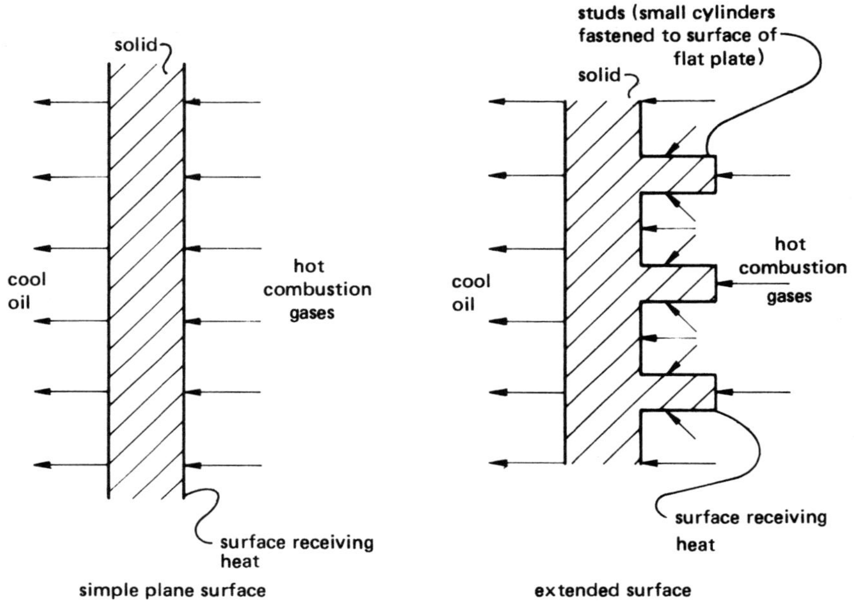


Fig. 4.5 – Plane and extended surfaces.

Consider the single stud (small cylinder) shown in Fig. 4.6.

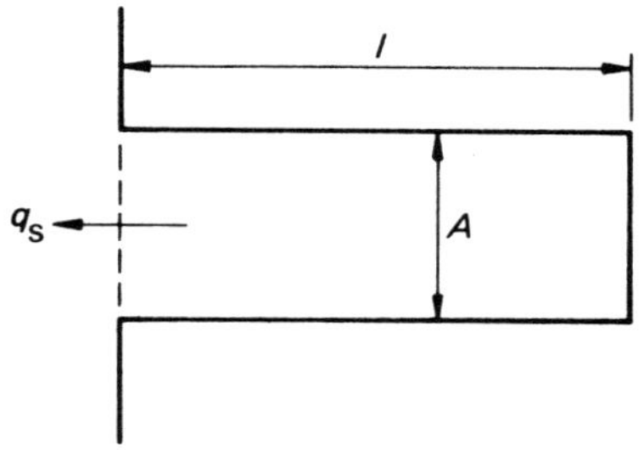


Fig. 4.6 – Extended surface as a stud.

The rate of heat transfer ($q$) through the base of one stud may be obtained from the following equation:

$$q_s = m\lambda A_s \tanh ml\,(\theta_g - \theta_1)$$

where

$$m = \sqrt{\frac{hC_s}{\lambda A_s}} = \sqrt{\frac{2h}{\lambda r}}$$

$C_s$ = circumference of stud (m) = $2\pi r$, $h$ = convective heat-transfer coefficient (see Chapter 5) (W/m$^2$.K), $\theta_1$ = temperature of metal at base of stud (K), $\theta_g$ = temperature of hot combustion gasses (K), tanh = a mathematical function with values as given in Table 4.1 (–).

**Table 4.1 – Values of** ***ml*** **and** tanh *ml*.†

| | | | | | | | | | |
|---|---|---|---|---|---|---|---|---|---|
| *ml* | 0 | 0.5 | 1 | 1.5 | 2 | 3 | 4 | 5 | 6 |
| tanh *ml* | 0 | 0.4621 | 0.7616 | 0.9052 | 0.9640 | 0.9951 | 0.9993 | 0.9999 | 1 |

† A complete set of values of tanh *ml* is available in a set of mathematical tables.

If the extended surface is in the form of a fin (Fig. 4.7), then

$$m = \sqrt{\frac{hC_f}{\lambda A_f}} = \sqrt{\frac{2h}{\lambda b}}$$

where cross-sectional area of fin, $A_f = bl$, and circumference of fin, $C_f = 2l$.

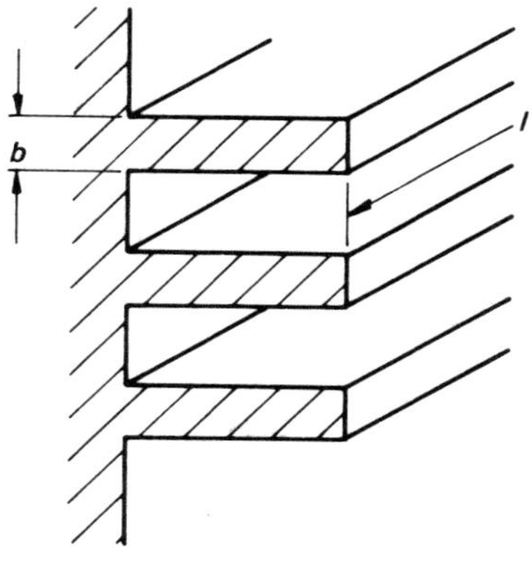


Fig. 4.7 – Extended surface as a fin.

It is only advantageous to use studs or fins where the rate of convective heat transfer to the metal surface is slow in relation to the conduction of heat through the plate.

The heat transfer through a wall fitted with studs is made up of two components: the heat flow through the studs and the heat flow through the areas of wall between the studs. If $A$ is the area of the wall, $A_s$ is the base area of a single stud and $p$ is the number of studs, then the heat flow through a wall with studs is given by:

$$\underbrace{Q_T}_{\text{total heat flow}} = \underbrace{pq_s}_{\text{heat flow through studs}} + \underbrace{h(A-pA_s)(\theta_g-\theta_1)}_{\text{heat flow through wall between studs}}\,.$$

The equivalent flow of heat through a wall without studs is given by:

$$Q = hA(\theta_g - \theta_1)\,.$$

This equation will be considered more fully in the next Chapter.

The principal use made of studs or fins is in gas-fired V-shaped fryers. In this type of fryer the oil in contained in a pan fitted with a V-shaped base (Fig. 4.8).

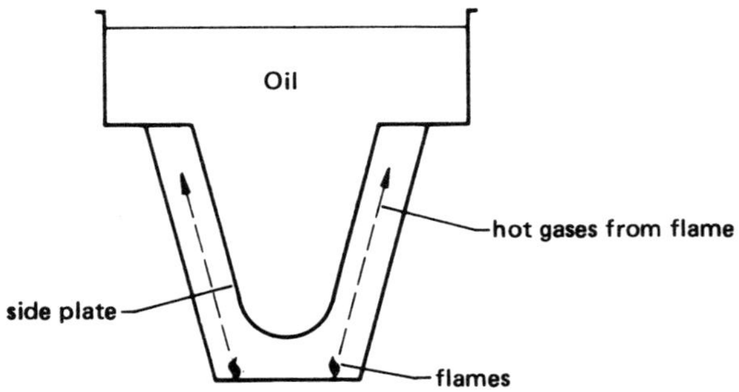


Fig. 4.8 – Cross-section of V-shaped fryer.

The reason for this shape will be considered in Chapter 5 when fryers are discussed in more detail. The hot combustion gases pass along one side of the side plates from the gases to the oil. It is evident from the diagram that the area of plate for heat transfer is very small. This area may be increased by fitting fins vertically downwards on the side plates or by fitting a bank of studs to the bottom and sides of the V-section. Either way the surface area is extended considerably. The V-section plate itself is made of thin metal (ensuring high thermal conductivity) whilst heat transfer from the gases to the plate is slow (see Chapter 5). In this case the use of fins/studs is advantageous in increasing the heat-transfer rate.

## 4.6 UNSTEADY-STATE CONDUCTION: DIFFUSIVITY

Unsteady-state conduction is concerned with the heating and cooling of materials (for example, foods) where there is a change of temperature with time. Thus virtually every cooking process represents an unsteady-state heat transfer problem. The temperature in the food during heating depends on two factors:

(i) position in the food – experience tells us that food is cooler in the middle than on the outside;
(ii) time from start of cooking (heating) – experience tells us that the longer the food is cooked the warmer it gets.

Chilling food is also an example of unsteady-state heat transfer.

The fact that temperature in the food is dependent on both these factors (position and time) makes the solution of unsteady-state heat transfer problems much more complex than steady-state problems.

In unsteady-state problems the aim is to calculate the temperature distribution (by position) in the food at different times and thereby define the degree of cooking/chilling. The basic equation we need to solve is a second-order differential equation, which for one-dimension only is:

$$\frac{\partial \theta}{\partial t} = \alpha \frac{\partial^2 \theta}{\partial x^2}$$

where $\theta$ = temperature, $x$ = distance, $t$ = time.

The term $\alpha$ is called the **thermal diffusivity** and is a property of the food, that is,

$$\alpha = \frac{\text{thermal conductivity}}{\text{heat stored}} = \frac{\lambda}{C_p \rho}$$

where $C_p$ = specific heat of food (J/kg.K), $\rho$ = density of food (kg/m$^3$).

The units of $\alpha$ are

$$\frac{\text{W kg.K m}^3}{\text{m.K J kg}} = \frac{\text{W.m}^2}{\text{J}} = \frac{\text{J.m}^2}{\text{s.J}} = \frac{\text{m}^2}{\text{s}} .$$

The definition of thermal diffusivity ($\alpha$) as thermal conductivity divided by heat stored implies that in unsteady-state conduction two factors are important:

(i) thermal conductivity, that is, how quickly or slowly heat is transferred from a hot region to a cold region of the food,
(ii) heat storage, that is, how much heat has to be stored in the food to increase its temperature.

This contrasts with steady-state heat transfer where only conduction is important. Heat storage is not important in this case because temperatures are not changing with time.

As given, the above differential equation implies a number of assumptions:

(i) heat transfer is in one direction only; the equation becomes more complex if two or three dimensions are considered;
(ii) the thermal diffusivity of the material is independant of temperature;
(iii) the material is homogeneous: its physical properties are uniform throughout the material;
(iv) no phase changes (melting, freezing, evaporation, condensation) take place in the food during heating/cooking.

Usually some or all of the above assumptions are not valid in any particular situation. This is certainly true when considering the heating/cooling of food. Indeed, rarely are any of the above assumptions correct as far as catering problems are concerned. To avoid these assumptions one must complicate the differential equation and resort to the computer to carry out the calculations involved (see Chapter 9). We will content ourselves here with fairly simple calculation methods. These methods are excellent for illustrating what is going on in the food and indicating the limitations of catering processes. It must not be forgotten however, that we have made the above assumptions and that in interpreting the results of our calculations these assumptions must be considered before arriving at any conclusions.

The two methods used to calculate the temperature distribution in food are fairly straightforward. The first method (Section 4.7) involves a graphical construction whilst the second method (Section 4.8) involves the use of published charts, derived from the basic differential equation using a computer solution.

## 4.7 UNSTEADY-STATE CONDUCTION: SCHMIDT'S METHOD

This method may be derived by considering the food to be sub-divided into a number of slabs and looking at the heat balance for one of these slabs. An internal slab is illustrated in Fig. 4.9.

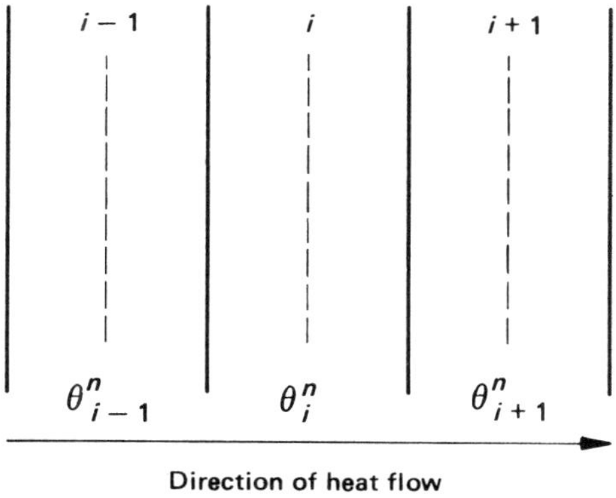


Fig. 4.9 – Unsteady-state conduction in an internal slab.

The heat balance for one slab over one time increment is:

$$\begin{bmatrix}\text{Gain in heat by}\\ \text{conduction through}\\ \text{surface of slab}\end{bmatrix} - \begin{bmatrix}\text{Loss in heat by}\\ \text{conduction through}\\ \text{other surface of slab}\end{bmatrix} = \begin{bmatrix}\text{Increase in}\\ \text{enthalpy of}\\ \text{the slab}\end{bmatrix}$$

This may be expressed in mathematical terms as:

$$\frac{\lambda A}{\Delta x}(\theta_{i-1}^n - \theta_i^n) - \frac{\lambda A}{\Delta x}(\theta_i^n - \theta_{i+1}^n) = \frac{A \Delta x C_p \rho}{\Delta t}(\theta_i^{n+1} - \theta_i^n)$$

| Rate of heat gain (W) | Rate of heat loss (W) | Rate of increase in enthalpy (W) |
|---|---|---|

where $\lambda$ = thermal conductivity of food (W/m.K), $\Delta x$ = thickness (m), $A$ = area through which heat is flowing ($m^2$), $C_p$ = specific heat of food (J/kg.K), $\rho$ = density of food ($kg/m^3$), $\Delta t$ = time interval(s).

In the heat balance we may delete $A$ and multiply through by $\Delta x/\lambda$. This gives:

$$\theta_{i-1}^n - \theta_i^n - \theta_i^n + \theta_{i+1}^n = \frac{\Delta x^2 C_p \rho}{\Delta t \lambda}(\theta_i^{n+1} - \theta_i^n).$$

The term

$$\frac{\Delta t \lambda}{\Delta x^2 C_p \rho}$$

simplifies to

$$\frac{\Delta t \alpha}{\Delta x^2}$$

where $\alpha$ = thermal diffusivity of food (Section 4.6).

Therefore we may re-write the equation as:

$$\theta_i^{n+1} - \theta_i^n = \frac{\Delta t \alpha}{\Delta x^2} (\theta_{i-1}^n + \theta_{i+1}^n - 2\theta_i^n) .$$

If $\Delta t\alpha/\Delta x^2$ is fixed at $\frac{1}{2}$ by the choice of suitable values of $\Delta x$ and $\Delta t$, the equation simplifies to:

$$\theta_i^{n+1} = \tfrac{1}{2} (\theta_{i-1}^n + \theta_{i+1}^n).$$

This equation can be expressed in words as: **The temperature in an incremental slab at the end on one time increment is equal to one half of the sum of the temperatures of the two adjacent slabs at the end of the previous time increment.** The equation may be used in the form of a graphical construction as illustrated in Fig. 4.10.

The food to be considered is divided up into a number of imaginary slices each slice being of thickness $\Delta x$. The starting time $t = 0$. It is assumed that at $t = 0$, the outside surface of the food is suddenly raised to some pre-determined temperature and is held at that temperature throughout the heating/cooling period whilst the remainder of the food remains at its original temperature. Heat flow is in one direction only. After one time interval ($t = 1$) the temperature distribution in the food is estimated by joining up alternate temperature points (marked 0) starting from the outer surfaces. In the example given we assume heating is in one direction but on to both surfaces of the food. The temperature distribution after the second time interval ($t = 2$) is obtained by joining up the points marked 1, again starting from the point 1 which is nearest the outside. This process is continued by joining up the points marked 2, then 3, and so on. We continue with the construction for as long as we wish but usually it is until the centre temperature reaches a particular value. The actual heating-up time is calculated from the number of time intervals ($\Delta t$'s) needed to achieve the centre-temperature required. The value of one $\Delta t$ is obtained from the formula:

$$\Delta t = \frac{(\Delta x)^2}{2\alpha} .$$

In the limit, if a large number of slices are used the disjointed lines of Fig. 4.10 would become curves.

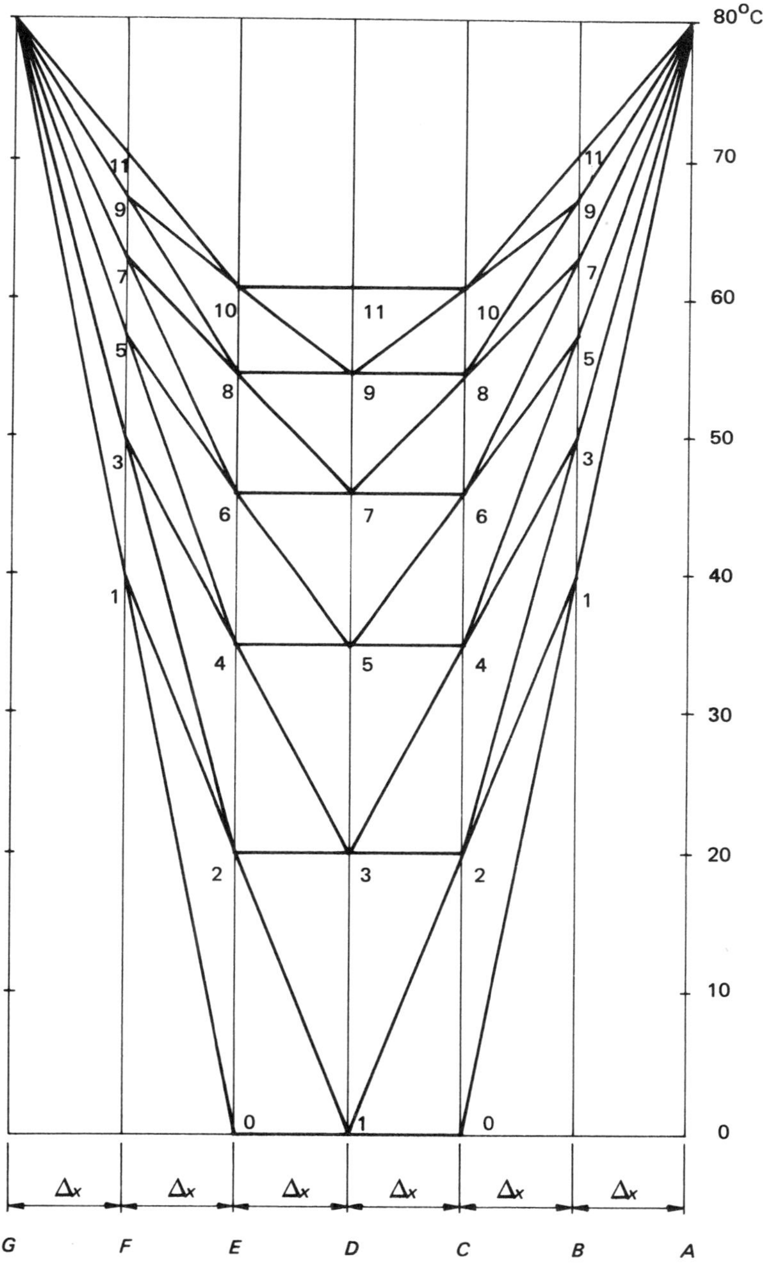


Fig. 4.10 – Schmidt's method: graphical form.

It should be evident from the example given that since one is only averaging temperatures between two points, it is not really necessary to carry out the graphical construction. It can be done more quickly and accurately using a table. The example used above is re-calculated in Table 4.2 using the tabular form. Thus 10 time intervals are required for the centre temperture to reach 60°C. This can be calculated in terms of time as with the graphical method.

**Table 4.2 – Schmidt's method: tabular form.**

| | $G$ | $F$ | $E$ | $D$ | $C$ | $B$ | $A$ |
|---|---|---|---|---|---|---|---|
| $t = 0$ | 80 | 0 | 0 | 0 | 0 | 0 | 80 |
| $t = 1$ | – | 40 | – | 0 | – | 40 | – |
| $t = 2$ | 80 | – | 20 | – | 20 | – | 80 |
| $t = 3$ | – | 50 | – | 20 | – | 50 | – |
| $t = 4$ | 80 | – | 35 | – | 35 | – | 80 |
| $t = 5$ | – | $57\frac{1}{2}$ | – | 35 | – | $57\frac{1}{2}$ | – |
| $t = 6$ | 80 | – | $46\frac{1}{4}$ | – | $46\frac{1}{4}$ | – | 80 |
| $t = 7$ | – | 63 | – | $46\frac{1}{4}$ | – | 63 | – |
| $t = 8$ | 80 | – | $54\frac{1}{2}$ | – | $54\frac{1}{2}$ | – | 80 |
| $t = 9$ | – | $67\frac{1}{4}$ | – | $54\frac{1}{2}$ | – | $67\frac{1}{4}$ | – |
| $t = 10$ | 80 | – | 61 | – | 61 | – | 80 |

The accuracy of both the graphical and tabular forms of Schmidt's method is dependent on how many slices are taken; the more slices the more accurate the solution. This method is a very simple finite-difference procedure which is the basis of most of the more-elaborate computer methods. These will be considered in Chapter 9.

If the temperature variation near the surface during the early stages of heating is required in more detail then a smaller value of $\Delta x$ is needed at the start. At subsequent stages of heating this can be changed to a larger value.

The temperature distribution at the start need not be uniform. The procedure in this case is to divide the solid up (as before) and use the temperatures where these slices cut the temperature distribution line, as illustrated in Fig. 4.11.

As Fig. 4.11 also inplies, it is not necessary for both sides of the solid to be suddenly raised to the same temperature. The application of Schmidt's method to cylinders and spheres has been described by Patton (1944).

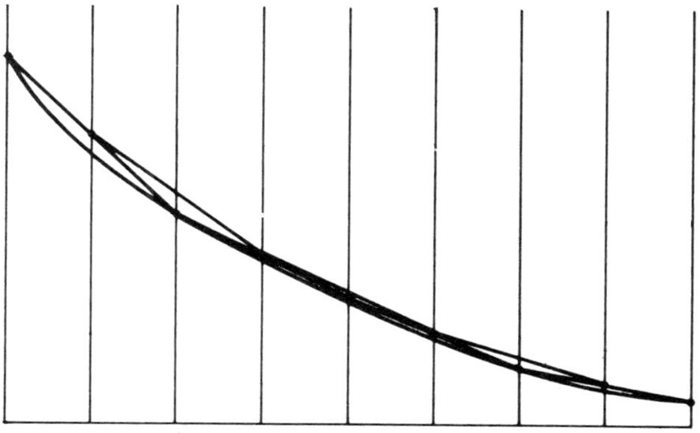

Fig. 4.11 – Schmidt's method: non-uniform initial temperature distribution.

So far, it has been assumed that the surface temperature is held constant. This is not the case when heating solids by convection because there is a thermal boundary layer on the surface. Although the concept of a thermal boundary layer is not discussed in detail until the next Chapter, it is convenient to show how it may be incorporated with Schmidt's method at this stage. The method involves adding to the solid another layer of thickness, $\Delta x^*$, which represents the thickness of the boundary layer. The modified construction is illustrated in Fig. 4.12.

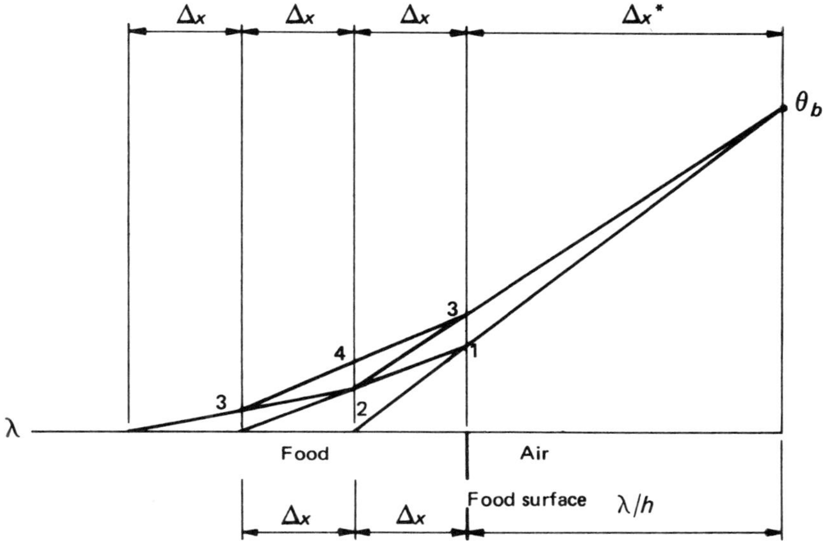


Fig. 4.12 – Schmidt's graphical method modified for convection.

The thickness of the boundary layer ($\Delta x^*$) is equal to the thermal conductivity of the boundary layer fluid divided by the convective-heat-transfer coefficient for the boundary layer ($\lambda/h$).

If the convection coefficient ($h$) changes with time, $\Delta x^*$ can be altered after the appropriate time intervals have elapsed. It can be seen, from Fig. 4.12, that the surface temperature of the food is actually rising. This is in contra-distinction to the fixed surface-temperature case (Fig. 4.10), where it was assumed that the surface of the food was suddenly increased to a specified temperature and remained constant at that temperature during the whole of the heating period.

## 4.8 UNSTEADY-STATE CONDUCTION: PUBLISHED CHARTS

These charts have been published in the literature by various workers. They represent the precise solution of the second-order differential equation for particular shapes of solid and include in their solution the assumptions set out earlier in Section 4.6. Unlike the Schmidt method, they do not assume that the surface of the food is suddenly raised to some particular temperature. They allow for the food to be immersed in a fluid (gas or liquid) at a constant temperature. As will be seen in the next chapter, the surface of the food will be at a different temperature to that of the bulk of the fluid (lower for heating and higher for cooling). The charts assume that the heat-transfer coefficient between the surroundings and the food remains constant throughout the heating period. This is rarely the case in practice. The charts are plotted in terms of four dimensionless groups which may be derived for this situation by dimensional analysis. The groups are:

$$\text{Fourier number, } Fo = \frac{\alpha t}{r_m^2},$$

$$\text{Biot number, } Bi = \frac{r_m h}{\lambda},$$

$$\text{Position ratio, } n = \frac{r}{r_m},$$

$$\text{Reduced temperature, } \textcircled{H} = \frac{\theta_s - \theta}{\theta_s - \theta_o}.$$

Where $\theta$ is the temperature of the solid at position $r$ at time $t$ (from the start of heating/cooling), $\theta_s$ = temperature of surroundings, $\theta_o$ = original uniform temperature of solid (at $t = 0$), $h$ = convective-heat-transfer coefficient between surroundings and the surface (see next chapter), $r_m$ = distance from the midplane to the surface (the half-width in the case of a slab and the radius in the case of a cylinder or sphere), $r$ = distance from the midplane to the point in the body under consideration. The values of $\alpha$ and $r_m$ are food properties and food dimensions respectively; $\theta_o$ is the initial temperature of the food which must be constant throughout the volume of the solid. $\theta_s$ is the temperature setting of the equipment and $h$ is known for the equipment from convective heat-transfer theory (see the next chapter).

There are several sets of charts available in journals and textbooks. The charts of Gurney and Lurie (1923) enable the temperature distribution throughout the food to be calculated (a range of values of $n$). The charts of Dalgleish and Ede (1965), on the other hand, give centre, surface and mean temperatures ($r = 0$, $r = r_m$ and $r_{\text{average}}$) but cover a wide range of values of the other dimensionless groups. Some of these charts are reproduced in Appendix B.

Usually, the calculations carried out using the charts of Dalgleish and Ede are of two types.

(1) The user wishes to calculate the centre or surface temperature of the food after a given time. In this case the Fourier and Biot numbers are known and may be used to obtain a value of the reduced temperature from the appropriate chart. Once the reduced temperature is obtained, the temperature $\theta$ may be calculated since the temperatures $\theta_s$ and $\theta_o$ are known. $\theta$ refers to the centre temperature if the centre temperature chart is used and to the surface temperature if the surface temperature chart is used.

(2) The user wishes to calculate the time required for the centre/surface temperature to reach a given value. In this case the reduced temperature and Biot number are known and therefore the Fourier number may be obtained from them using the appropriate chart. Once the Fourier number is known, the time may be calculated since the values of thermal diffusivity and radius ($r_m$) are known.

The charts themselves have been calculated for three geometries:

(a) infinite slabs (thickness $2r_m$);
(b) infinite cylinders (radius $r_m$);
(c) infinite spheres (radius $r_m$).

In practice not all objects can be treated as being of infinite extent because heat flow takes place in either two or three dimensions. The latter (finite objects) can be considered to be formed when two or three infinite objects intersect, for example, a finite cylinder is formed when a infinite cylinder intersects with a infinite slab, as shown in Fig. 4.13. The length of the finite cylinder is equal to the thickness of the infinite slab and has a radius equal to that of the infinite cylinder. Similarly, the intersection of three infinite slabs of the same thickness form a finite cube, and the intersection of three infinite slabs of different thickness form a finite rectangular solid (or slab). Thus two- and three-dimensional problems can be solved for a limited number of special cases using the charts for infinite solids.

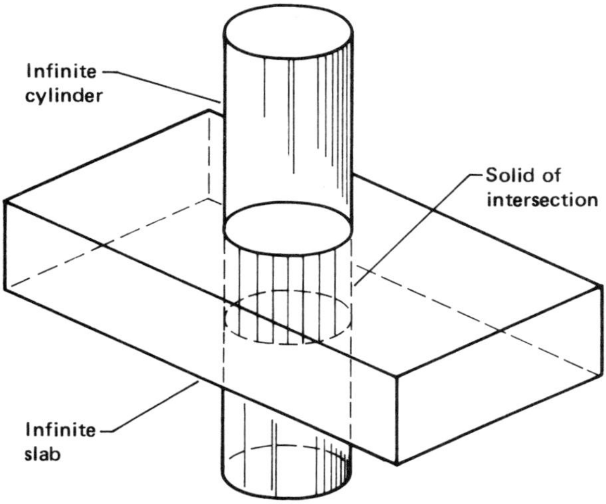


Fig. 4.13 – Finite cylinder formed from infinite cylinder and infinite slab.

The temperature in the finite solid may be calculated from the product of the reduced temperatures Ⓗ of the infinite objects, for example, in the case of a finite cylinder (Fig. 4.13).

$$Ⓗ_{fc} = Ⓗ_{is} \times Ⓗ_{ic} ,$$

where $fc$ = finite cylinder, $is$ = infinite slab, $ic$ = infinite cylinder.

The relationship for a finite rectangular solid is:

$$Ⓗ_{frs} = (Ⓗ_{is})_1 \times (Ⓗ_{is})_2 \times (Ⓗ_{is})_3$$

where 1, 2 or 3 refer to the three intersecting infinite slabs.

When employed in this way, the charts can be used to calculate centre or surface temperatures for cylinders, cubes and brick-shaped solids. They cannot be used however to calculate the time it takes to achieve a desired centre or surface temperature.

The chart method's principal advantages over the Schmidt method are that it is less time consuming and usually more accurate (unless the tabular method is computerised). The advantage of the Schmidt method is that it illustrates the process of unsteady-state conduction and can be used for a non-uniform initial temperature distribution.

In Section 4.6, it was pointed out that unsteady-state conduction involves two processes: thermal conduction and thermal storage. Compared with many other materials, solid foods are poor conductors of heat and this represents one of the major problems of heating food. It is a slow process to transfer heat from the outside surface of the food to its centre. We can overcome this problem, to some extent, by microwave heating and this will be discussed in Chapter 7. If frozen foods are heated, or when foods are frozen, the problem is exacerbated by the need to conduct into the food latent heat to melt the ice as well as to increase the food temperature. The calculation procedures so far described are not applicable to frozen foods; other methods will be considered in Chapter 8.

We describe only a few items of equipment in this Chapter because the processes by which heat arrives at the surface of the food need to be considered before most types of equipment can be discussed in any detail. This we will do in Chapters 5 and 6 and so will pave the way for a discussion of most other types of cooking equipment.

Since thermal conduction through food is slow, we would expect the food to heat up more quickly (as measured say by its centre temperature) if the food is thin rather than thick. Similarly, if the temperature the food is exposed to is raised, we would expect the rate of heating to accelerate because raising this temperature increases the (temperature) driving force for heat conduction (see Section 4.2). This may be illustrated using Schmidt's graphical method. Consider a food pack initially at 20°C with a thermal diffusivity $\alpha = 0.25\ \text{mm}^2/\text{s}$. If the surface of the pack is suddenly raised to 100°C (six slices taken in the construction) then it may be shown that the effect of pack thickness on heating-up time (defined as the time for the centre to reach 60°C) is as follows:

| Pack thickness | Heating-up time (to 60°C) |
|---|---|
| 24 mm | 192 seconds |
| 18 mm | 108 seconds |
| 12 mm | 48 seconds |
| 6 mm | 12 seconds |

The relationship between heating-up time and pack thickness is not linear; the results show that the thinner the pack the shorter the heating time and the higher the throughput of food from the equipment. The customer however, may object to thin packs if these are steaks or hamburgers.

The effect of the heating temperature on heating-up time (heating to 60°C at the centre) for a 24 mm thick pack gives results as follows:

| Heating temperature | Heating-up time (to 60°C) |
|---|---|
| 100°C | 192 seconds |
| 120°C | 147 seconds |
| 140°C | 122 seconds |
| 160°C | 106 seconds |

The relationship is non-linear again; the results show that increasing the heating temperature reduces the heating-up time. If the heating temperature is increased to too high a value however, chemical and physical changes may take place in the food which may make it unacceptable to the customer. In addition, the reduction in heating time caused by increasing the heating temperature decreases as the heating temperature rises. A rise of 20°C from 100 to 120°C reduces the heating time by 45 seconds, whereas the same rise from 140 to 160°C reduces the time by only 16 seconds. Similar calculations give heating-up times for different centre temperatures (food pack 24 mm thick, heating temperature 100°C):

| Centre temperature | Heating-up time |
|---|---|
| 40°C | 96 seconds |
| 60°C | 192 seconds |
| 80°C | 352 seconds |

Again there is a non-linear relationship and low centre temperatures are advantageous in terms of equipment throughput. From a microbiological and aesthetic point of view however, low centre temperatures may be unacceptable.

There are no reports in the literature of the use of the Schmidt method for calculating heating times for food and the accuracy of this method is therefor not known. Dickerson and Read (1968) measured the centre temperature of a can containing chicken *à la* king heated in steam. The temperature progression was also calculated using charts which are the Schmidt-method equivalent of the Dalgleish charts, that is, the surface temperature of the container was assumed constant at the steam temperature throughout the calculation. The

Biot number measures the temperature gradient between the heating medium and the food (can) surface and therefore the Schmidt method and the method used here are special cases of Biot number from the more general Dalgleish charts, that is, the Biot number is very large (steam temperature = surface temperature). Good agreement (within 2°C) between the experimental and calculated temperatures was obtained as shown in Fig. 4.14. The thermal diffusivity of the chicken *à la* king was measured and this value used in the calculation. The can was treated as an infinte cylinder intersecting with an infinite slab for the calculations. The diagram below shows that the surface temperature of the food did not, in the experiment, meet the assumption of the calculation procedure (that is, instantaneously reach 99°C – the steam temperature – and then remain constant).

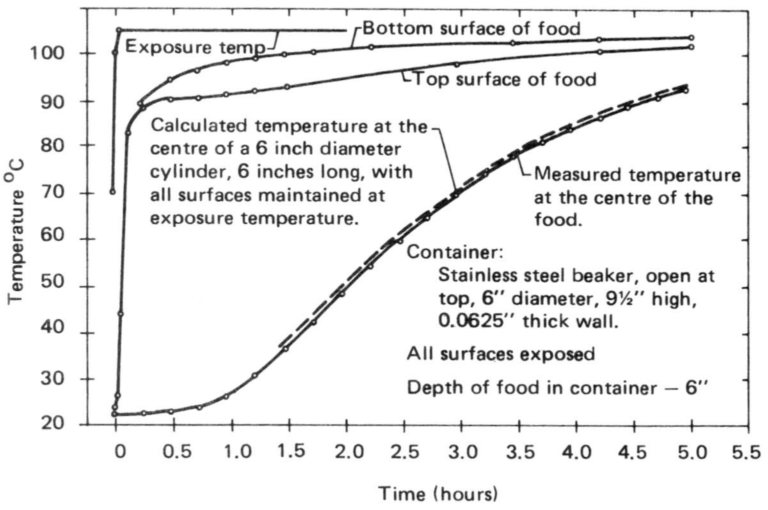


Fig. 4.14 – Heat transfer test on chicken *à la* king in steam chamber (ambient pressure).

Dagerskog and Bengtsson (1974) measured temperatures inside meat patties during double-sided contact frying. The results for various plate temperatures are shown in Fig. 4.15. The surface temperature measurements do not conform to the Schmidt assumption, confirming the finding of Dickerson and Read. The centre temperatures confirm the general shape of those of Dickerson and Read. The significant feature of the surface temperature curves, particularly for the higher surface temperature experiments, is the dip in the curve after 2 to 2.5 minutes. This cooling effect was attributed to liquid pressed out during the onset of shrinkage of the patties.

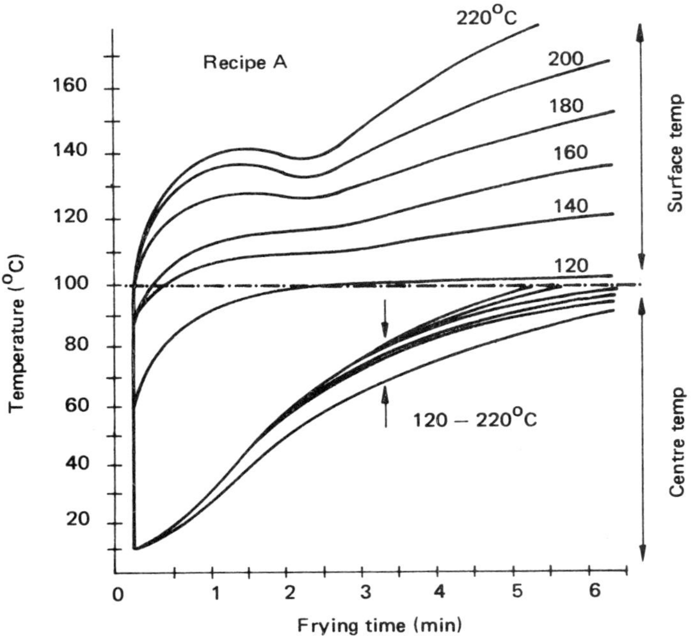


Fig. 4.15 – Influence of pan temperatures between 120–220°C on product surface and centre temperature.

## 4.9 GRIDDLES/FRY PLATES

Griddles/fry plates approximate to the type of heating problem which may be solved by using Schmidt's method. Heating solids by direct contact with a hot surface is utilised in the contact grill (griddle grill) and the griddle (plate) or fry plate. The former have two heating surfaces with the food sandwiched between them whilst the latter have a single heating plate. The heated surfaces are either perfectly flat or grooved. Griddle plates (single plates) are heated with either gas (Fig. 4.16) or electricity whilst contact (two plate) grills are electrically heated.

A drain channel and collection tray are normally provided to collect liquid cooking losses. A variety of products (steaks, hamburgers, eggs, bacon, chops, liver, pancakes, mushrooms, tomatoes and so on) are cooked by this method. The griddle grill is suitable for thicker food items such as steaks and hamburgers. The griddle plates are made up of thick metal to prevent distortion when they are heated (up to 300°C). Gas burners/electrical elements are situated below the plate in an enclosed box. Rapid heat-up and even temperature distribution are design requirements. Minimum plate temperatures are 220 to 230°C. These

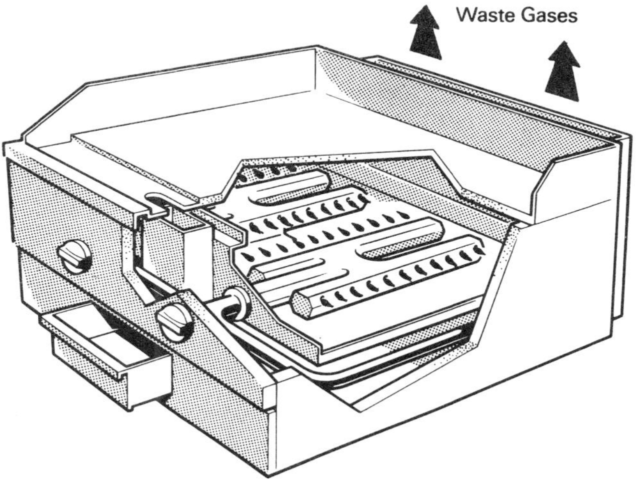


Fig. 4.16 – Countertop griddle.

temperatures would produce some drying of moisture from the food surface, a factor which is not taken into account in the Schmidt method. This will be discussed in Chapters 9 and 10.

An early American Gas Association (1950) report discussed the combustion aspects of gas-heated griddle design. This report is of interest because it stressed the considerations which need to be taken into account in the design of a relatively simple appliance.

**(a) Removal of combustion products (ventilation)**

The area of the ventilation openings must not be too large, otherwise the combustion products escape before they have given up their heat; on the other hand it must not be too small, otherwise they will tend to 'smother' the flames. Both situations lead to reduced efficiency. A properly designed ventilation system should permit burner operation with just sufficient excess air to ensure good combustion under all normal operating conditions (Chapter 2). Best heat distribution is obtained by using slits (for ventilation) in the rear side plate of the box, located at a sufficient height above the burner to allow the combustion products to escape from the combustion zone.

**(b) Air supply**

Openings for the combustion air are below the burners in the side walls. They must provide both the primary air (air mixed with gas before it reaches the flame) and also secondary air (air mixed into flame). Preferably the openings should be circular to prevent blockage by dust and lint (fluff, and so on); long narrow slots are easily clogged. They should also be located so as to give an even distribution of air to the flames.

**(c) Heat distribution**

Even distribution of heat is achieved by spacing the burners regularly under the plate and by including baffles to direct the flow of hot combustion products. The burners themselves consist of rows of small flames, itself an aid to even heat distribution. Even heat distribution is aided by having a griddle plate of high heat capacity (large volume) and by adding fins or ribs to its underside. The ribs/fins help to direct the flow of hot gases and increase the heat-absorbing area (Section 4.5).

**(d) Burner-plate distance**

The burner should be at the correct distance from the plate, otherwise, if it is too near, incomplete combustion could result. On the other hand, if it is too far away, heat transfer is poor and may give uneven distribution. The best location of the burner (giving best thermal efficiency consistent with good combustion) will vary with each griddle design and must be determined by performance tests (using a water load).

A later report of the American Gas Association (1958) reported results of tests carried out on contemporary equipment but with increased heat input rates. Combustion products' losses were in the range 34 to 42% (with satisfactory combustion as measured by carbon monoxide in the flue gases) and hot start thermal efficiencies (measured using a water load) in the range 49 to 52%. A peak efficiency was obtained at a particular heat input, the efficiency falling below the peak at heat inputs above and below the optimum.

Higher heat input rates (cold start) also reduced the time to heat the plate from room temperature to 232°C and the time to heat the water load through 78°C. The use of thermostatic control (cf. manual control) was found to be both feasible and desirable. It was found in this case that higher heat inputs did not increase the cooking speed of hamburgers because the full input rating of the burner was not needed to keep the plate temperature at the value controlled by the thermostat.

Tests were then carried out with experimental bar burners (see Chapter 11) giving heat inputs several times as high as those of the burners used in the tests just described. It was found that the heating speeds of the contemporary equipment could be increased by more than 2.5 times their then (1958) heating speed by increasing heat input rates without appreciable loss in water heating efficiency. Temperature uniformity over the plate was also improved. The most important factors in achieving these improvements were found to be the location of burners, the use of heat reflector plates between the burners, control of the air supply and prevention of leakage along the fry-top edges and selective flue inlet baffling. The new design gave 94% of the fry-top surface within ±28°C of the 232°C thermostat setting (cf. 86% with a contemporary range) and 80% within ±14°C (cf. 77% with a contemporary range). The maximum variation in spot temperatures observed (maximum minus minimum) was 44°C. Start-up time to 232°C was 13 minutes (cf. 18 minutes with an improved contemporary range or 29 minutes with the original contemporary range). Thermal efficiency was 42%, the same as that of the original contemporary range but less than that of the improved contemporary range.

A newer form of griddle plate is used for making pancakes. It consists of a heated rotating cylinder. The liquid pancake mixture is fed onto the cylinder in a thin layer and cooks in direct contact with it. They are automatically peeled off the cylinder and stacked in a pile.

Fischer (1977) has described a continuous grill (Fig. 4.17). The food is transported along two heated grilling plates by pusher-bars. It is 'flipped' over between the plates so that it is heated from both sides. Each plate is heated from below by radiation.

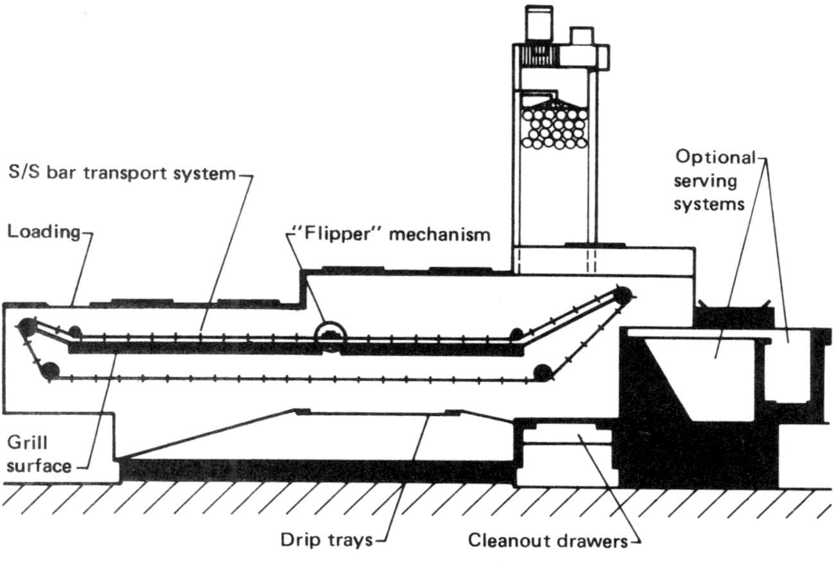


Fig. 4.17 – Continuous grill.

## 4.10 HEAT PIPES

In Section 4.8 we discussed the problems met in heating solid foods rapidly because of their low thermal conductivity. Microwaves were mentioned as a partial solution to this problem and will be considered in Chapter 7. An alternative to microwave heating, which is also much cheaper, is to insert a rod of metal (thermal pin) down the centre of, say, a joint of meat. Heat may then be conducted down the rod, thereby releasing heat within the meat itself. The situation is represented in Fig. 4.18. The enlarged end of the rod receives heat from the oven in the same way as the outside of the meat. This end is blackened so that it absorbs a larger proportion of the radiant energy falling on it (see Chapter 6). The metal rod is a good conductor of heat and therefore this heat rapidly passes down the rod and is released inside the joint, supplementing the heat falling on the meat surface. Similar devices have also been used for the rapid baking of potatoes.

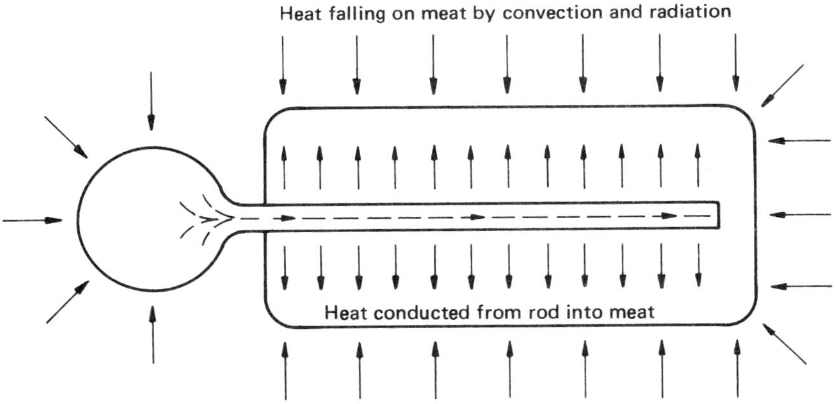


Fig. 4.18 – Metal rod for conducting heat to centre of meat.

A more effective device than a simple metal rod for this purpose is a heat pipe. This is constructed as shown in Fig. 4.19.

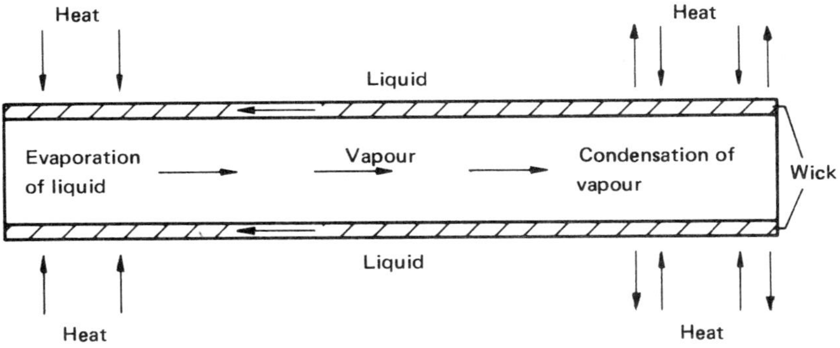


Fig. 4.19 – Heat pipe.

The heat pipe is a sealed tube lined with a wick and containing a small quantity of water, alcohol or other liquid (but no air). The liquid receives heat from the oven and is vapourised. The vapour travels down the tube. The other end of the tube (inserted in the meat) is cooler and so the vapour condenses back to liquid, giving up its latent heat. The liquid returns to the other end of the tube *via* the wick. As we shall see in Chapter 5, the rate of heat transfer from a condensing liquid is very high. This means that the heat pipe is a more effective means of transferring heat into the centre of the joint than a solid rod of metal. Herbert and Irving (1974) tested a heat pipe with joints of beef, pork and lamb and reductions in cooking times of 24 to 42% were observed. Reductions with chickens were small, probably because of the arrangement of meat on the carcase. In terms of consumer acceptability however, the use of a heat pipe was not recommended for pork.

Corfield (1975) and Basiulis (1973) discussed the development of a griddle based on a heat pipe for transferring the heat from natural gas flames to the surface of the griddle plate. The heating-up time to 204°C was less than 10 minutes and the variation in temperature over the plate was within 3°C. The evenness of temperature distribution is attributable to the vapour in the heat pipe condensing at a specific temperature all over its under-surface. This performance is superior to that set in national standards for both gas and electric griddles. Hurley and Miskolczy (1975) have described the development of a gas-fired griddle, an oven (but not of the forced-convection type), a deep-fat fryer and a water heater involving the use of heat pipes. These designs use a fan to deliver air to the burner and the flames are used under impingement conditions, which enhances the heat-transfer rate from the flames to the heated surface (heat pipe) as will be seen in the next Chapter. The heat pipe then conveys the heat to the surface of the griddle, plate, oven walls, oil container and water container respectively. The griddle heated up to 204°C in 7 minutes and the surface temperature variation was less than 2°C. At 81% the combustion system efficiency was twice that of a conventional griddle. The oven's combustion system had an efficiency of 80%. It maintains (it is claimed) constant and uniform temperature in the oven cavity over a range of 93 to 232°C, temperature uniformity is independent of the position of the food in the oven, high humidity is maintained, the need for a large fan is eliminated and the reconstitution time of frozen foods is independent of position in the oven.

The fryer was 77% efficient, used 2.10 MJ/kg and fried raw potatoes at a rate of 38.1 kg/h. It was claimed that there was also a lower oil breakdown than in conventional fryers. This can be achieved because the high heat inputs achieved using a heat pipe dispenses with the need of high fryer-wall temperatures.

## 4.11 NOMENCLATURE

| | |
|---|---|
| $b$ | thickness (m) |
| $h$ | convective heat-transfer coefficient (W/m$^2$.K) |
| $l$ | thickness or length (m) |
| $m$ | $\sqrt{\frac{hC}{\lambda A}}$ (1/m) |
| $n$ | position ratio (–) |
| $p$ | number of studs (–) |
| $q$ | rate of heat flow (W) |
| $q_s$ | rate of heat flow through a stud (W) |
| $r$ | radius (m) |
| $t$ | time (s) |
| $\Delta t$ | time interval (s) |
| $x, \Delta x$ | distance (m) |
| $\Delta x^*$ | thickness of a boundary layer (m) |
| $A$ | area (m$^2$) |
| $Bi$ | Biot number (–) |
| $C$ | circumference (m) |
| $C_p$ | specific heat (J/kg.K) |
| $Fo$ | Fourier number (–) |
| $Q$ | heat flow (W) |
| $R$ | resistance to heat flow (K/W) |
| $\alpha$ | thermal diffusivity (m$^2$/s) |
| $\theta$ | temperature (°C or K) |
| H | reduced temperature (–) |
| $\lambda$ | thermal conductivity (W/m.K) |
| $\rho$ | density (kg/m$^3$) |

## 4.12 PROBLEMS

**(1)** (a) The inner wall surface of a cold store is to be maintained at −18°C whilst the outside wall surface temperature is at 22°C. The walls of the store are made of felted slab mineral wool sandwiched between sheets of 0.5 mm thick stainless steel.

Calculate the quantity of heat which would need to be removed from the store for the following thicknesses of mineral wool:

(i) 25 mm,
(ii) 50 mm,
(iii) 100 mm.

The total surface of the cold store is 50 m$^2$. The thermal conductivity of stainless steel is 26.4 J/m.s.K and that of felted slab mineral wool is 25.8 mJ/m.s.K.

(b) How would you decide the actual thickness of insulation to use?

(2) (a) Calculate the percentage reduction in the heat loss through the structure of an oven operating at 170°C on the inner wall surface (outside surface temperature is 20°C) if the thickness of insulation is increased from 12.5 mm to 25 mm. The total surface area of the oven is 4 $m^2$ and the thermal conductivity of the insulating material is 25 mJ/m.s.K.

(b) What is the value of this heat saving if the oven is operated for 4 h each morning and 6 h each evening, 6 days/week, 50 weeks/year and the cost of energy is 0.5 p/MJ?

(c) If the cost of providing the extra insulation was £100, would you be prepared to pay this extra sum? Discuss the factors involved in arriving at your decision.

(3) Calculate the effect on the rate of heat transfer of using 2500 studs per $m^2$ on a plane plate. The studs are each 5 mm diameter and 10 mm long, plate temperature = 150°C, combustion gases temperature = 1250°C, thermal conductivity of metal studs = 25 W/m.K, convective heat-transfer coefficient = 15 W/$m^2$.K (to studs and flat plate).

(4) Use Schmidt's method to calculate the time for a hamburger centre temperature to reach 80°C, 60°C and 40°C by both the graphical method and the tabulation method.

Data: Hamburgers 24 mm, 18 mm, 12 mm and 6 mm thick (divide into 6 sections), initial temperature is 0°C, outer surface is raised to 240°C, 220°C and 200°C instantaneously and held there (on both sides).

$$\lambda = 1.06 \text{ W/m.K}$$
$$C_p = 3.76 \text{ kJ/kg.K}$$
$$\rho = 960 \text{ kg/m}^3$$

List the assumptions made in the calculations and discuss the results.

(5) A 2 kg joint of beef, initially at 15°C, is palced in an oven at 175°C. Assuming the heat-transfer coefficient ($h$) is 15 W/$m^2$.K, estimate the time required for the centre of the joint to reach 95°C. What is the surface temperature at the end of cooking? (Use Dalgleish charts).

Data: $\lambda = 1.0 \text{ W/m.K}$,
$C_p = 4 \text{ kJ/kg.K}$,
$\rho = 1000 \text{ kg/m}^3$.

Assume the beef is spherical in shape ($V = \frac{4}{3}\pi r^3$).

Discuss the results.

(6) A tray of chicken *à la* king at 60°C is placed in a refrigerator, the ambient temperature of which is 4°C. The tray is 0.2 m wide × 0.3 m long × 0.1 m deep. The heat transfer coefficient in the refrigerator is 6 W/m$^2$.K. The thermal diffusivity of the food is 0.15 mm$^2$/s and its thermal conductivity is 0.52 W/m.K. What is the centre temperature of the food after 5 h in the refrigerator? (Use Dalgleish charts and consider the tray as a finite slab formed from three intersecting infinite slabs.)

Comment on the result.

## 4.13 REFERENCES

American Gas Association (1950), Research Report No. 1150.

American Gas Association (1958), Research Report No. 1284.

Basiulis, A. (1973), Trans. ASME, Paper No. 73-WA/HT-16P 11-15.

Corfield, D. B. (1975), *Activities Report*, **27** (1), 52-57.

Dagerskog, M. and Bengtsson, N. E. (1974), *Lebensm.-Wiss.u. Technol.*, **7**, 202-207.

Dalgleish, N. and Ede, A. J. (1965), Ministry of Technology, National Engineering Laboratory, NEL Report No. 192.

Dickerson, R. W. and Read, R. B. (1968), *Food Technology*, **22**, 37-39, 49, 51, 52.

Fischer, S. (1977), paper in Glew, G. (ed.), *Catering equipment and systems design*, Applied Science Publishers.

Gurney, H. P. and Lurie, J. (1923), *Ind. Eng. Chem.*, **15**, 1170 ff.

Herbert, B. and Irving, R. (1974), Home Economics, November, 12-14.

Hurley, J. R. and Miskolczy, G. (1975), Paper to 10th IECEC Proceedings.

Keniston, L. (1977), *HCIMA Jnl.*, December, 11, 13-15, 17.

Patton, T. C. (1944), *Ind. Eng. Chem.*, **36**, 990-996.

CHAPTER 5

# Convection

## 5.1 INTRODUCTION

Heat transfer by convection takes place because of the movement of a fluid. Part of the fluid may be heated up, and this warmer fluid moves and gives up its heat by mixing with cooler fluid. Conversely, part of the fluid may be cooled down; this cooler fluid moves and absorbs heat from the warmer fluid thereby cooling the latter down. If the fluid moves of its own volition, that is without assistance of stirring either by hand or mechanically, then we have **natural convection**. If the fluid moves as a result of manual or mechanical stirring then we have **forced convection**. In forced convection situations there is always a certain amount of natural convection but often the contribution of natural convection to the total heat transfer is so small that it may be ignored. This is strictly the meaning of forced convection. In some instances the contributions of natural convection and forced convection are both significant and both have to be considered.

Fluid movement in convection situations may be either **laminar** or **turbulent**. Natural convection can be so great that the natural circulation currents generate a tubulent flow. Forced convection can take place at both low and high flow rates, that is under both laminar and turbulent conditions. Hence convection situations may be classified as:

natural convection – laminar flow
natural convection – tubulent flow
forced convection – laminar flow
forced convection – turbulent flow.

It is also possible to have simultaneous forced and natural convection under both laminar and turbulent flow conditions. There is clearly a strong relationship between the fluid flow situation and convective heat transfer. This will become more apparent in the remainder of this chapter.

Convection takes place in all catering equipment where temperatures above or below room temperature occur, although it is more important in some particular instances, for example, forced-convection ovens and blast freezers, than others. Heat transfer during both boiling and condensation will also be considered in this chapter. These are both convection heat-transfer situations and can lead to very high heat-transfer rates.

## 5.2 THE THERMAL BOUNDARY LAYER

### 5.2.1 Heating solids with fluids

In order to heat or cool a fluid, the heat normally passes from/to a solid surface into/from the fluid. We know (Section 3.8) that the fluid moving near to a solid surface is slowed down because of its influence. This causes a boundary layer in fluid flow. There is a similar **thermal boundary layer** in convection heating/cooling. In this case there is a temperature change in the fluid near to the solid surface as shown in Fig. 5.1.

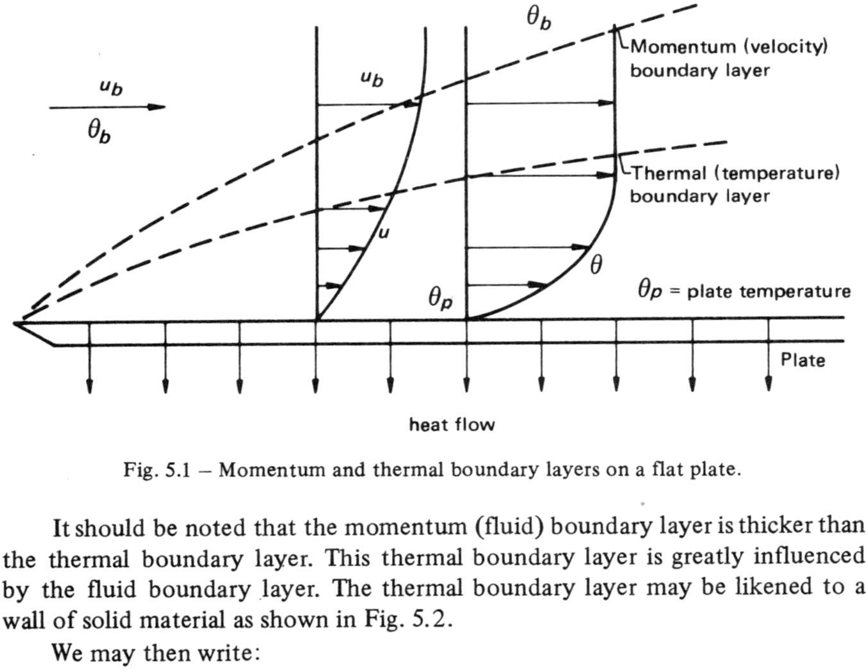


Fig. 5.1 – Momentum and thermal boundary layers on a flat plate.

It should be noted that the momentum (fluid) boundary layer is thicker than the thermal boundary layer. This thermal boundary layer is greatly influenced by the fluid boundary layer. The thermal boundary layer may be likened to a wall of solid material as shown in Fig. 5.2.

We may then write:

$$q = \frac{\lambda A(\theta_b - \theta_p)}{l}$$

where $\theta_b$ = 'bulk' temperature of the fluid (°C or K), $\theta_p$ = temperature of the plate (°C or K), $A$ = area of plate ($m^2$), $\lambda$ = 'thermal conductivity' of the boundary layer (W/m.K), $l$ = 'thickness' of the boundary layer (m).

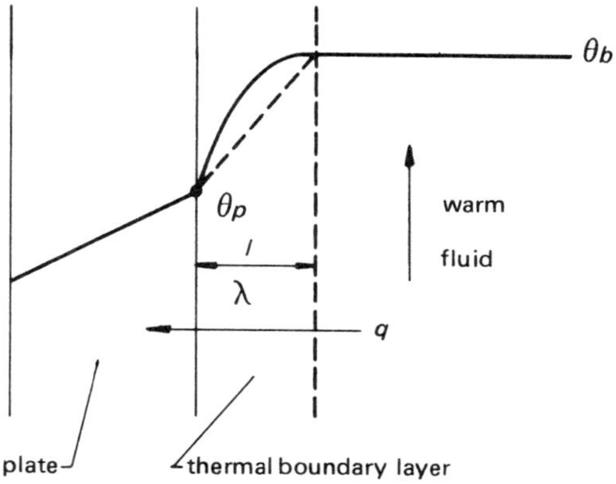


Fig. 5.2 – Boundary layer for a fluid heating a solid.

Because it is difficult to measure the thickness of the thermal boundary layer (they are difficult to define experimentally and they are often very thin) it is usual to write:

$$q = hA(\theta_b - \theta_p)$$

where $h = \lambda/l$ and is called the convective-heat-transfer coefficient (W/m$^2$.K).

Re-arranging:

$$q = \frac{(\theta_b - \theta_p)}{\left(\dfrac{1}{hA}\right)} = \frac{(\theta_b - \theta_p)}{R}$$

where $R = (1/hA)$. This is analogous to

$$q = \frac{(\theta_1 - \theta_2)}{\left(\dfrac{l}{\lambda A}\right)} = \frac{(\theta_1 - \theta_2)}{R}$$

where $R = (l/\lambda A)$ (Section 4.2). In other words we again have the analogy with Ohm's law:

$$\text{flow} = \frac{\text{driving force}}{\text{resistance}}$$

(Section 4.2).

If the flow is intially laminar then, with a smooth plate, it will remain laminar. A turbulent flow will first produce a laminar boundary layer which will give way to a turbulent boundary layer at a Reynolds number of 500 000. The Reynolds number in this case is defined as $(lu\rho/\mu)$ where $l$ is the distance from the edge of plate, provided the plate is smooth and the leading edge (starting edge) is rounded off. On the other hand, if the mainstream is disturbed before it reaches the surface, or the leading edge is abrupt, then most of the boundary layer will be turbulent. The transition from a laminar to a turbulent boundary layer in practice can take place over a wide range of Reynolds number. This was discussed for the fluid boundary layer in Section 3.6.

Fig. 5.3 shows the variation of the convective-heat-transfer coefficient ($h$) and the thickness of the thermal boundary layer ($y$) with distance ($l$) from the leading edge for a turbulent flow (velocity, $u$ = 3 m/s).

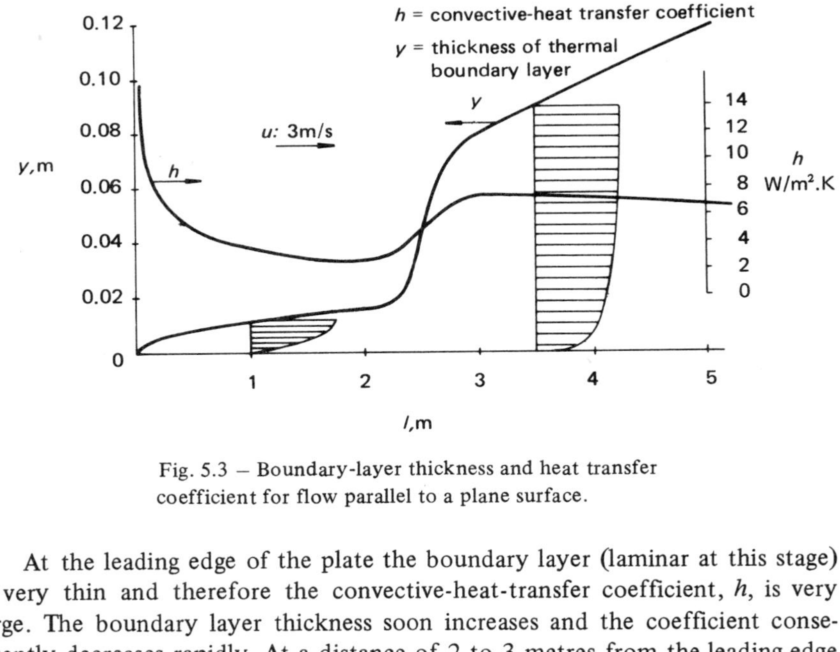


Fig. 5.3 – Boundary-layer thickness and heat transfer coefficient for flow parallel to a plane surface.

At the leading edge of the plate the boundary layer (laminar at this stage) is very thin and therefore the convective-heat-transfer coefficient, $h$, is very large. The boundary layer thickness soon increases and the coefficient consequently decreases rapidly. At a distance of 2 to 3 metres from the leading edge the flow changes from laminar to turbulent, the boundary layer thickness decreases (as a result of the turbulence) and the coefficient increases and thereafter remains approximately constant.

Fig. 5.3 indicates that, at a leading edge, the heat-transfer rate is very high and this could cause burning of food at such a position. This happens when chips are heated in a forced-convection oven: they burn, or brown relatively too much, at the sharp corners because of this effect.

If a solid food is heated by convection, the appropriate equation is, as we have seen:

$$q = \frac{\lambda A(\theta_b - \theta_p)}{l} .$$

Each term in this equation affects the rate of heat transfer ($q$). The thermal conductivity of the boundary layer ($\lambda$) is a property of the fluid and cannot be altered because the heating medium is fixed by other considerations. The area of food surface ($A$) exposed to the air should be as high as possible for high heat-transfer rates, although normally other factors such as portion/pack size and customer preferences affect it. For fast rates of heating the ratio of surface area to volume of the food should be high because this maximises the quantity of heat flowing into the food and, as was seen in Section 4.9, minimises the distance over which heat has to be conducted.

The bulk temperature of the fluid ($\theta_b$) is the oven setting and this may be increased to increase the rate of heat transfer. This will cause an increase in the food surface temperature (Section 4.7) which may be unacceptable. The reason for this increase in food-surface temperature is as follows. The rate at which heat arrives at the surface of the food by convection is balanced by the rate at which it leaves by conduction. If the rate of arrival increases, the rate of leaving must also increase. The rate of conduction can only increase if the conduction driving force is increased; this is caused by an increase in surface temperature. In turn, the rise in surface temperature will reduce the convection driving force ($\theta_b - \theta_p$), partially offsetting the advantage of the higher oven temperature. This need to balance the rates of heat transfer will cause a non-linear relationship between oven temperature and heating-up rate for the food.

The surface temperature of the food ($\theta_p$) should be low for high heat-transfer rates. At the start of heating this is normally the case but as heating proceeds the surface temperature will rise, having an adverse effect on the heat-transfer rate. Thus the rate of heat transfer by convection will fall during the heating period (other things being equal).

The thickness of the boundary layer ($l$) should be as small as possible to increase the heat-transfer rate. If this happens, the food surface temperature will rise. The food surface temperature is such as to balance the rate of heat arriving at the surface with the rate at which heat is conducted inwards into the food. If the convection rate increases, then the conduction rate will also need to increase to re-strike a balance between the two. This can only be achieved if the food-surface temperature rises to increase the driving force for conduction. The effect of this will also be to decrease ($\theta_b - \theta_p$) and hence the convection driving force will fall, thereby partially counter-acting the effect of the thinner boundary layer. This need to balance the heat-transfer rates will cause a non-linear relationship between boundary-layer thickness and heating-up rate for the food.

The boundary-layer thickness may be reduced by a number of methods including:

(i) increasing the velocity of fluid near the surface;
(ii) impinging the fluid directly onto the surface rather than allowing it to flow parallel to the surface;
(iii) physically removing the boundary layer by using scraper blades;
(iv) disrupting the boundary layer by allowing a liquid near the surface to boil;
(v) disrupting the boundary layer by allowing a vapour to condense on the surface.

The practical usefulness of these methods is discussed later in the chapter.

### 5.2.2 Heating liquids with fluids

Suppose now we are using hot combustion gases to heat oil in a deep-fat fryer. The gases are separated from the oil by means of a metal tube or wall, so the situation may be represented as in Fig. 5.4.

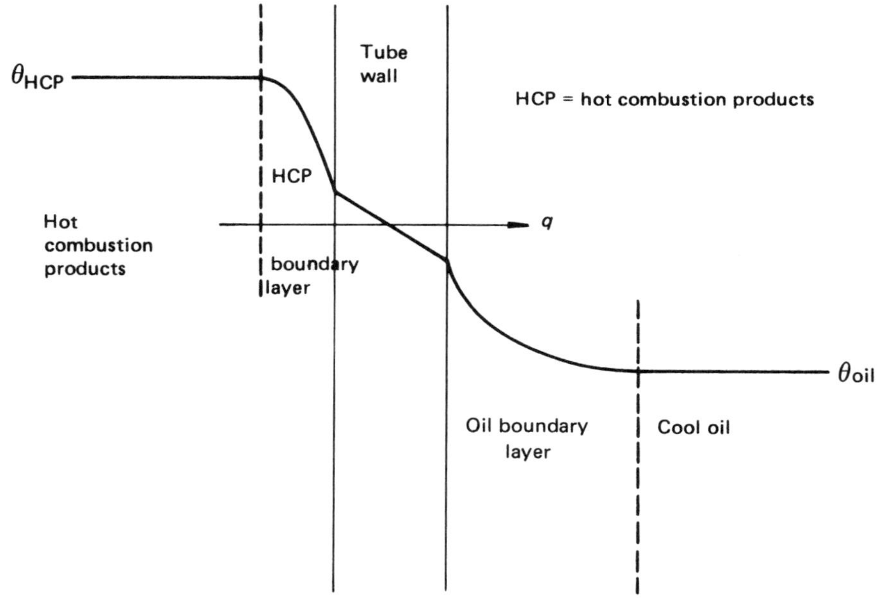


Fig. 5.4 – Boundary layers for a fluid heating another fluid.

This is analogous to a composite wall and we may write (Section 4.3):

$$q = \frac{(\theta_{HCP} - \theta_{OIL})}{R_{HCP} + R_{TUBE} + R_{OIL}}$$

where

$$R_{HCP} = \left(\frac{1}{h_{HCP}A}\right)$$

$$R_{TUBE} = \left(\frac{l}{\lambda A}\right)$$

$$R_{OIL} = \left(\frac{1}{h_{OIL}A}\right).$$

Here $h_{HCP}$ = convective coefficient for HCP, $A$ = surface area of tube, $l$ = thicknes of tube wall, $\lambda$ = thermal conductivity of tube material, $h_{OIL}$ = convective coefficient for oil, $\theta_{HCP}$ = temperature in bulk of hot combustion products, and $\theta_{OIL}$ = temperature in bulk of oil.

The overall resistance to heat flow is $R_{HCP} + R_{TUBE} + R_{OIL}$. The electrical analogy for this situation is illustrated in Fig. 5.5. This is the same as the composite wall situation of Section 4.3, except that two boundary layers are present (replacing two of the solid wall components). In most cases the resistance of the tube is negligible compared with the resistances of the two boundary layers.

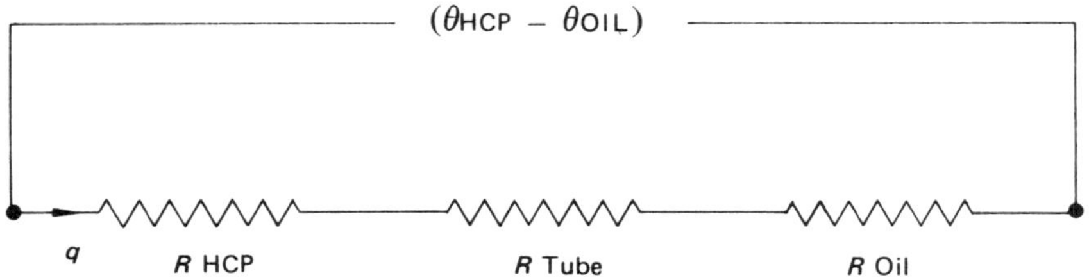


Fig. 5.5 – Electrical analogy for a fluid heating another fluid.

When a solid is heated by convection only one boundary layer exists. If we are heating a liquid with a fluid however, there are two boundary layers. Both of these offer a resistance to heat flow and either or both may be reduced in thickness by one of the methods listed at the end of the previous section. Which possibility do we aim at? The total resistance to heat flow is:

$$R_{HCP} + R_{TUBE} + R_{OIL} .$$

The metal is thin and (relatively) a good conductor, therefore $R_{TUBE} \approx 0$ compared with $R_{HCP}$ and $R_{OIL}$. For the other two resistances, three possibilities exist:

(i) $R_{HCP}$ is much greater than $R_{OIL}$: in this case we attempt to reduce $R_{HCP}$;
(ii) $R_{OIL}$ is much greater than $R_{HCP}$: in this case we attempt to reduce $R_{OIL}$;
(iii) $R_{HCP}$ and $R_{OIL}$ of similar magnitude: in these cases we attempt to reduce both.

In the first two possibilities, one of the boundary layers is so large compared with the other resistances that it controls the rate of heat transfer and is often referred to as the **controlling resistance**.

In the case of the two boundary layers the driving force for convective heat transfer was seen to be $(\theta_{HCP} - \theta_{OIL})$. If this difference is increased, then the heat-transfer rate will also increase. It should be noted however, that $\theta_{OIL}$ is not the highest temperature to which the oil is subjected. The oil in contact with the metal tube is at a higher temperature because of the boundary layer. This is important when some liquid foods are heated. If the food can be damaged by excessive temperatures, then it is the tube surface temperature which is important rather than the bulk temperature. Examples of this are excessive degradation of frying oil and 'burning' milk on the bottom of a pan. The liquid near the surface is continually exchanged with the bulk liquid and hence all the fluid is adversely affected by thermal degradation in the boundary layer. If water is heated this consideration is not important because water is not decomposed or degraded but boils. If the rate of heat transfer to boiling water is increased, rather than an increase in temperature taking place, it boils more rapidly with more water being lost to the kitchen. This results in excessive energy consumption and a steamy kitchen, neither of which are desirable. It does not increase the rate of heating of the potatoes, say, in the pan.

We have already mentioned that the metal is usually thin and of high thermal conductivity and has therefore negligible resistance. However, factors such as mechanical strength, inertness to the food and ease of manufacture tend to over-ride thermal considerations.

When heating liquids with fluids, the surface area for heat transfer between the two is governed by the need for a metal to contain the liquid. This area is not governed by customer considerations as when solid food is heated. It is therefore possible to increase the surface area for heat-transfer in this case. The use of studs and fins for this purpose was discussed in Section 4.5.

## 5.3 LOCAL AND AVERAGE CONVECTION COEFFICIENTS

If the reader refers back to Fig. 5.3 he will observe that the boundary layer increases in thickness with distance from the starting edge of the plate. Therefore at points near to the starting edge the resistance to heat-transfer is small ($h$ is high) whilst further away the resistance increases ($h$ is small). There is then a 'local' convective coefficient at each point on the plate. The value of this coefficient changes with distance from the starting edge. Usually, however, we are not concerned with these local values because we want to calculate the total heat transfer through the plate. In this case we use an average coefficient for the whole plate (which is of course derived from all the local values). It is important to be certain, when carrying out calculations using coefficients with data taken from the literature, which of the two are quoted.

At a certain distance from the starting edge of the plate, both the fluid-flow and thermal boundary layers reach certain thicknesses which then remain constant over the remainder of the plate. They are said to become 'fully developed'. It is important to distinguish areas where the boundary layers are developing from those where they are fully developed because, as just mentioned for a plate, the value of the coefficient is changing during the development of the thermal boundary layer.

An example of the change in local convective coefficient around a cylinder is shown in Fig. 5.6. Each heavy line in the diagram represents the variation in the coefficient around the cylinder for a particular velocity of the fluid approaching the cylinder. The value of the coefficient at any point is its radial distance from the cylinder surface. As the velocity increases, the coefficient becomes larger, that is, the line is further away from the cylinder surface.

It is clear from this figure that the boundary layer, in the case of a cylinder, never becomes 'fully established'. It also shows that local coefficients are often rather impracticable and in practice average coefficients are better. Nevertheless, the distribution of the local coefficient around a solid may explain certain observed effects. This is particularly true in catering situations where the equipment is small and edge/entry effects are usually important.

One further example of the variations in velocity ($\mu$), temperature ($\theta$) and convection coefficient ($h$) is given in Fig. 5.7. This is for the natural convection of heat from a hot vertical plate. In this case (hot plate, cool fluid, as on the outside surface of a cooking appliance) the warm air film produces flows upwards.

The outside of a refrigerated cabinet represents the case of a colder surface and warmer fluid. In this situation the fluid film flows downwards.

In Sections 4.3 and 4.4 we implied that the heat loss through a wall, say of an oven, was governed by the thermal conductivity of the wall materials only. It should be apparent now that this was an over-simplification. In practice, an oven contains a boundary layer of hot gases on its inside surfaces and also a boundary layer of fluid (air) on its outside surfaces. The resistance to heat flow of these two boundary layers is not inconsiderable, relative to the resistances of the solid materials of the oven wall. These boundary layers therefore offer added resistance to heat flow and so reduce the heat loss from the oven. In this case their presence is beneficial. The converse is true for the food inside the oven. Heat is being transferred by convection from the hot combustion products to the food surface. Unfortunately there is a boundary layer of these around the food offering a resistance to the flow of heat. This will then slow down the rate of heating-up of the food, which is a disadvantageous effect.

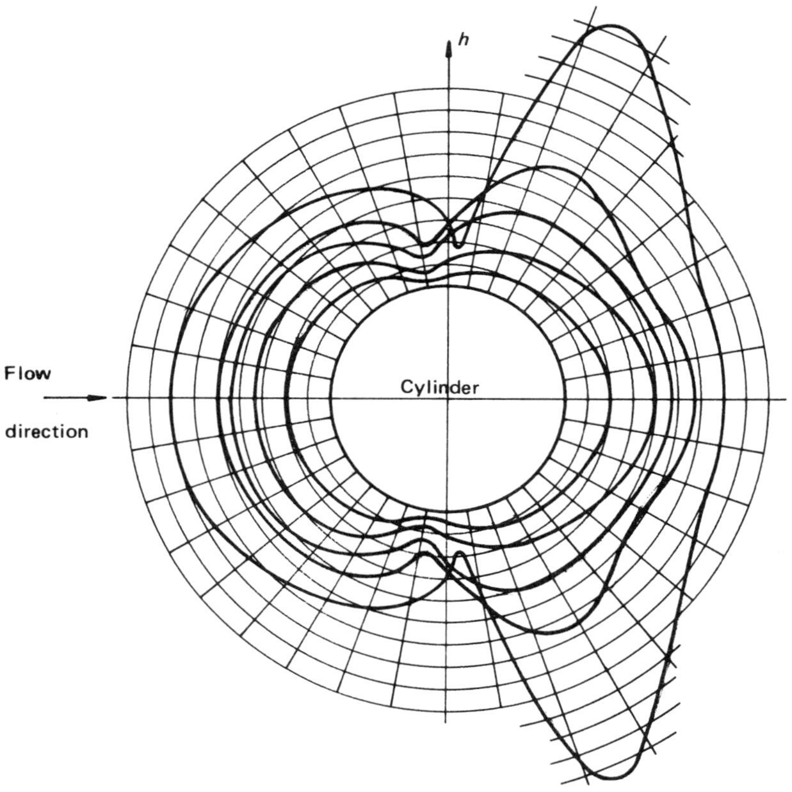


Fig. 5.6 – Variation of local convective coefficients around a cylinder.

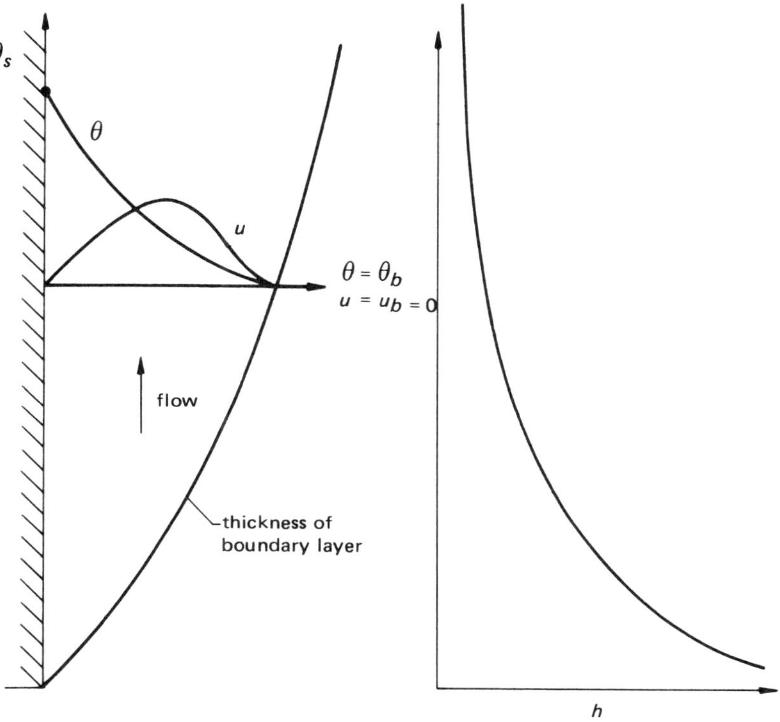


Fig. 5.7 – Variation of local convective coefficient on a vertical plate.

## 5.4 DIMENSIONAL ANALYSIS OF CONVECTION

In the previous section we introduced the concept of convective-heat-transfer coefficients but did not state how values of these could be obtained under any given set of circumstances. It has also been implied that there is a relationship between heat-transfer by convection and the fluid flow situation. In order to explore this relationship between fluid flow and convective heat transfer we need to resort to the use of dimensional analysis (as for fluid flow in Section 3.7).

### 5.4.1 Forced convection

For forced convection, experience tells us that the following variables would be important. As in Section 3.7, each variable is given a symbol and its dimensions expressed in terms of mass ($M$), length ($L$), time ($T$), temperature ($\theta$) and heat ($H$).

| Variable | Symbol | Dimensions |
|---|---|---|
| heat flow rate | $q$ | $H/L^2 T$ |
| characteristic length | $l$ | $L$ |
| temperature difference | $\Delta\theta$ | $\theta$ |
| fluid velocity | $u$ | $L/T$ |
| fluid denisty | $\rho$ | $M/L^3$ |
| fluid viscosity | $\mu$ | $M/L T$ |
| fluid specific heat | $C_p$ | $H/M\theta$ |
| fluid thermal conductivity | $\lambda$ | $H/L T\theta$ |

There are eight variables and five primary quantities ($H, L, \theta, T, M$) and therefore there are three dimensionless groups. We may now write:

$$[q]^1 [l]^\alpha [\Delta\theta]^\beta [u]^\gamma [\rho]^\delta [\mu]^\epsilon [C_p]^\iota [\lambda]^\kappa = [1]^\circ$$

where [ ] means 'the dimensions of'. Each variable is raised to some power ($\alpha$, $\beta$, $\gamma$, and so on) with one variable raised to the power of 1. If we insert the dimensions of each variable we get:

$$[HL^{-2}T^{-1}]^1 [L]^\alpha [\theta]^\beta [LT^{-1}]^\gamma [ML^{-3}]^\delta [ML^{-1}T^{-1}]^\epsilon [HM^{-1}\theta^{-1}]^\iota [HL^{-1}T^{-1}\theta^{-1}]^\kappa = [1]^\circ .$$

Equating the indices of the primary quantities gives:

condition on $H$: $1 + \iota + \kappa = 0$;
condition on $L$: $-2 + \alpha + \gamma - 3\delta - \epsilon - \kappa = 0$;
condition on $\theta$: $\beta - \iota - \kappa = 0$;
condition on $T$: $-1 - \gamma - \epsilon - \kappa = 0$;
condition on $M$: $\delta + \epsilon - \iota = 0$.

There are five equations and seven unknowns and therefore the equations can be solved in terms of two unknowns. Solving in terms of $\delta$ and $\iota$:

from $H$: $\kappa = -1 - \iota$;
from $M$: $\epsilon = -\delta + \iota$;
from $\theta$: $\beta = \iota + \kappa = \iota - 1 - \iota = -1$;
from $T$: $\gamma = -1 - \epsilon - \kappa = -1 + \delta - \iota + 1 + \iota = \delta$;
from $L$: 

$$\begin{aligned}\alpha &= 2 - \gamma + 3\delta + \epsilon + \kappa \\ &= 2 - \delta + 3\delta - \delta + \iota - 1 - \iota \\ &= 1 + \delta.\end{aligned}$$

Substituting back into the original equation and collecting up variables raised to the same power gives:

$$\left[\frac{lu\rho}{\mu}\right]^{\delta}\left[\frac{\mu C_p}{\lambda}\right]^{\iota}\left[\frac{ql}{\Delta\theta\,\lambda}\right]^{1} = [1]^{o}.$$

All the groups of variables in the square brackets are dimensionless and constitute dimensionless groups.

This equation may be written as:

$$\left(\frac{ql}{\Delta\theta\,\lambda}\right) = \phi''\left(\frac{lu\rho}{\mu},\ \frac{\mu C_p}{\lambda}\right)$$

| Nusselt number ($Nu$) | Reynolds number ($Re$) | Prandtl number ($Pr$) |
|---|---|---|

where $\phi''$ means 'a function of'. It is usual to write the Nusselt number as $hl/\lambda$ because $h = q/\Delta\theta$.

Hence the Nusselt number ($Nu$) contains the convective film coefficient, the Reynolds number ($Re$) characterises the flow situation, as we saw in Chapter 3, the Prandtl number ($Pr$) contains only terms which are properties of the fluid and hence characterises the fluid. It is therefore possible to carry out experiments using different flow situations (characterised by $Re$) and different fluids (characterised by $Pr$) and measure the effect of these on the Nusselt number ($Nu$). These may be correlated together. The equations obtained are often of the following type:

$$Nu = W(Re)^{X}(Pr)^{Y}$$

where $W$, $X$ and $Y$ are constants for any particular geometrical situation, that is, a cylinder, a sphere, flow inside a tube, flow across the outside of a tube and so on. Each geometrical shape is characterised by a 'characteristic dimension' ($l$) and it is this dimension that is used in calculating the Reynolds and Nusselt numbers. It is important to ascertain what the 'characteristic dimension' is when using correlations from the literature although usually the standard practice is fairly obvious, for example, the diameter of a cylinder is its characteristic length.

Thus, once experimental data become available such that $W$, $X$ and $Y$ can be determined for any particular geometry, there is no need to carry out any further experimental work. The correlation will apply to other fluids, flow rates and values of the Nusselt number for that particular geometry. The only limitation of this is that the correlations should only be used within the range of values of Nusselt, Reynolds and Prandtl numbers for which $W$, $X$ and $Y$ have been estimated. One should also note whether or not the values of $W$, $X$ and $Y$ apply to local or mean values of the convective coefficient, as noted in Section 5.3. Sources of convection correlations are given in Appendix C.

### 5.4.2 Natural convection

In natural convection, the variables taken into account are heat flow rate, characteristic length, temperature difference, fluid density, fluid viscosity, fluid specific heat and fluid thermal conductivity. These are all as in the case of forced convection but there is also a term $(\beta g)$ which is used instead of fluid velocity $(u)$. The reason for this is that in natural convection the **buoyancy** of the fluid is important rather than the velocity at which it moves. Fluid velocity is induced by external means whereas the 'buoyancy' is a result of the natural convection situation. The term $\beta$ is the coefficient of cubical expansion of the fluid (dimension $\theta^{-1}$) while $g$ is the accelaration caused by gravity (dimension $LT^{-2}$).

A dimensional analysis of natural convection gives rise to the following relationship:

$$Nu = W(Gr)^X (Pr)^Y$$

where the values of $W$, $X$ and $Y$ in natural convection are different to those in forced convection under similar circumstances. The dimensionless group, $Gr$, is the Grashof number and is defined as:

$$\frac{(\beta g)\Delta\theta\, l^3 \rho^2}{\mu^2}\,.$$

The Prandtl number is nearly the same (0.72 for air, 0.75 for carbon dioxide and 0.94 for water vapour) for all gases over a wide range of temperature and therefore a value for $(Pr)^Y$ is often included in the value of $W$.

## 5.5 BOILING AND CONDENSING FLUIDS

In catering, liquids are boiled as a means of either heating them for the customer or transferring heat to a solid object, say potatoes. We also use condensing steam as a means of transferring heat to foods in equipment such as steamers, boiling pans and *Bain Maries* (wet type).

### 5.5.1 Boiling

The best way to understand the phenomena of boiling is to make reference to the 'boiling curve' shown in Fig. 5.8. The heat transfer coefficient ($h$) is plotted against temperature excess, that is the temperature of the heating surface in excess of the temperature of the bulk of the water.

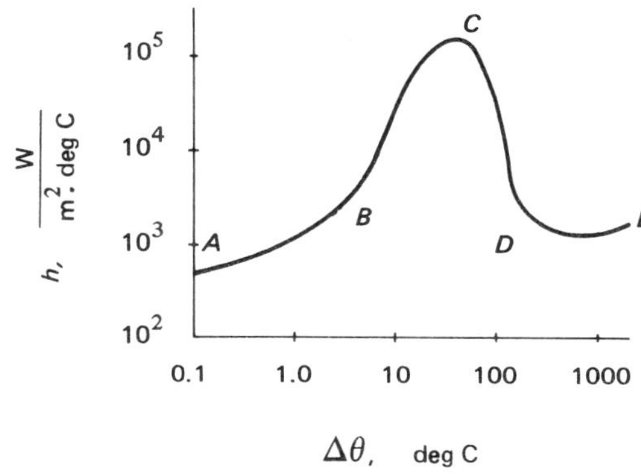


Fig. 5.8 – Boiling heat transfer coefficients.

Consider the heating of a pan of water on either a gas or electric boiling table. Heat is transferred by conduction through the pan bottom to the water. Essentially the pan bottom acts as a hot surface above the boiling point of the water. Initially the heat from the hot surface (pan bottom) heats the water and this heat is transferred to the water by natural convection. The water near the hot surface is slightly superheated (raised to a temperature just above the boiling point) and when it rises and reaches the surface by natural convection, it evaporates. This accounts for steam rising from the surface of the water before it appears to be boiling (bubbling). (Initially the water is de-superheated before it reaches the surface.) This first stage is shown as stage $A$ to $B$ in Fig. 5.8 and is called **interface evaporation** or the pre-boiling stage. In region $B$ to $C$, bubbles begin to form on the pan bottom; they break away from there when they are large enough and then are dissipated in the water before reaching the surface, that is, the steam condenses back to water. The collapse of the bubble is accompanied by a slight noise, the 'singing' of a kettle. This is the start of **nucleate boiling**.

Bubbles then form more rapidly so that they reach the surface and are dissipated as steam (towards $C$). If bubbles are formed so rapidly that they cover the whole of the pan bottom they form what is, in effect, a stėam boundary layer and this reduces the rate of heat transfer ($C$ to $D$). This region is not reached in catering equipment and, in any case, should be avoided from a heat-transfer point of view because of the insulating effect of the steam layer. Reference to Fig. 5.8 shows that the heat-transfer coefficient increases as the pan of water passes from $A$ to $B$ to $C$. $A$ to $B$ represents natural convection whilst $B$ to $C$ represents heat transfer under boiling conditions, the latter giving considerably higher heat-transfer rates than the former.

In nucleate boiling, the bubbles are created on the pan bottom by the expansion of air dissolved in the water or by steam bubbles formed on the small cavities in the surface of the pan. The bubbles grow, the size depending on the surface tension at the steam-water interface, the temperature and the pressure. Each bubble may do one of three things (depending on the temperature excess):

(i) it may collapse on the pan bottom (where it is being formed);
(ii) it may expand and detach itself from the pan bottom, only to be dissipated in the water above it, or,
(iii) it may expand and detach itself and then rise to the surface where it is dissipated as steam.

Heat-transfer rates are very high during nucleate boiling (on the water side) because of the intense agitation caused by the bubbles leaving the pan surface and also because heat is carried away from the surface as latent heat in the vapour and some or all of it is then released into the water above the surface.

The roughness and type of material of the pan base are very important factors in determining bubble formation. Increasing the pressure, as in a pressure cooker, increases the heat-transfer coefficient. The phenomenon described here, where the water is stationary except for stirring by bubbles, is termed **pool boiling**.

### 5.5.2 Condensation

If the temperature of superheated steam is reduced, it will become saturated and then **supersaturated**. This condition in unstable and condensation will occur. It is necessary to have a site where nucleation of droplets of water can take place and this is provided by the solid surfaces to which it is desired to transfer the heat in the steam. Once condensation starts it will continue at a high rate, provided the latent heat released is removed and that a plentiful supply of steam to the surface is maintained. If the condensate (water) collects on the surface, it forms a barrier (boundary layer) to further heat transfer and therefore it must be allowed to drain away quickly. The rate of condensation is limited by the rate at which condensate can drain from the surface. The rate of drainage depends on the slope and size of the surface.

A further factor is the mode of condensation. During **filmwise** condensation the water condensate forms a film over the whole surface whilst during **dropwise** condensation, drops of water form on the solid surface. Dropwise condensation gives much higher heat-transfer coefficients than filmwise condensation but it is very difficult to achieve in practice. Therefore filmwise condensation is normally used. Heat transfer by condensation occurs when cold food is placed in a steamer. Once the temperature of the food surface rises above the condensation temperature however, condensation ceases and heat transfer takes place by convection (from the steam).

Several other factors affect heat transfer during condensation. If the temperature of the surface is hotter than the saturation temperature of the steam (boiling point) no condensation will occur and heat transfer will be by normal convection. This will result in very much lower heat-transfer coefficients than if condensation were occurring. Air present in the steam (called 'non-condenable' gas) will accumulate near the heat-transfer surface. If this air is not 'bled off' and a film of air forms on the surface, heat transfer can be retarded quite significantly. This air is always present when catering equipment is started up. Steam also always contains air, as the water used in the production of steam contains dissolved air. The temperature at which condensation occurs depends on the saturation temperature, which is governed by the pressure of the steam.

## 5.6 APPLICATIONS OF CONVECTION HEATING

Summarising the earlier discussion of this chapter, the main control over convection heating rates were seen to be increasing the fluid temperature, thereby increasing the driving force and reducing the resistance to heat-transfer by decreasing the thickness of the boundary layer. It is evident how the first of these may be achieved, but how do we achieve a reduction in the thickness of the boundary layer? The earlier sections of this chapter and Section 3.9 (impinging jets) suggest the following possibilities:

(a) increasing the velocity of the fluid (or Reynolds number),
(b) using impinging jets to break up the flow along a flat surface;
(c) physically removing the boundary layer by means of a scraper blade (for viscous liquids);
(d) breaking up a liquid film on a surface by heating the liquid to such an extent that it boils, the bubbles of vapour breaking up the liquid film;
(e) breaking up a gas boundary layer on a surface by allowing gas (strictly vapour) to condense on the surface, the liquid drops breaking up the gas film.

We will now discuss how these possibilities are exploited in catering-equipment design.

### 5.6.1 Forced-convection ovens

Forced-convection-oven design makes use of the effect that increasing the fluid velocity reduces the thickness of the boundary layer. The flow situations in the two types of convection oven were described briefly in Section 3.9. An example of a convection oven fitted with an interior fan is shown in Fig. 5.9.

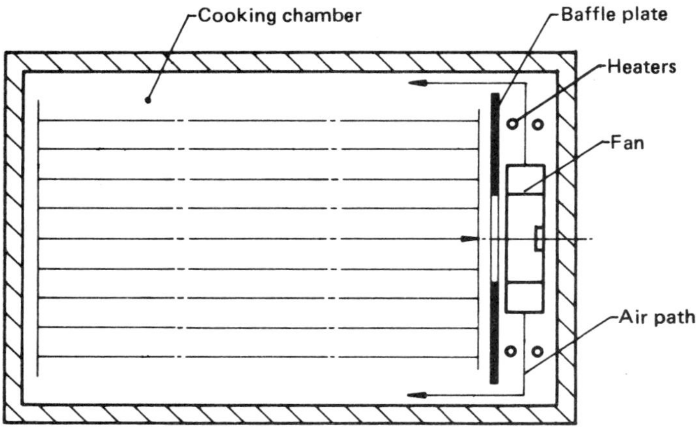


Fig. 5.9 – Forced-convection oven with interior fan.

The fan is situated at one end of the cooking space behind a baffle plate. Air enters the fan through a circular hole in the baffle plate and is delivered to the cooking space around its four sides. Electric heating elements are situated behind the plate so that the air heats up as it passes over them and before it reaches the cooking space. Circulation of the air in this way increases the velocity of the air as it passes over the food. This reduces the thickness of the boundary layer and hence increases the rate at which convected heat arrives at the food surface. In turn this increases the rate of heating-up of the food.

Convection ovens fitted with fans external to the oven space require a system of ducts to direct the air to and from the cooking space. One possible arrangement is shown in Fig. 5.10. The ducts must be designed to produce an even flow (one with equal velocity) over the food from the top to the bottom of the oven otherwise uneven heating rates will occur. There is also more air flowing in the vertical ducts at the bottom than at the top, hence the reduction in duct width from top to bottom. Heating elements are placed in the air passage after the fan to heat up the air.

Gas-fired convection ovens are of different design to the electrically-heated type because of the need to incorporate stable flames. They may be of the externally-heated type (the indirect type) in which the hot combustion products from the flame pass over pipes whilst the air is drawn through the pipes into the oven space. In this way heat is exchanged from the hot combustion products to the cooler air. As this heat-exchange process involves two boundary layers (hot combustion products and air), there is therefore a loss of efficiency because of the layer of metal separating the hot combustion products and the circulating air in the oven. The other type of gas-fired convection oven avoids this heat-exchange process by introducing the hot combustion products directly into the circulating air by means of a throat or venturi as shown in Fig. 5.11.

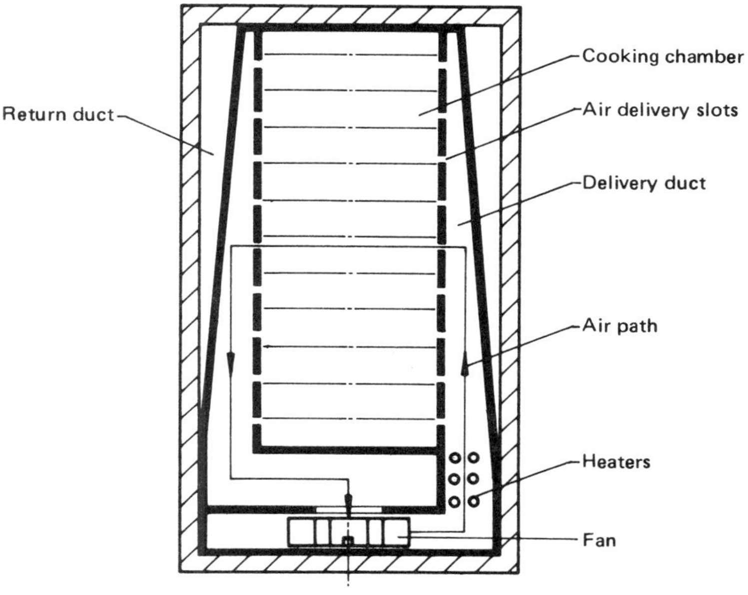


Fig. 5.10 – Forced-convection oven with external fan.

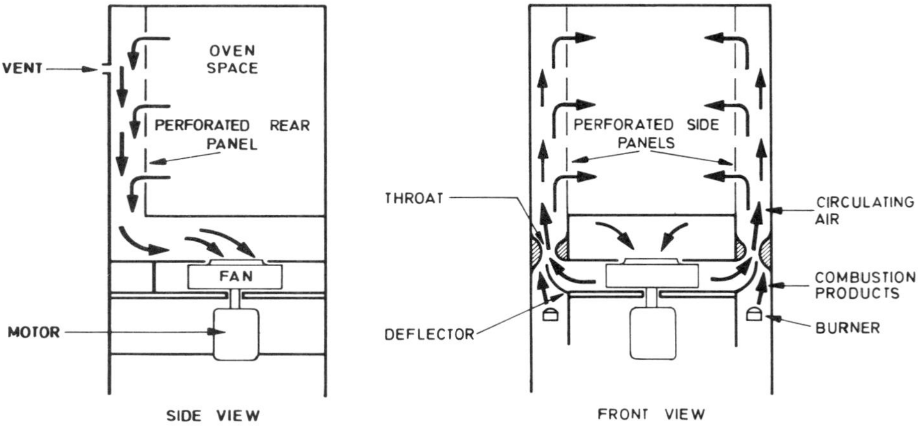


Fig. 5.11 – Directly heated, gas forced-convection oven.

This design eliminates the heat-exchange stage, is more efficient than the externally-heated type and heats up more quickly.

The detailed design of convection ovens varies very much from manufacturer to manufacturer as each has his own ideas about how the air flow is to be controlled. Successful design is as much an art as a science at the present because we do not have the knowledge to design the oven from first principles or, if we did, it would involve complex computer calculations which would be extremely costly. The design of the air flow and heating system is in itself more complicated than calculating the heat-transfer from the moving air to the food. In general the flow of air in many designs of convection oven approaches that of the flow of air over a flat plate.

Forced-convection ovens give a more even distibution of temperature than conventional ovens because of the use of battle plates in them to distribute the air flow more evenly. They are used for prime cooking and for end-cooking/re-heating of frozen foods. A wide range of sizes is available.

The American Gas Association has published a series of studies on forced-convection ovens: Hampel and Neu (1956), Hampel (1960), Spinell (1960 and 1963) and Perry and Stas (1967). The work reported is largely concerned with the empirical aspects of design and cooking tests rather than the application and verification of heat-transfer theory.

Spinell (1963) used the chart method (similar to the Dalgleish charts but of more limited range) to predict the heating times of a pack of frozen escalloped chicken and noodles. The lower specific heat of the frozen material and the latent heat required to melt the product were ignored, the specific heat above the freezing point being used in the calculations. An assumed heat-transfer coefficient of 21.0 W/m$^2$.K was used. The specific heat of the food was calculated on the basis of the recipe and the thermal conductivity (0.55 W/m.K) was estimated from that of water (0.64) and raw beef (0.53). The density of the food was determined experimentally. The food was initially at $-15^\circ$C. The following comparison between calculated and measured times was obtained.

| Heating to centre temperature of | Calculated time | Experimental time |
|---|---|---|
| 60°C | 16.8 minutes | 18.7 minutes |
| 71°C | 19.1 minutes | 20.0 minutes |
| 82°C | 22.1 minutes | 21.7 minutes |

Changing the thickness of the pack produced the following results for a final centre temperature of 82°C.

| Pack thickness | Calculated time | Experimental time |
|---|---|---|
| 19.0 mm | 15.4 minutes | 14.1 minutes |
| 25.4 mm | 22.1 minutes | 21.7 minutes |
| 31.8 mm | 29.0 minutes | 28.9 minutes |
| 38.1 mm | 35.2 minutes | 36.0 minutes |

The above calculations assumed the 127 mm long by 102 mm wide by 25 mm thick slab was infinite. Assuming is was made from three infinite intersecting slabs, the heating time was shown to be only slightly less than that obtained by making the infinite slab assumption. In view of the assumptions made, agreement between calculated and experimental times recorded above are very good.

Walker and Glew (1969) used Dalgleish charts to compare calculated and experimental heating-up times for frozen potato packs. The packs were heated from 0°C to 80°C. The thermal diffusivity of potato was calculated from published data and the convective-heat-transfer coefficient was determined experimentally (14.2 W/m$^2$.K) from the evaporation rate of water contained in half-full lidded aluminium trays in the oven. Their data are reproduced in Fig. 5.12.

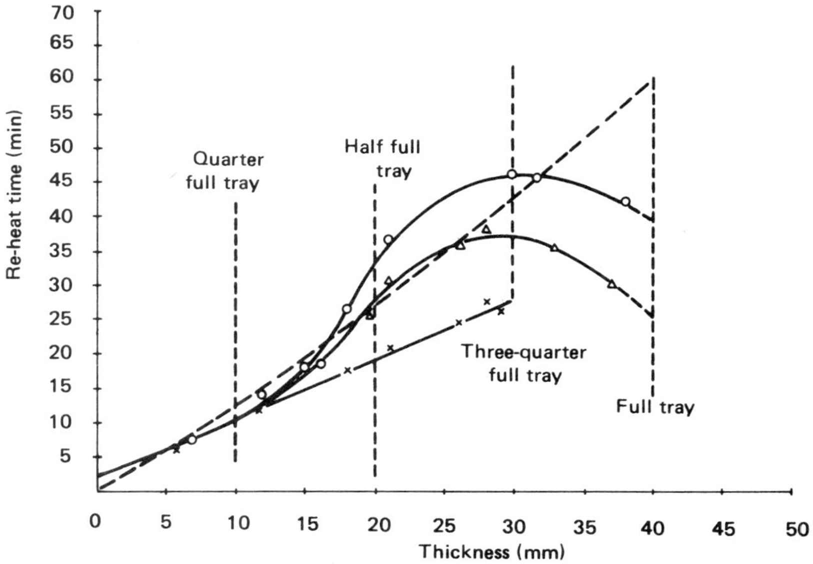


Fig. 5.12 – Variation of food reheat time with pack thickness. X = water, Δ = thin potato mix, O = thick potato mix. The thick broken curve indicates the theoretical curve for thick potato mix.

They considered that for depths of potato around the half-full mark the theoretical curve was in quite good agreement with the experimental times considering the assumptions made. They postulated that for fuller trays, the difference between experimental and calculated results was caused by the air space within the tray being much thinner so that the heat transferred through the air space approached that by conduction alone. Consequently the convective coefficient is reduced and calculated times are then too short. For full or almost full trays, the potato touched the lid increasing the value of the convective coefficient so that calculated times were too long.

Dorney and Glew (1974) tested eleven forced-convection ovens. They measured the air velocity across the product containers with the ovens fully loaded. There were large differences between the highest and lowest velocities occuring in each oven: the difference varied from 73 to 206% of mean velocity over the eleven ovens. Corresponding figures for temperature were 9 to 35%. The velocity data in particular show the difficulties of design. They plotted mean reheat times (tests carried out under identical conditions) against mean air velocity for each oven (Fig. 5.13).

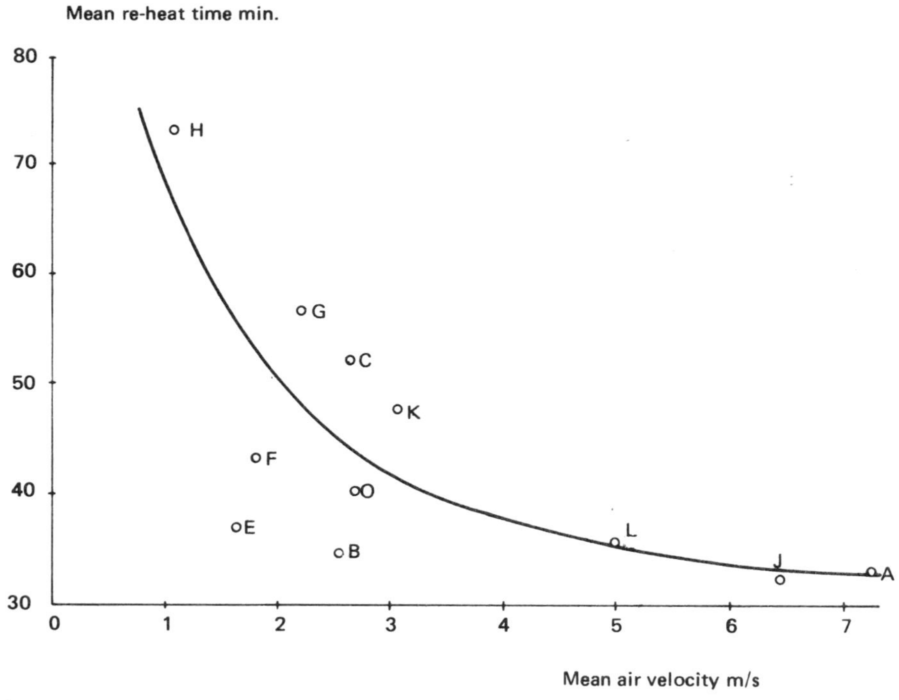


Fig. 5.13 – Mean reheat time v mean air velocity.

These results confirm the advantage of higher air velocities for shorter heating times. They also demonstrated that higher temperature differences are caused, to some extent at least, by high velocity variations. The use of Dalgleish charts has been illustrated by Milson and Kirk (1977). They demonstrated the effect of increasing oven-air temperature and convective-heat-transfer coefficient on the time a food pack takes to reach a centre-pack temperature of 80°C and the corresponding surface temperature (Table 5.1).

**Table 5.1 – Predictions for a forced-convection oven.**

| Oven air temperature (°C) | Convective-heat-transfer coefficient ($W/m^2.K$) | Time to reach centre-pack temperature of 80°C (minutes) | Surface temperature of pack when centre temperature reaches 80°C (°C) |
|---|---|---|---|
| 180 | 5 | 83 | 83 |
| 180 | 15 | 32 | 88 |
| 180 | 30 | 23 | 110 |
| 150 | 15 | 42 | 87 |
| 180 | 15 | 32 | 88 |
| 210 | 15 | 26 | 90 |
| 240 | 15 | 22 | 96 |

These data show clearly the advantages of using high convective coefficients (high air velocities) and high oven-air temperatures. The relationships between these two variables and heating time are not linear and a law of diminishing returns operates. Surface temperature also increases as heating time diminishes and in one case in the table it exceeds 100°C. This would not happen in practice because once its temperature reaches 100°C, moisture would start to evaporate from the food suface and the temperature would remain at 100°C whilst moisture is present at the surface (see Chapter 10 for further discussion of this). The reason for this 'false' prediction is that the charts used do not allow for the possibility of a phase change, such as the evaporation of water. A similar situation could arise if air below 0°C is used to cool food and the surface begins to freeze. Care needs to be exercised therefore when using these charts; the user must be conscious of the assumptions made in deriving them. The above predictions show that there are limits to oven-air temperature and convective-heat-transfer rate above which the food may be damaged from the point of view of the customer. These limits will depend on the thermal properties of the food and its size and shape.

### 5.6.2 Liquid heating

Equipment used to heat liquids in catering may be classified under a number of headings.

#### 5.6.2.1 *Liquid in a container which is independent of the heat source*

**Open-ring gas burners** allow the flame to impinge on the under-side of the container (usually a pan) holding the liquid. Three types of equipment employ such burners and these are illustrated in Figs. 5.14, 5.15 and 5.16.

Fig. 5.14 – Open-top boiling table fitted above an oven as part of a range.

Fig. 5.15 – Open-top boiling table independent of an oven.

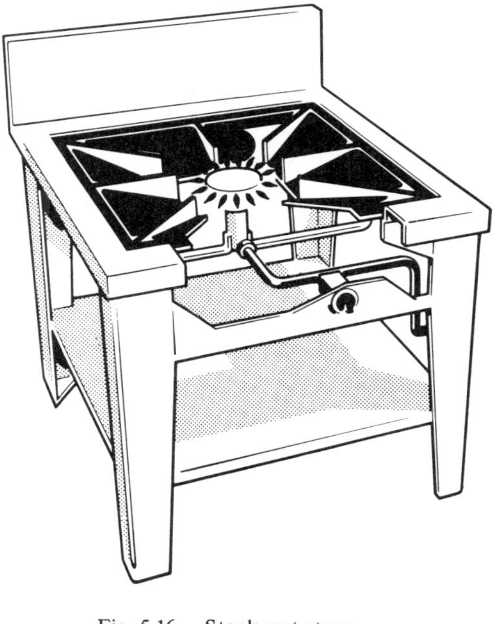

Fig. 5.16 – Stock-pot stove.

The open-ring gas burner(s) are mounted in a metal framework, which also includes pan supports. Spillage trays are fitted under the burners in the first two examples illustrated. Control of liquid heating is by means of a gas tap, manually operated.

The heat-transfer situations with this type of liquid heating involve:

(i) convective heat transfer from the flame to the base and sides of the container (outer surface);
(ii) conduction through the container base and walls;
(iii) convection from the container base (inner surface) to the liquid food.

If a solid is being heated by the liquid, for example boiling potatoes, or a liquid containing solid particles is being heated, for example heating a vegetable soup, then two additional heat transfer situations are involved:

(iv) convection from liquid to surface of solid food (particles);
(v) unsteady-state conduction from surface of solid food (particles) inwards.

It is evident that there are many resistances to the heat-transfer and hence we would not expect this type of equipment to be very efficient.

The American Gas Association (1955) tested a (then) contemporary open-top range and showed how heating-up times of stock-pots could be reduced by using higher output burners, with negligible loss in efficiency, satisfactory combustion and satisfactory control. The tests carried out showed that as the diameter of a stock-pot on a burner increases, so does the thermal efficiency, until a maximum is reached after which a further increase in diameter leads to a falling off in efficiency. The depth of water in a pot does not, apparently, affect thermal efficiency.

These observations indicate clearly the importance of adjusting the size of the flame to the size of the stock-pot (pan) being heated if maximum thermal efficiency is to be attained. The use of relatively large flames with small pans is, thermally, inefficient. Further work used a booster burner (air supplied with a fan) with an open-top range to reduce heating-up time by as much as 49% with a thermal efficiency of 42–45%, as reported by Marn and Rowland (1966).

Chigier and Milson (1973) have reviewed the research on the heat transfer caused by flames impinging on plates. This is the same as, or approximates to, a pan heated on an open-ring burner. The effect of burner nozzle to plate distance on the distribution of heat transfer rate (flux) is illustrated in Fig. 5.17, taken from work by Buhr (1969).

This diagram shows that if the burner to plate distance is sufficiently small ($h/d < 29.5$ in this case, $h$ = burner nozzle to plate distance, $d$ = burner diameter) then the maximum heat flux does not occur on the burner axis. The reviewers interpreted this to mean that if maximum heat-transfer rates and maximum heat fluxes are to be achieved, complete or nearly complete combustion

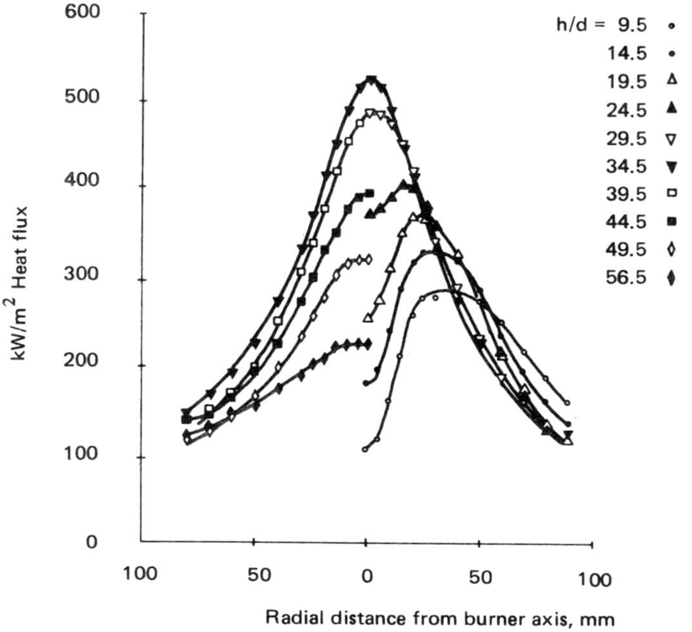


Fig. 5.17 – Distribution of heat flux through a water-cooled plate. Town-gas-air flame, $d$ = 6 mm, $Re$ = 15 700, air/gas 1.3/1.

should be achieved before the flame impinges on the surface. Higher heat-transfer rates are achieved by using burners with higher gas/air exit velocities (as would be expected).

Keniston (1977) has drawn attention to the excessive waste of energy apparent with gas-fired open-top ranges. A simple mechanical sensor (see Chapter 11) was devised which would shut off the gas supply to a burner 30 seconds after a pan has been taken off. This time delay allows the cook to change a pan or agitate a container away from the flame without the need to restart the burner.

**Electrically-heated solid tops** have elements situated under a metal plate. The solid top is supported on a framework and is itself strong enough to support the pan. Several sizes are available and they are used in appropriate combinations, often but not always mounted above an oven. Control of each is by means of an appropriate, manually-operated controller (Chapter 11). An alternative to this is a flattened electrical element, formed in the shape of a spiral. These are only suitable for light applications.

Heat transfer in this case is similar to that with open-ring gas burners except that the convective-heat-transfer resistance between the flame and the base of the container is eliminated. In its place however there is a 'contact resistance' to heat-transfer between the solid top and the base of the pan, Hampel (1960) has discussed this. If two metal surfaces in contact are examined with a sufficiently large magnification, each surface appears to consist of peaks and valleys with occasional points of contact when peaks on each surface coincide. Between these points of contact is an air film which effectively acts as the controlling resistance to heat transfer (rather than conduction through the areas of metal contact). Hence this film between the plates in contact acts as a barrier to heat transfer.

**Gas-fired solid tops** (Fig. 5.18) have a flame situated underneath the centre of a solid top (with a small hole in the middle). The location of the flame under the centre gives a high temperature at the centre (bull's eye) for rapid heating with lower temperatures away from the centre for simmering. The rate of heating is controlled by the actual location of the pan on the solid top rather than by the manual-operation of a gas tap.

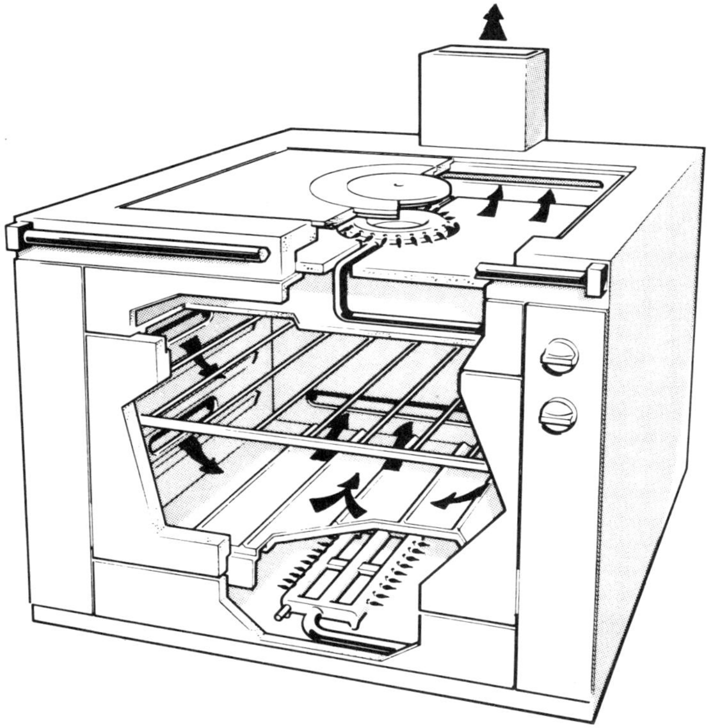

Fig. 5.18 – Solid top fitted to a range.

An evenly-heated version of a gas-fired solid top is also a possibility. From a heat-transfer point of view, a gas-fired solid top may be considered as either an open-top together with a contact resistance (between solid top and container) and conduction through the solid top or as an electric solid top plus a convection film between the gases and the base of the solid top. Either way, there are more resistances to heat-transfer than by the other two methods.

Jacobson (1956) investigated the design of gas-fired solid tops with uniform heat distribution. He showed that it was possible to double the heat input rates of (then) contemporary units, although a fall in thermal efficiency accompanied this. This latter loss was felt to be justified in view of the decrease in heating-up time. An experiment was set up to determine the optimum heat-input rate and the corresponding flue area. The size of the flue was shown to have an effect on thermal efficiency and heating-up time. Jacobson (1956) also investigated the design of gas-fired solid tops with a non-uniform heat distribution. He showed that it was possible to decrease heating-up times to one half of the (then) contemporary value by increasing the heat input and without affecting the 60% hot-start thermal efficiency. Again flue size was shown to affect efficiency.

In recent years continuous boilers have been developed (Fischer, 1977). Two of the types available are illustrated in Figs. 5.19 and 5.20. The former conveys the food in buckets whilst the latter uses blades to carry it through the water. Cooking times can be varied over a wide range and these machines handle many kinds of products.

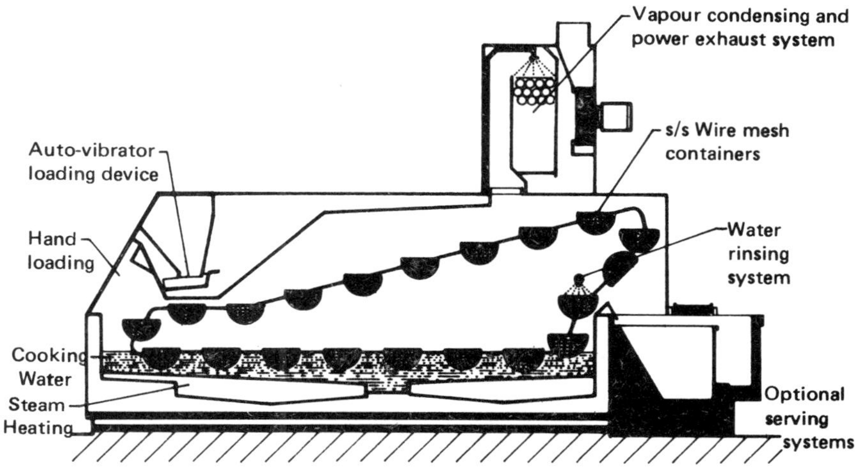


Fig. 5.19 – Continuous boiler: food conveyed in buckets.

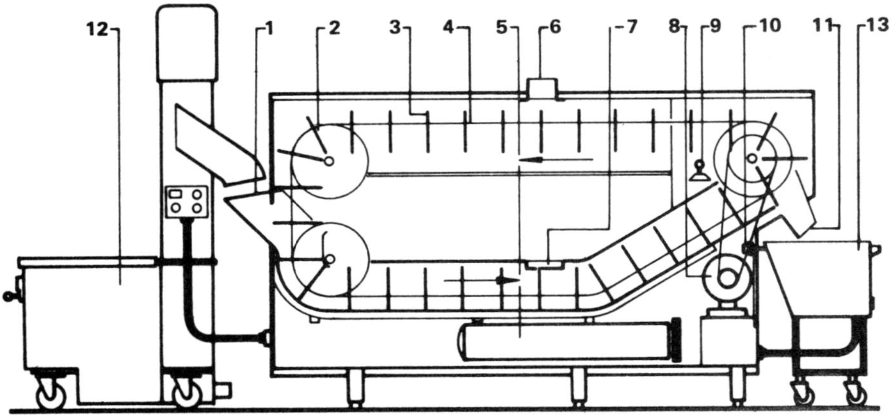


1 Filling funnel
2 Chain wheel
3 Slats
4 Transport chains
5 Counter stream (calorifier) unit
6 Vapour discharge ducts
7 Overflow
8 Drive motor
9 Spray device
10 Safety switch
11 Discharge
12 Automatic loading device/ universal conveyor
13 Heated standard distribution trolley

Fig. 5.20 – Continuous boiler: food conveyed by blades.

**5.6.2.2** *Liquid in a container which has a heat source incorporated into its design*
Boiling pans make up this category of liquid-heating device. Several types are made. Directly heated or single pan units are heated by a gas flame, electrical elements or steam. In this case the heat is applied directly to the container holding the liquid (although the flame does not impinge directly on the metal surface). Viscous liquids can easily be damaged by this method of heating because the container temperature may be quite high if control is inadequate and 'hot spots' occur.

In the case of indirectly heated or jacketed pans (Fig. 5.21), heat is applied to a metal surface which is separated from the container surface by a gap filled with water (gas and electric heating only). The water is sealed in this jacket and is under partial vacuum. Because the heat is first transferred to water and is then transferred to the food, hot spots may be eliminated. Steam is formed in the jacket. The water may be contained in an unsealed jacket instead. An alternative electric version of this type has elements mounted on a plate about 15–20 cm below the containing pan.

Directly heated pans fitted with a removeable inner pan (porringer) also enable liquids to be heated indirectly (Fig. 5.22).

Taps are fitted to allow the boiling pan to be filled with water and with the larger sizes a drain tap is also incorporated for ease of emptying. Some

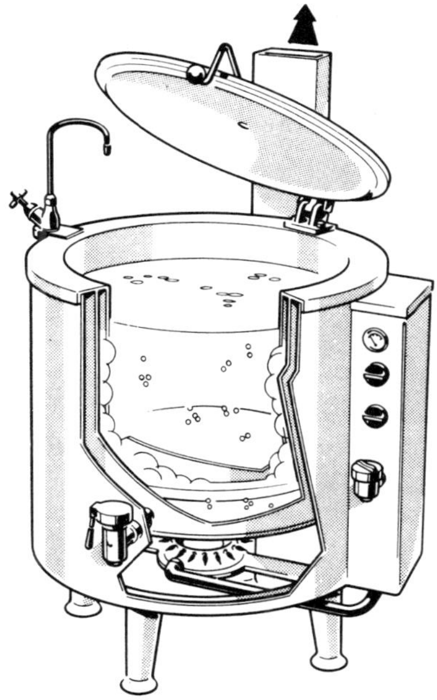

Fig. 5.21 – Boiling pan (jacketed).

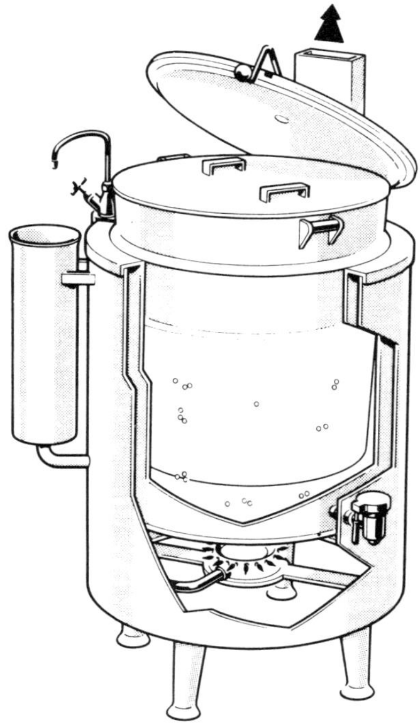

Fig. 5.22 – Boiling pan (dual purpose).

boiling pans are designed to tilt for emptying. The jacketed pan is used for heating custards, milk puddings, porridge, and other viscous liquids that burn easily, whilst the directly-heated pan is used for cooking root vegetables and green-stuffs, boiling meat and making soups. No research/design studies on boiling pans have been reported in the literature.

The heat-transfer processes involved are conduction through the metal container walls and convection on either side of these walls (except where electric elements are used for heating). If heating is by steam, then the gas boundary layer is broken up by allowing the steam to condense on the vertical surface of the pan. This breaking-up of the boundary layer increases the rate of heat transfer considerably. If the steam does not condense, a (steam) boundary layer with a high resistance to heat flow is formed in the normal way. However, if the steam condenses, a layer of liquid water forms on the surface, and considerable quantities of heat are released at the wall during condensation. The main design requirement is that the water drains away from the surface as quickly as possible, otherwise the hot water will form its own boundary layer. This method can therefore only be used on vertical surfaces where natural drainage can take place. The use of steam is advantageous in that its temperature depends only on its pressure. Hot spots on the pan wall do not occur because all the steam is at the same pressure, and hence the same temperature.

Two major problems exist when viscous liquids are heated in boiling pans. First, high local temperatures on the metal heat-transfer surface may damage the food by subjecting the layer of food near to the surface to too high a temperature. Then, with very viscous liquids, the liquid in the heated boundary layer will only move very sluggishly. Therefore the mixing of warmed liquid near the wall with the bulk liquid which is cooler is a relatively slow process.

In small-scale cooking, these problems are overcome by the semi-continuous, manual stirring of the liquid. This mixes the boundary layer liquid with the bulk of the fluid, but it is not feasible for large-scale operations except on an occasional basis. If the liquid is not too viscous continuous mechanical agitation would be an improvement on manual stirring, but with additional capital costs for the stirrer and motor added to the cost of the equipment. With more viscous liquids, more drastic action is required. The method used to disturb the boundary layer in this case is to scrape physically the boundary layer off the surface and mix it back into the bulk fluid. This may be achieved in one of two ways. First, the traditional boiling pan can be fitted with scraper blades moving over the inside base of the pan, with heating from the underside of the base. The scraper blades must touch the surface for the process to be successful, otherwise a thin layer of liquid will be left behind.

The second method is used for cooling sauces in, say, cook-freeze installations and could also be used for heating purposes. This method uses scraper blades inside a vertical cylinder. Heat is transferred though the walls of the cylinder and the liquid to be cooled/heated is continuously pumped through the cylinder. This equipment is costly in relation to other catering equipment and for small volumes cleaning is excessive. However, for large throughputs it is probably the only way of achieving sufficiently rapid cooling.

A further problem with boiling pans is the large surface area of food, causing a high heat loss to the atmosphere by steam evaporation and convection from the free surface. This heat loss may be reduced by placing a lid on the pan, thereby reducing the evaporation loss and convection (the steam condenses on the lid and returns to the bulk of the liquid and the air between the food surface and lid acts, to some extent, as an insulating layer). A further means of reducing this heat loss is to place one or more layers of hollow plastic balls (whose softening point is at least 120 to 145°C) on top of the liquid food. Tests carried out on an open tank of water at 90°C, insulated on the sides and bottom, showed that the amount of heat required to maintain this temperature was reduced by 69.5% by using one layer of such balls and by 75.5% by using a double layer of them (Capricorn Industrial Services Ltd., undated). At 80°C, evaporation of water is reduced by 88% with one layer of balls and 89% with two layers. They also reduce splashing and smells, fumes and steam emissions. They have been employed successfully in a water boiler used for making tea and with heated water tanks used for reconstituting prepared foods in pouches. However, caterers do not seem to have made widespread use of this development to date.

### 5.6.2.3 *Water boilers*

Boiling water is required for making tea and coffee (independently of the boiler) and steam is required for heating and mixing beverages. Coffee-making equipment where the production of hot water and then coffee are integrated into a single unit also uses hot water.

**Bulk boilers** (Fig. 5.23) are the simplest type of boiler, and are available with a range of features. The simple urn is simply a pan of water heated by immersion heaters or with a series of small gas jets in an enclosed jacket underneath the water pan. A tap is fitted to enable the water to be drawn off and a lid permits manual filling. A slightly more elaborate type is insulated and fitted with a gauge glass and thermometer, so that it is possible to see (by looking at the gauge glass) how full the boiler is.

Jacketed urns are available for heating milk. In this case a water jacket surrounds the milk, as so to prevent burning. Large boilers connected to the water mains and with overflow connections are also available. The American Gas Association (1948) published a study of factors affecting the design of coffee urns, with particular reference to ventilation, air supply, burner location, burner design and controls.

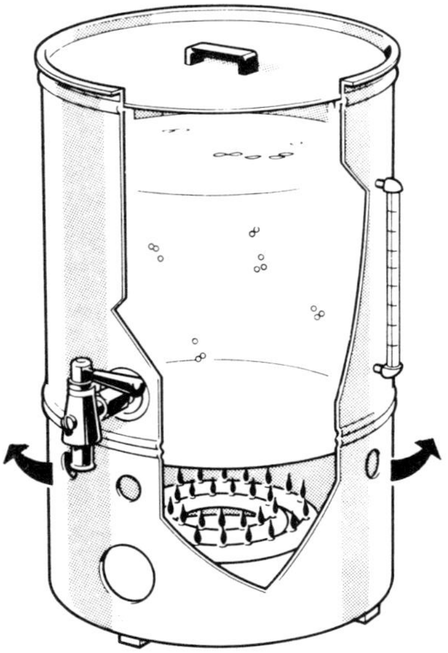

Fig. 5.23 – Bulk water boiler.

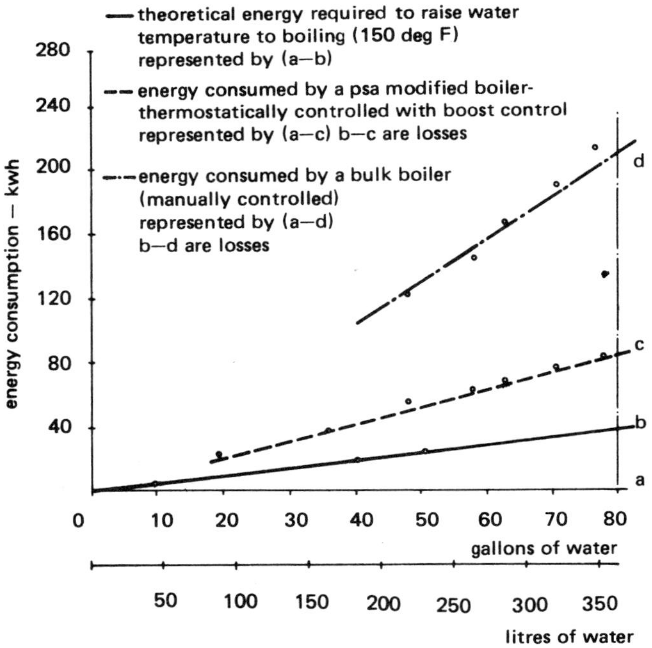


Fig. 5.24 – Energy savings obtained in bulk boiler by fitting thermostat.

Keniston (1977) reported the results of fitting a thermostat with boost control to a manually-controlled electric bulk boiler. This device allows the water temperature to be controlled at 98°C when no boiling water is required, but when it is, a switch is operated to over-ride the thermostat and reconnect the heater elements to raise the water rapidly to its boiling point. The tests (Fig. 5.24) showed that a 60% energy saving could be achieved by fitting this control, with the added advantage of a marked reduction in maintenance. A gas-operated version is also available.

**Expansion or overflow boilers** (Fig. 5.25) are based on the principle that water expands when it is heated. The tap/valve through which the boiling water is removed from the boiler is situated 20 to 50 mm above the cold-water level. This ensures that water can only be drawn off when it is boiling and so has expanded upwards into an expansion chamber and hence to the draw-off level. A separate tap may be fitted lower down so that hot (not boiling) water may also be drawn off. This type of boiler is fitted with its own cold-water tank, where the level of water is controlled by a ball valve. The gas tap and draw-off tap are usually linked together so that when water is drawn off, the gas valve is automatically opened.

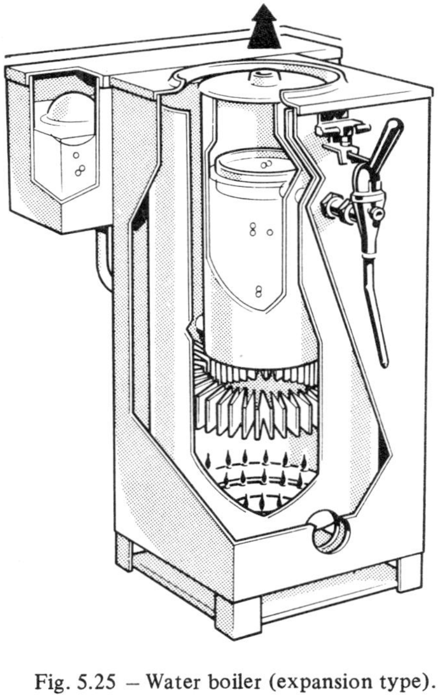

Fig. 5.25 – Water boiler (expansion type).

Some types have a reserve chamber which allows higher delivery rates for short time periods. Cold water is admitted to this chamber jacket before it enters the main chamber and provides a reservoir of heated water. Little preheating takes place once the initial reserve is used. The boiling tank is of small capacity and can be readily descaled. Rapid boiling is achieved by providing a large heating surface to a small amount of water. To further this end a crennelated heat-exchange surface is used. This type of boiler is mounted on a counter.

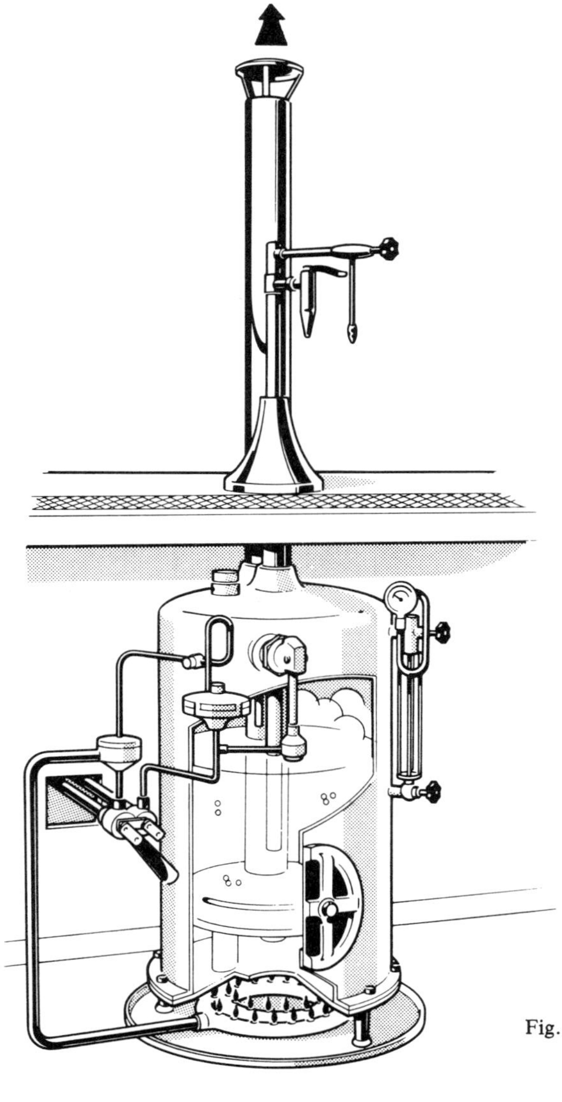

Fig. 5.26 – Water boiler (pressure type).

**Pressure boilers** (Fig. 5.26) are housed under a counter top and heat and store water at about 105°C at a pressure up to 51.7 kN/m². In this type the water offtake is below the water level. When the outlet tap is opened, boiling water is forced out by the internal pressure. When the water tap is closed, the pressure builds up again until it reaches a pre-set level at which the gas supply is shut off by a control (pressure stat). This boiler may also be used to supply steam for steam injection. The boiler is below counter level (so making its appearance unimportant), but water and steam are delivered above counter level because of the pressure in the boiler. The water-feed cistern must be fitted at a height above the boiler such that the static pressure is greater than the boiler pressure. The flow of water into the boiler is controlled by a ball valve either inside the boiler or in the balancing tank when this is fitted. The major disadvantage of this type is that it must be dismantled for descaling, a skilled operation. It is subject to periodic inspection and insurance.

Electrical models, with re-siting of the elements, have largely overcome the scale problem. The build-up of scale (in these models) on the elements and in the inlet pipes has been reduced considerably. Scale only builds up on the base of the boiler, where its presence is relatively unimportant.

**Café sets.** These consist of either an expansion boiler or pressure boiler together with one or more storage urns for the brewing and storage of coffee and for the heating and storage of milk. Similarly, side urns can be provided so that tea can be brewed and stored for quick service. A café set, incorporating a pressure boiler and coffee infuser and storage urn, is shown in Fig. 5.27.

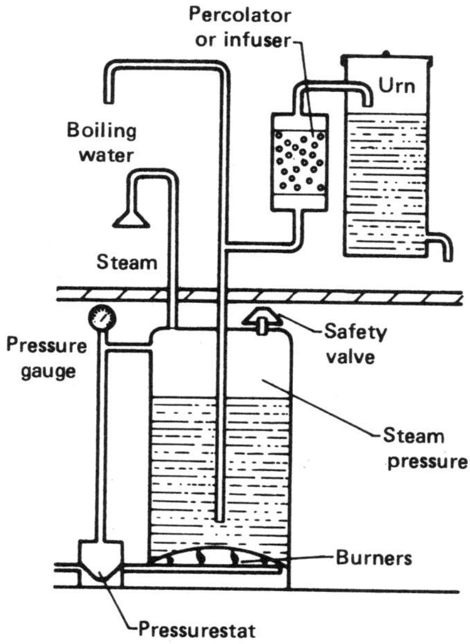


Fig. 5.27 – Café set.

### 5.6.3 Steaming ovens and steamers

As their name indicates, steaming ovens and steamers use steam as the heat-transfer fluid. In units operating at atmospheric pressure the steam is at 100°C, whilst for those operating above atmospheric pressure the temperature of the steam depends on the pressure generated in the unit:

| | | |
|---|---|---|
| 0 Pa | 100°C | |
| 34.4 Pa | 108°C | (note: $1\,kN/m^2 = 1\,Pa$ |
| 68.9 Pa | 115°C | where Pa = Pascal) |
| 103.4 Pa | 121°C | |

The higher pressures and hence temperatures give a larger temperature difference (driving force) between the steam and the food surface, so aiding convection. Some steam may also penetrate into the food, giving some internal heating. However, the steam in contact with the colder food will give rise to heat transfer under condensing conditions and this will enhance the heat-transfer rate to the food surface. Because the water in the food is at the same pressure as the steam, its boiling point is also elevated and this accelerates the rate of the chemical reactions which constitute the cooking process. Since the steam atmosphere is 'wet' (see Chapter 10), drying out of the food will not occur even if the temperature is above 100°C. Leaching of soluble constituents is also avoided (cf. boiling). Steamers working at atmospheric pressure (atmospheric steamers) have a tray of water in the bottom of the oven, manually filled or filled automatically from an external feed cistern as shown in Fig. 5.28. The water is heated by gas or electricity and steam escapes through an open vent, which ensures that the pressure in the oven remains atmospheric.

Pressure steamers operate in the pressure range 4.4 Pa to 103.4 Pa. In the larger types the cistern is inside the oven (Fig. 5.29) and the pressure inside the oven is controlled either by a thermostat (controlling the water temperature) or by a pressure stat (controlling the pressure). A trap must be fitted to allow the escape of water and grease without steam blowing out through it. The pressure steamer cooks more quickly (see above) but is more expensive than the atmospheric type.

If a centralised steam supply is available then this may be used directly in steamers provided it is clean. Steamers are available as one large chamber or, say, three chambers which may be linked together. Smaller steamers are now being used with the advantage that they heat small quantities of food within a few minutes. These may be used to obtain a better match between demand and production, thereby reducing warm-holding time. These either have an external boiler (Fig. 5.30) incorporated so as to make a single unit, or the steam may be generated inside the steamer (Fig. 5.31) by spraying water onto its walls, which are heated by electrical elements.

Smaller steamers may also be fitted with a row or several rows of steam jets (Fig. 5.30). These jets of steam blow onto the food surface, blowing away

Fig. 5.28 – Steam oven heavy duty (atmospheric).

Fig. 5.29 – Steam oven (pressurised).

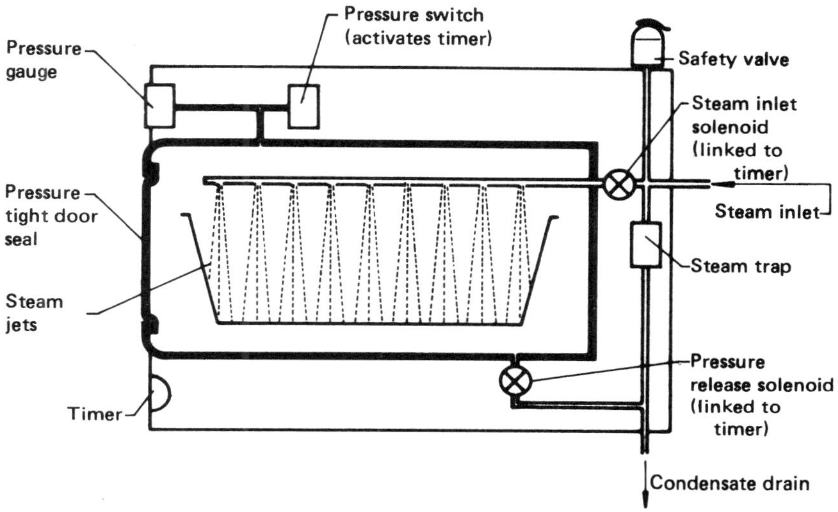


Fig. 5.30 – Steamer with external steam generation and steam jets.

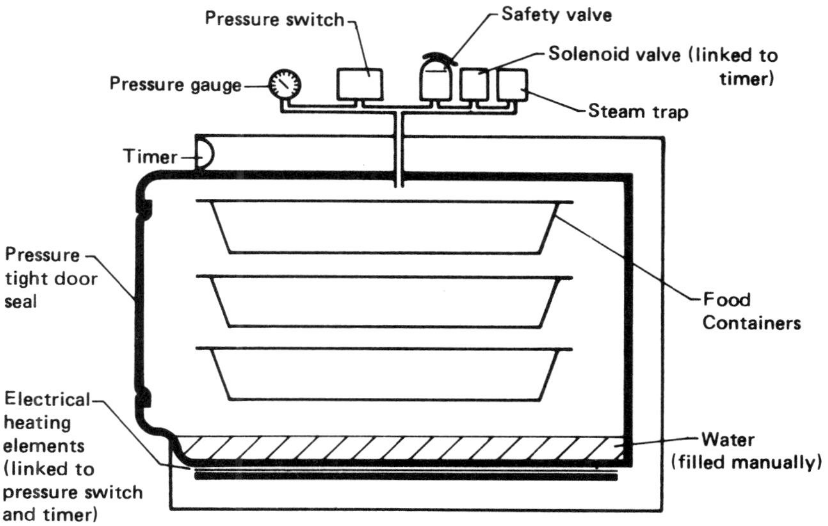


Fig. 5.31 – Steamer with internal steam generation.

condensed moisture and breaking up the boundary layer as they impinge on the food. If water remains on the food surface it will form a boundary layer and impair the condensing steam situation. Similarly, if air from the steam and food (steam and food often contain small proportions of air) builds up on the food surface, an air boundary layer will form. The advantage of having steam jets in the steamer is that water and/or air boundary layers tend to be removed by the action of the steam jet on the food surface. Another means of removing air, and condensate is to have a continuous flow of new air-free steam passing over the food surface with continuous evacuation of steam from the cooking vessel. This is claimed to increase the convective-heat-transfer coefficient three-fold compared with a pressure steamer.

It should be noted that steamers operating above atmospheric pressure require compulsory inspection and insurance. Pressure steamers should be fitted with a safety device which locks the door when the steam is turned on: this protects operatives from scalding. Doors must be effectively sealed to prevent leakage which would cause a deterioration in working conditions and a waste of energy.

Continuous pressure cookers are now available as shown in Fig. 5.32. The trays of food are loaded and unloaded automatically and are conveyed through the equipment on an inclined roller conveyor by the force of gravity. A water spray is used to achieve rapid steam condensation at the end of cooking and before unloading.

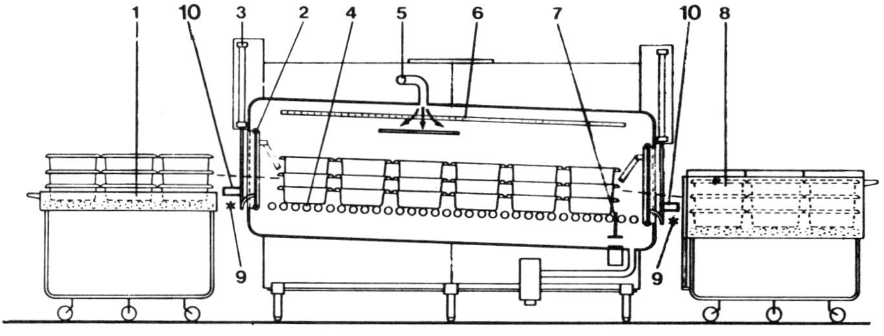


(1) Feed Trolley (2) Door
(3) Door drive (4) Roller Conveyor
(5) Steam inlet (6) Water Spray
(7) Container Delivery Stop (8) Delivery Trolley with "Gastro-Norm" containers
(9) Additional Roller Conveyors 3'3" long can be supplied (10) Security device

Fig. 5.32 – Continuous steamer.

Ling *et al.* (1974 and 1976) have commented that, in theory, one would expect higher heat-transfer rates using condensing steam ($h \approx 5.67$ to $17.01\ \text{kW/m}^2.\text{K}$) than with immersion in hot water ($h \approx 0.28$ to $1.70\ \text{kW/m}^2.\text{K}$). Nevertheless, heating times are less with water blanching than with steam blanching. They attributed this effect to the formation of an air film or an air-water foam on the food surface caused by gas evolved from the intercellular gas in the food. Experimental observations indicated that most of the gas left the inside of the food particle in 10 to 20s, and was held as a foam on the surface until it was swept away by the steam. After 50 to 60s of heating, the effect of the gas disappeared. They also showed that the velocity of the steam did not affect the heat-transfer rate in the range 0.03 to 0.79 m/s but did give a small increase at velocities over 30.5 m/s. This work did not consider the effect of air contained in the steam nor the effect of the steam velocity on removing the condensed water from the food surface. However, the authors pointed out that current theories showed that steam velocity has a small effect on heat transfer, though the effect of any turbulence of the liquid is marked. The effect of steam velocity is confirmed by the authors' results. From a practical point of view, they suggested that the intercellular air could be removed by de-gassing the food under vacuum before steam heating. This should lead to a reduction in the heating time.

### 5.6.4 Fryers

**Brat pans or shallow tilting fryers** (Fig. 5.33) have a variety of uses, not all of which may be considered as frying because they can also be used to boil and stew. Essentially they are shallow containers, the bases of which have large surface areas and are heated by gas flames or electrical elements. They may be used to heat liquids and solids suspended in a liquid (oil or water) or they may be used as griddles on which solid foods are heated directly or with a minimum of liquid (fat) present.

They are also useful where a combination of these processes is being used, such as cooking a roux followed by cooking the sauce, or frying meat and vegetables followed by simmering (braising). The heat-transfer processes in such equipment have been considered earlier. No design studies of brat pans have been published in the literature.

Brat pans can be tilted by means of either a lever or hand-wheel, so that they may be quickly emptied. A non-tilting brat pan is also marketed, showing a considerable cost-saving as a result of eliminating the tilting gear.

**Deep fat fryers** are available in three main designs. In the V-shaped pan, heat-transfer is from the combustion products through the sides of the V-shape. The

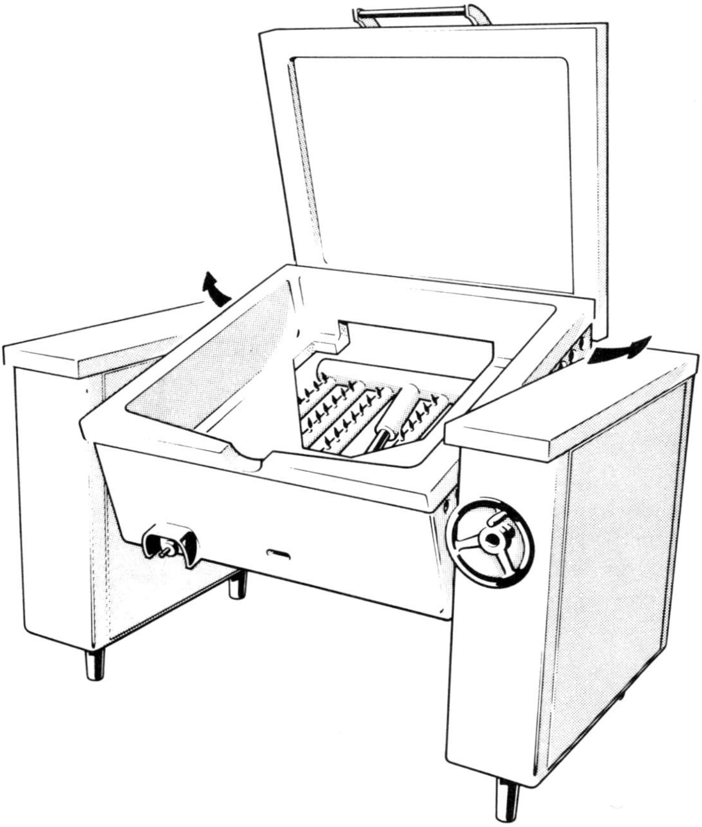

Fig. 5.33 – Brat pan.

flames are situated above a 'cool zone' in the base of the V (Fig. 5.34). The flat-bottom-pan design is heated with electrical elements immersed in the oil or burners firing into tubes immersed in the oil (Fig. 5.35). The third design is a flat-bottom pan with heating from underneath its base. No 'cool zone' is possible with with this type.

In the first two designs, there is a 'cool zone' of oil below the heat source. Particles of food sink into this cool zone and collect there for removal since there are no convection currents capable of re-circulating them. Because the temperature is relatively low in the cool zone, these waste food particles do not char there, spoil the oil or contaminate the flavour of other foods. A drain is built into the cool zone so that these particles can be run off with the oil. The mixture is filtered to remove the particles before the oil is returned to the fryer.

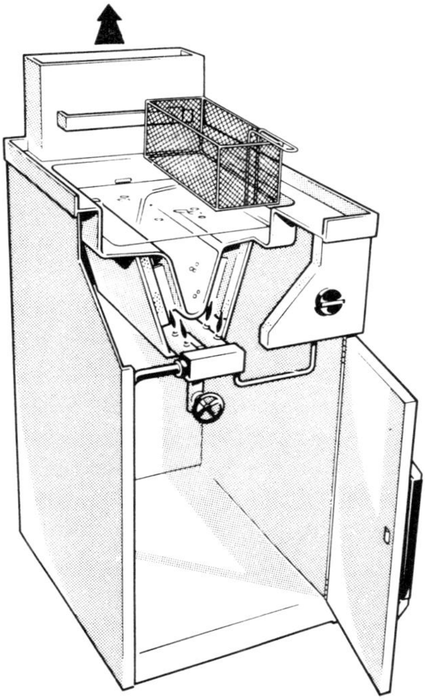

Fig. 5.34 – 'V' pan fryer.

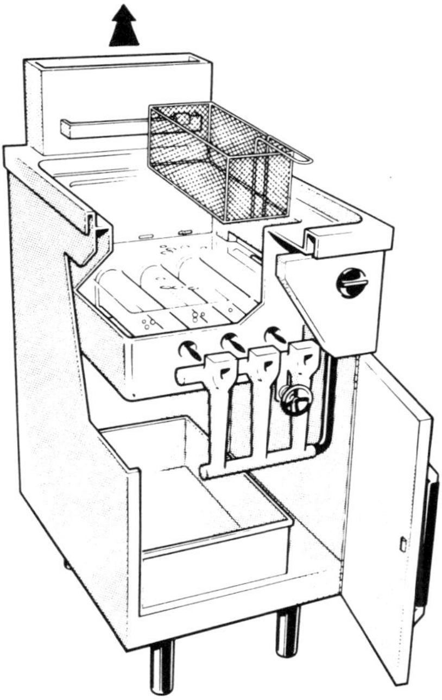

Fig. 5.35 – Immersion tube fryer.

It is important that all the oil in the fryer (except that in the cool zone) is at the same temperature. If temperature variations exist, then the temperature of some of the oil may be too high, with consequent damage to the oil. The fryer must be designed so that the oil circulates around the fryer (except the cool zone) and so maintains a constant temperature throughout its volume. The oil/air interface area should be as low as possible to minimise oil oxidation.

The sensitivity of the oil to temperature limits the maximum permissible temperature of the metal surface in contact with the oil. Normally, an increase in this temperature would be used to achieve an increase in heat-transfer rate. An alternative possibility is to increase the surface area of the heating surface and this may be achieved, for example, by using electrical elements of high surface area.

The advantage of the deep-fat fryer, from a heat-transfer point of view, is that the liquid is at a more or less constant temperature throughout its volume. When a solid food is immersed in it therefore, even heating over the entire food surface is achieved, giving overall even surface browing. The heat-transfer processes taking place during deep-fat frying are complex because the oil temperature is well above that of the boiling point of water. Water (as steam) is evolved from the food surface and so the food is covered with a steam/oil foam for at least part of the frying period (see previous section). Oil absorption also takes place on the food surface. Consequently we have mass-transfer processes taking place at the same time as the heat-transfer processes.

Continuous fryers have been described by Fischer (1977). The food is carried through the fryer on a conveyor belt. Some types unload at the opposite end to the loading station, whereas others turn the food over and unload at the loading end (Fig. 5.36). Different manufacturers have different means of turning the food over. Continuous oil filtration is required with these units (see Section 3.12).

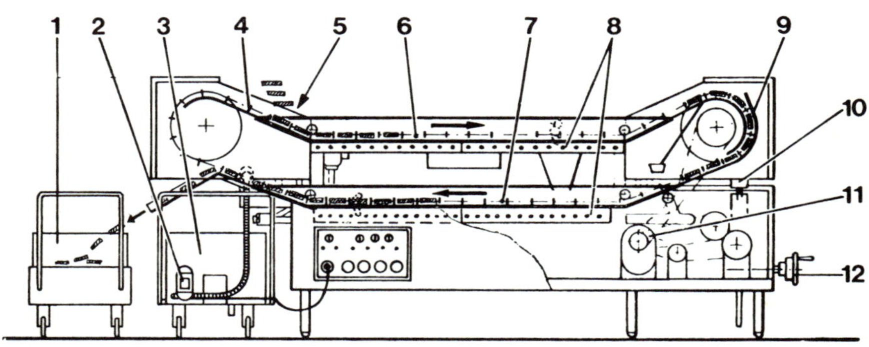


(1) Delivery Trolley (2) Oil Circulating Pump (3) Oil Receiving Tank Trolley (4) Conveyor (5) Feed Zone (6) Top Frying Pan (7) Bottom Frying Pan (8) Heaters (9)Turnover with Guide Plate

Fig. 5.36 – Continuous fryer.

The **pressure fryer** combines the features of the deep-fat fryer with those of the pressure steamer. The pressure fryer consists of a deep pan of oil, heated by electrical immersion elements. The food is placed in the hot oil and the fryer lid is fastened on. The evolution of steam from the food on immersion in the oil causes the presssure inside the fryer to build up as with a steamer. A valve allows steam to escape and yet maintains the pre-set pressure in the fryer. The evolution of steam is suppressed by the higher pressure (higher boiling point of water at higher pressure), and in turn this causes better contact between the oil and the food surface, thereby increasing the rate of heat transfer and reducing the loss of moisture. Turbulence is created in the fat which tends to roll food around. The condensate is removed through a valve and must not be allowed to mix back with the oil. Safety valves must be used.

**Deep-fat fryer studies** have been reported by the American Gas Association (Siskin, R. A. *et al.*, 1952) and these are worthy of detailed attention. A study of available fats and oils led the authors to conclude that heat-transfer surfaces adjacent to the oil/fat could be allowed to reach 343 to 371°C without damage. This temperature then defines the rate at which heat can be passed through the gas and oil boundary layers (the metal-wall resistance was assumed negligible). Oil-side heat-transfer coefficients were measured for horizontal, vertical and 45° inclined plane walls and for tubular elements. The results showed that the highest coefficients were obtained for a 45° inclined plane wall, then horizontal surfaces, with vertical surfaces giving the lowest coefficients. In all cases, the heat-transfer coefficient increased as fat temperature increased.

The range of coefficients for all the experiments was 370 to 587 $W/m^2.K$. They considered that an average value of 499 $W/m^2.K$ was representative of all shape factors and surface and oil temperatures. These results were obtained with no food being fried. If the oil was stirred during heating, the heat-transfer coefficient could be increased to as much as 1761 $W/m^2.K$. Stirring reduces the temperature difference through the oil film, at constant heat throughput. It should be used therefore to obtain an increase in heat-transfer rate through the oil boundary layer without overheating of the oil taking place. This will increase the capacity of the fryer. If food is fried in the oil, oil agitation and circulation is increased by steam evolution from the food. Tests with potato frying showed that this effect increased the oil-side coefficient by about the same extent as stirring under the same conditions, that is to 875 to 951 $W/m^2.K$. Thus, in this case, stirring would be of no particular benefit. If little bubbling (caused by steam evolution) takes place, then stirring may be advantageous. However, the enhanced coefficient obtained during frying is accompanied by a proportionate decrease in the wall temerature. Hence the two effects presumably cancel each other out. A design figure of 440 $W/m^2.K$ was chosen to cover any and all designs of fryer.

Gas-side coefficients were determined experimentally in three experimental fryers. Values were found to be in the range 35 to 56 $W/m^2.K$ or about 41 $W/m^2.K$ on average. This value relates to an inclined flat wall and is independent of angle of inclination, shape of combustion chamber, burner design and aeration. It also applies to baffled immersion tubes of 12 to 16 mm inside diameter and 47 to 71 mm length. It is clear therefore that the gas-side coefficient is controlling as it is only one-tenth of the oil-side coefficient. Hence one method of increasing the heat flow from the combustion products to the oil is to increase the surface area on the gas-side by the use of fins (or studs) as discussed in Section 4.5. Alternatively higher combustion-product temperatures may be used, provided hot spots are avoided.

The temperature of the oil in the fryer is not constant during the frying period. As soon as the food enters the oil, surface moisture is evaporated to steam. The heat for this is provided by the oil, causing a drop in oil temperature. In turn, the thermostat (immersed in the oil) operates the burner which starts to inject heat into the oil. Eventually the rate of heat input to the oil exceeds the rate of heat loss from the oil to the food and the oil temperature rises again. In the oil-cooling period, the potatoes absorb heat (typically) at three times the rate it is supplied from the burners. During the temperature-recovery stage, the heat absorption rate of the potatoes may be one fifth or less of the heat-input rate. Effectively the hot oil at the start of frying acts as a heat reservoir and, the higher the oil/food ratio, the lower the temperature drop. The higher the usable heat input rate, the faster the temperature-recovery rate and the lower the temperature drop. The usable heat-input rate equals the burner-heat rate minus flue-loss rate minus jacket- and surface-heat losses rates.

The prototype fryer developed had the following characteristics:

| | |
|---|---|
| mass of oil in fryer | 14.0 kg |
| oil/food ratio | 5.85:1 |
| cycle time | 4.25 minutes |
| usable heat | 0.97 kW/kg |
| production rate | 2.5 kg/h.kg oil |
| average fat temperature during cycle (initial and final temperatures 174°C, lowest 149°C) | 163°C |
| $CO_2$ in combustion products | 8.0% |
| thermal efficiency | 66.4% . |

A later report (Dewerth and Komora, 1965) reported the development of a prototype counter-top deep-fat fryer. No heat-transfer data were included in this report, but a thermal efficiency of 75% was attained. A gas-fired infra-red burner, an oil container (kettle) with a flat bottom and fins on the gas-side and uncovered solid-sided baskets were found to be advantageous features of the design.

More recently a new Swiss design has been described (Anon., 1977). The efficiency of the gas-fired version is claimed to be 82%, because the combustion products leave at 255°C (compared with 371°C for the AGA 1952 prototype). This is achieved by using deep ribbing on the gas-side. In the electrical version, elements are cast in light alloy and their surface temperature does not exceed 205°C when maintaining an oil temperature of 180°C. The other main design feature of this fryer appears to be shallow ribbing on the oil-side which causes the oil to move upwards in a rotary motion.

## 5.7 NOMENCLATURE

| | |
|---|---|
| $h$ | convective-heat-transfer coefficient ($W/m^2.K$) |
| $l$ | thickness or length (m) |
| $q$ | rate of heat flow (W) |
| $u$ | velocity (m/s) |
| $y$ | thickness of boundary layer (m) |
| $A$ | area ($m^2$) |
| $C_p$ | specific heat (J/kg.K) |
| $Gr$ | Grashof number (–) |
| $Nu$ | Nusselt number (–) |
| $Pr$ | Prandtl number (–) |
| $R$ | resistance to heat flow (W/K) |
| $Re$ | Reynolds number (–) |
| $\beta$ | coefficient of cubical expansion of a fluid (1/K) |
| $\theta$ | temperature (°C or K) |
| $\Delta\theta$ | temperature difference (K) |
| $\lambda$ | thermal conductivity (W/m.K) |
| $\mu$ | dynamic/absolute viscosity (kg/m.s) |
| $\rho$ | density ($kg/m^3$) |

## 5.8 PROBLEMS

(1) Use Dalgleish charts to calculate the time taken for the centre of a food pack to reach 80°C. Investigate the effect of changes in heat-transfer coefficient and warm air temperature on this time.

Data: Food pack is 24 mm thick (assume infinite slab). Initial temperature of pack is 0°C.

$$\lambda = 1.06\ W/m.K$$
$$C_p = 3.76\ kJ/kg.K$$
$$\rho = 960\ kg/m^3.$$

(a) Use convective coefficients ($h$) of $5\ W/m^2.K$, $15\ W/m^2.K$ and $30\ W/m^2.K$ all at an oven air temperature of 180°C.

(b) Use oven air temperatures of 150°C, 180°C, 210°C and 240°C all with a convective coefficient ($h$) of 15 W/m².K.
(c) Using the $Fo$ and $Bi$ values from (a) and (b) calculate the corresponding surface temperatures when the centre is at 80°C.
(d) Present the results graphically and comment on them.

(2) It is desired to estimate the convective-heat-transfer coefficient in a forced-air oven. The food packs have a cross-section (rectangular) of 20 cm width × 4 cm thick,

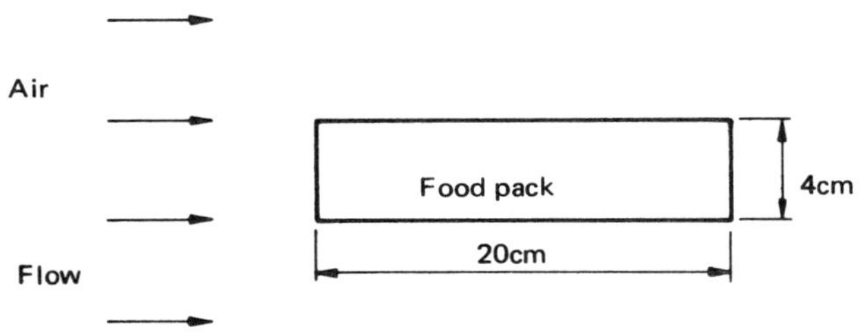


The following equation is recommended for calculating the average convective-heat-transfer coefficient over a flat plate in laminar flow, that is $Re < 500\,000$:

$$Nu = 0.664\, Re^{0.5}\, Pr^{0.33}.$$

The length of plate in the direction of flow is its characteristic dimension.
The properties of air under the conditions in the oven may be taken as:

$\rho = 0.6\ \text{kg/m}^3$
$\mu = 30\ \text{mg/m.s}$
$\lambda = 40\ \text{mW/m.K}$
$Pr = 0.68$.

Calculate the convective coefficient at the following average air velocities:

(a) 1 m/s;
(b) 4 m/s;
(c) 8 m/s;
(d) 12 m/s.

Comment on the data obtained using a graph.

(3) The interior of a refrigerator, having inside dimensions of 0.5 m wide × 0.5 m deep × 1.5 m high, is controlled at 4°C. The walls are constructed of two 3 mm thick mild steel sheets separated by 75 mm thick glass wool insulation. If the average convective-heat-transfer coefficients at the inner and outer surfaces are 2 W/m².K and 5 W/m².K respectively, estimate the

rate at which heat must be removed from the interior of the refrigerator to maintain the controlled temperature (4°C) if the kitchen is at 20°C. The thermal conductivity of glass wool is 25 mJ/m.s.K. Calculate the temperature of the outer surface of the refrigerator.

How important are the surface resistances in this case?

(4) Experimental work in jacketed vessels agitated with horizontal paddles attached to a central vertical shaft have been correlated by the following equation:

$$\frac{hD}{\lambda}\left(\frac{C_p\mu}{\lambda}\right)^{-\frac{1}{3}}\left(\frac{\mu_s}{\mu}\right)^{0.14} = 0.37\left(\frac{L^2N\rho}{\mu}\right)^{\frac{2}{3}}$$

where $h$ = convective coefficient on the food side, $D$ = diameter of vessel (0.1 m), $\lambda$ = thermal conductivity of the food (0.27 W/m.K), $C_p$ = specific heat of the food (2.4 kJ/kg.K), $\mu$ = viscosity of the food at the bulk temperature (1.34 kg/m.s), $\mu_s$ = viscosity of food at the surface temperature (1.34 kg/m.s), $\rho$ = density of the food (1262 kg/m$^3$), $L$ = length of the paddle (1.0 m) and $N$ = rate of revolution of the paddle (revolution/s).

Using the equation and data given:

(a) Derive a simple numerical relationship between $h$ and $N$.
(b) Evaluate the effect of $N$ on $h$ for values of $N$ = 1, 0.1 and 0.01 revolutions/s.
(c) Present the results of (b) graphically and comment on them.
(d) What factors would you consider before using a paddle device in a boiling pan?
(e) If the steam-side coefficient is 250 W/m$^2$.K, what paddle speed would you suggest be used?

## 5.9 REFERENCES

American Gas Association (1948), Research Report 1165.

American Gas Association (1955), Research Report 1239.

Anon. (1977), *Caterer & Hotelkeeper*, December 8th, 14.

Buhr, E. (1969), Dissertation, Technische Hochschule Aachen (cited by Chigier and Milson, loc. cit.).

Capricorn Industrial Services Ltd. (undated), *Technical data on Allpass.*

Chigier, N. A. and Milson, A. (1973), *La Rivista dei Combustibili*, **27** (4–5), 137–146.

Deworth, D. W. and Komora, L. J. (1965), AGA Research Report 1393.

Fischer, S. (1977), Paper in Glew, G. (ed.), *Catering equipment and systems design*. Applied Science Publishers.

Hampel, T. E. and Neu, R. F. (1956), AGA Research Bulletin 73.
Hampel, T. E. (1960), AGA Research Bulletin 81.
Jacobson, R. A. (1956), AGA Research Report 1257.
Jacobson, R. A. (1956), AGA Research Report 1263.
Keniston, L. (1977), *HCIMA Jnl.*, December, 11, 13–15, 17.
Ling, C.-C. A. *et al.* (1974), *Journal of Food Science*, **39**, 692–695.
Ling, C.-C. A. *et al.* (1976), *Journal of Food Science*, **41**, 1345–1348.
Marn, W. L. and Rowland, D. Y. (1966), AGA Research Report No. 1426.
Milson, A. and Kirk, D. (1977), Paper in Glew, G. (ed.), loc. cit.
Perry, E. H. and Stas, A. S. (1967), AGA Research Report 1430.
Siskin, R. A. *et al.* (1952), AGA Research Bulletin No. 66.
Spinell, D. M. (1960), AGA Research Bulletin 85.
Spinell, D. M. (1963), AGA Research Bulletin 98.
Walker, R. B. and Glew, G. (1969), *J. Fd. Technol.*, **4**, 307–318.

CHAPTER 6

# Radiation

## 6.1 INTRODUCTION

The transfer of heat by radiation involves a transfer of energy from one surface, at a higher temperature, to a second surface, at a lower temperature. The energy is transferred by electromagnetic waves. However, not all electromagnetic waves can be used to transfer heat. We are familiar with many forms of electromagnetic energy; these include the waves used to transmit radio and television pictures, radar waves, the waves used to heat foods in microwave ovens, infra-red rays, visible light, ultra-violet rays, the waves used by doctors to produce x-ray pictures and the dangerous gamma rays associated with nuclear reactions. These waves all belong to the same family but have different wavelengths ($\lambda$) and frequencies ($f$), as shown in Fig. 6.1. As its name implies, an electromagnetic

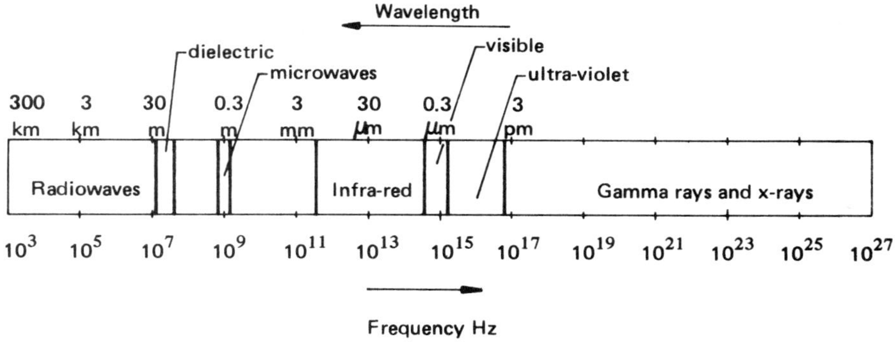


Fig. 6.1 – Different forms of radiation.

wave consists of two parts: a magnetic part and an electric part. As a wave travels through space (electromagnetic waves can travel through a vacuum) or through air, the strengths of the electric and magnetic fields vary with time, as shown in Fig. 6.2.

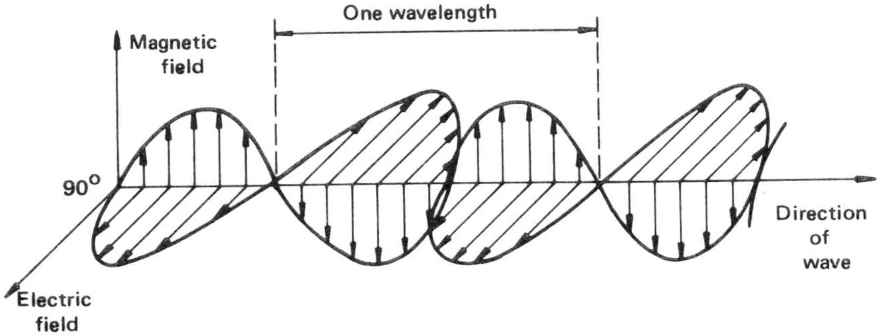


Fig. 6.2 – An electromagnetic wave.

All of these waves travel at the same speed, that of the speed of light ($c$), which is 300 million m/s. The relationship between frequency and wavelength is given by the equation:

$$f = \frac{c}{\lambda} .$$

We can now look at the relationship between all of the different kinds of electromagnetic waves. To do this we look at what is known as the electromagnetic spectrum. In Fig. 6.1, looking from left to right, the wavelength of the wave is decreasing and the frequency increasing.

In the next chapter we will concentrate on using the microwave region of the electromagnetic spectrum to heat foods. The remainder of this chapter is concerned with the area of the spectrum involving waves given off by materials by virtue of their temperature, that is thermal radiation. This covers (Fig. 6.3) roughly waves of wavelength 0.1 to 100 $\mu$m ($\mu$ = one millionth). A part of this range of wavelengths includes visible radiation.

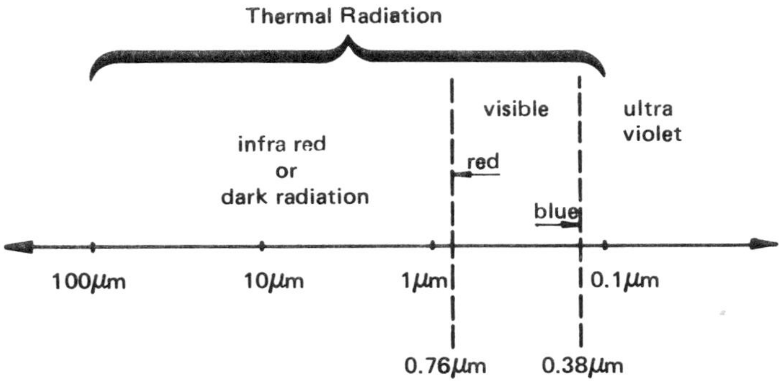


Fig. 6.3 – Thermal radiation.

## 6.2 CONCEPTS AND LAWS OF RADIATION

All bodies above absolute zero temperature (0 K or −273°C) are constantly loosing heat to their surroundings because they radiate energy. At the same time they will be receiving heat from their surroundings. When a body and its surroundings are both at the same temperature they both radiate and recieve heat at the same rate, even though their temperatures remain constant. If a body receives more radiation than it radiates then its temperature rises. Conversely, if a body radiates more heat than it receives then its temperature falls. Thermal radiation behaves in a very similar way to light in that it travels in straight lines and a surface must be able to 'see' another surface before it can receive radiation from it. When thermal radiation falls on a surface a proportion of the radiation will be reflected (Fig. 6.4). Of the remaining radiation, which actually passes through the surface, some will be absorbed by the body and the remainder will pass straight through it (transmitted radiation). The proportions of reflected radiation, absorbed radiation and transmitted radiation vary from one material to another.

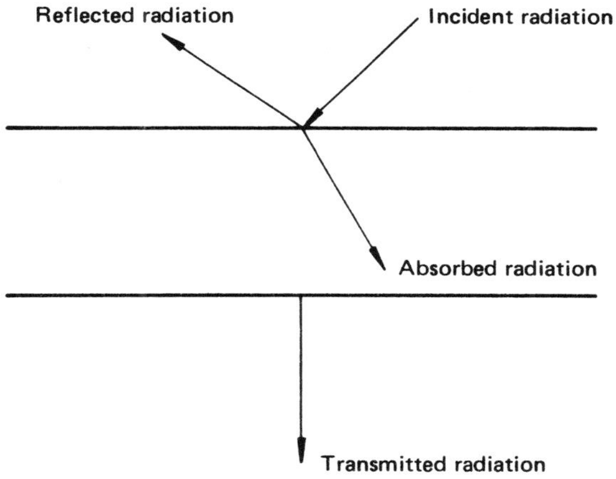


Fig. 6.4 – Radiation striking a surface.

The fraction of the radiant energy incident on (falling on) the surface which is absorbed defines the **absorbtivity** ($\alpha$) of the surface. The fraction incident on the surface which is reflected defines its **reflectivity** ($\rho$). The fraction which is transmitted defines the term **transmissivity** ($\tau$). Radiant energy falling on a surface is absorbed, reflected or transmitted and therefore:

$$\alpha + \rho + \tau = 1 .$$

An important concept in radiation is that of a **black surface**. This surface absorbs all the radiant energy incident on it. It does not transmit or reflect any of the incident radiation, that is for a black surface, $\alpha = 1$, $\rho = 0$ and $\tau = 0$. Conversely a black body is a perfect emitter (radiator) of heat, that is, it emits the maximum amount of heat possible for any surface. The term black surface is used because light is also a form of energy (Fig. 6.1) and if all energy, including light energy, is absorbed by such a surface, then it will appear black. In the case of thermal radiation however, some surfaces are almost black even though they are not black in appearance, for example snow or a white wall. This concept of a black surface leads to the definition of the **emissivity** of a surface ($\epsilon$) which is defined as the ratio of the energy actually emitted by a surface to the energy which would be emitted if the surface were black, that is for a black surface, $\epsilon = 1$. A grey surface is defined as that having an emissivity of less than 1, the emissivity being independent of temperature and wavelength. An opaque surface either absorbs or reflects all the radiant energy falling on it. It does not transmit any of the incident radiation, that is for an opaque surface, $\tau = 0$. An opaque surface usually absorbs the radiation within a very thin surface layer, less than 1 mm thick.

In addition to these concepts, five laws of radiation are important.

**Kirchoff's law** states that the (monochromatic) emissivity for a surface is equal to its (monochromatic) absorptivity provided the system is in thermodynamic equilibrium, that is all the surfaces are at the same temperature. It may be shown from Kirchoff's law that total emissivity and absorptivity are also equal for thermodynamic equilibrium. For black and grey bodies not in thermodynamic equilibrium (surface and surroundings at different temperatures) the equality also holds. For other surfaces, the deviation from this law increases as the temperature difference between the surface and its surroundings increases.

The **Stefan-Boltzmann law** defines the amount of radiant energy emitted from a black surface:

$$e = \sigma\theta^4$$

where $e$ = (total) emissive power of the surface ($W/m^2$), $\theta$ = absolute temperature of surface (K), $\sigma$ = Stefan-Boltzmann constant ($56.7\ nW/m^2.K^4$, where $n = 10^{-9}$ (nano)).

In the case of a grey surface, the emissive power of the surface is less than that for a black surface by a factor ($\epsilon$) measuring the greyness of the surface:

$$e = \epsilon\sigma\theta^4 .$$

**Planck's distribution law** defines the relationship between the emissive power of a black surface as a function of the wavelength of the radiation and the temperature of the surface. It can be expressed as a mathematical equation but is best shown as a diagram (Fig. 6.5).

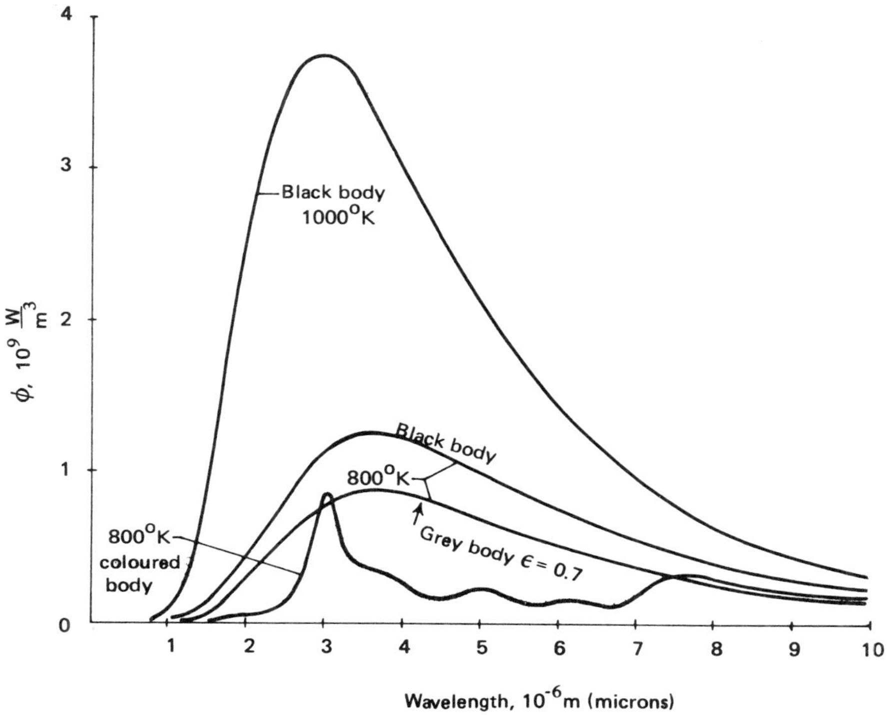


Fig. 6.5 – Planck's law.

When a surface such as that of an electric heating element becomes hotter it is apparent that the nature of the radiation changes in two ways. In the first place the amount of energy radiated increases, as shown by the fact that it browns toast more quickly. Secondly the radiation becomes visible and as the element becomes hotter still the nature of this visible radiation changes. As the temperature increases the radiation changes from a dull red glow, through a bright red to white (as in the case of an element in an electric light bulb).

The monochromatic emissive power (in Fig. 6.5) is the amount of radiation emitted at any one particular wavelength. The total emissive power of the Stefan-Boltzmann law is the amount of radiation emitted over the whole range of wavelengths, that is, it is the area under the curves of the Planck diagram. Fig. 6.5 also shows that the emissive power for a grey surface is less than that for a black one at the same temperature over the whole range of wavelength. This total emissivity of a grey surface as used in the Stefan-Boltzmann equation implies a constant monochromatic emissivity for the grey surface over the whole range of wavelength. If the monochromatic emissivity of the surface varies with wavelength, then the surface is **coloured** (as opposed to grey). If emission of thermal radiation is only in selected wavelength bands then the surface is a **selective emitter.** The emissivities of coloured or selective emitters vary with temperature.

The Planck law diagram shows that the wavelength at which monochromatic emissive power is at a maximum varies with temperature. **Wien's displacement law** gives this maximum in the form of a simple equation:

$$\lambda_{max}\theta = 2.9\,\text{mm.K}$$

where $\lambda_{max}$ is the wavelength at which maximum emissive power occurs for a temperature $\theta$. The variation of $\lambda_{max}$ with $\theta$ is shown in Fig. 6.6. Below temperatures of about 500°C, practically the whole of the emission is in waves too long to be recognised as light. Above 500°C the first visible dull red light appears, changing with further increases in temperature. At about 900°C the colour is cherry red, at 1100°C orange red and above 1400°C white.

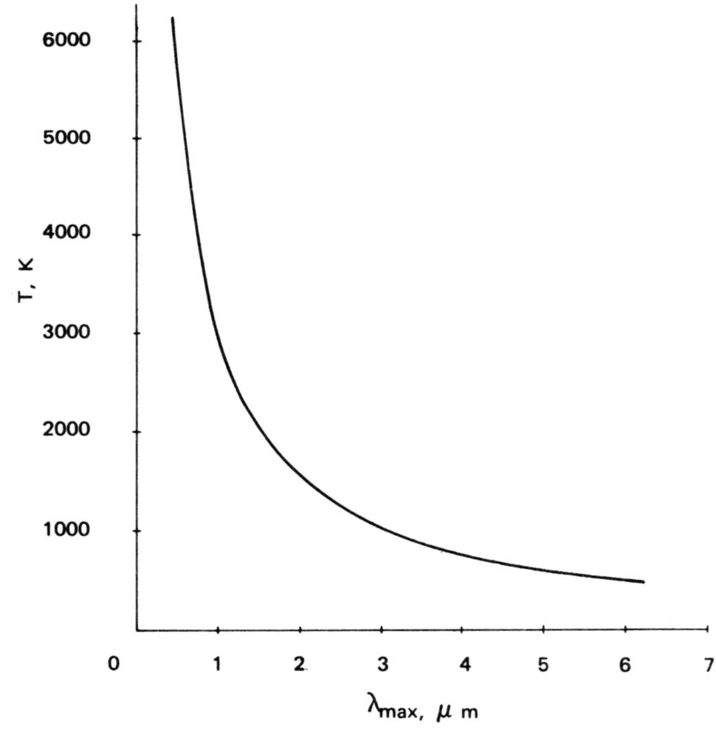


Fig. 6.6 – Wien's displacement law.

When thermal radiation hits the surface of food some of it is reflected but some starts to be absorbed. This absorption causes the density of the radiation to diminish rapidly from the surface of the food to the inside. The **Bouger-Lambert equation** expresses this absorption:

$$q_x = q_o e^{-Sx}$$

where $q_o$ = radiation flux at the surface (W/m$^2$), $S$ = extinction coefficient (1/m), $q_x$ = radiation flux (W/m$^2$) at a depth $x$ (m) from the surface.

The value of the extinction coefficient depends on the wavelength of the radiation and the density and chemical composition of the food. In Chapter 7 we shall see that there is an equivalent equation for microwave radiation. However, the difference in wavelength for the two types of radiation means that thermal (infra-red) radiation is absorbed very near to the food surface whereas microwave radiation penetrates much further into the food. Bengtsson *et al.* (1977) have given the data recorded in Table 6.1 for penetration depths in lean meats.

**Table 6.1 – Penetration depths in lean meats.**

| Wavelength of radiation | Temperature (°C) | Penetration depth (mm) |
|---|---|---|
| 2450 MHz (microwave) | +20 | 9.8 |
| | −20 | 87.0 |
| 915 MHz (microwave) | +20 | 18.5 |
| | −20 | 140.0 |
| 1.25 μm (infra-red) | +20 | 1.25 |
| 2–4 μm (infra-red) | +20 | 0.5 |

These data show clearly that the depth of penetration of microwaves is much greater than that of infra-red (thermal) radiation. These data also indicate that in the infra-red region, penetration is greater the shorter the wavelength.

## 6.3 RADIATION IN CATERING EQUIPMENT

We can now look at the way in which thermal radiation is used in catering equipment. As we have seen, all bodies emit and absorb thermal radiation. In the case of food in an oven, in general the gases in the oven, together with the walls of the oven and any fixtures and fittings, will be at a higher temperature than the food. This temperature difference between the gases/walls and the food surface will cause heat-transfer by radiation to occur. The amount of heat transfer by radiation in an oven will depend upon a number of factors: the emissivity of the oven walls, the emissivity of the gases in the oven, the emissivity of the food surface, the relative temperatures of the oven walls, food and gases in the oven and the geometry of the food and oven.

The higher the emissivity of the surfaces which make up the oven walls, the greater will be the radiant power from these surfaces and, given that the walls are at a higher temperature than the food in the oven, this will result in higher rates of heat transfer. Emissivity values depend very much on the surface conditions and a bright, polished surface will have a lower value than a dull or dirty surface.

In the case of gases in the oven we need to consider air, water vapour and, for direct-fired gas ovens, carbon dioxide and nitrogen. Oxygen and nitrogen do not emit significant amounts of radiation and are transparent to it, so they do not radiate any heat to the food, nor do they interfere in the radiation of heat from the oven walls to the food. On the other hand, water vapour and carbon dioxide are significant radiators of energy to the food (and to the oven walls).

Very little is known about the emissivity and absorbtivity of foods but most foods with high water contents have high emissivities. This is responsible for the unique heating characteristics of food in equipment such as grills, where the predominant mode of heat-transfer is by thermal radiation. We have seen earlier that radiation incident on the surface of a material can be reflected, absorbed or transmitted and we have also seen that a material with a high emissivity absorbs a larger proportion of this incident energy (Kirchoff's law). As the thermal radiation travels from the surface of the food towards its centre, the energy in the wave is absorbed, the absorbed energy being converted to heat. Because of the high absorbtivity, the incident wave cannot penetrate far into the food and all of its energy is converted into heat just below the surface. This results in the very intense surface heating which we associate with grilling cooking methods. The intense surface heat then travels into the centre of the food by conduction, which is a relatively slow process.

The geometry of the food and oven are important in a number of ways. Heat-transfer by thermal radiation can only take place if the food can 'see' the radiating surface. Consequently in a well-filled oven, a food item placed centrally in the oven will be shielded from thermal radiation by the foods surrounding it. This will be apparent from its slower rate of cooking and by less surface browning. Similarly, if a plate of food in an oven has a metal cover, this cover is opaque to thermal radiation and the waves cannot pass through it. The thermal radiation heats the metal cover which, when it becomes hot, emits thermal radiation which is then absorbed by the food. If the cover is at a lower temperature than the heating element, the radiation will be less intense.

The numerical treatment of radiation heat transfer in catering equipment is more complicated than the treatment of conduction and convection. This largely arises because all the surfaces are involved in emitting, absorbing or reflecting the radiant enrgy. However, the electrical analogy introduced in Sections 4.2, 4.3 and 5.2 can be extended to deal with simpler problems involving heat-transfer by radiation.

## 6.4 RADIATION INTERCHANGE BETWEEN SURFACES

In this section we consider first the radiant interchange between two parallel surfaces separated by a non-absorbent gas (for example air, but not combsution products). The situation is similar to that occurring with a grill. The effect of introducing an absorbing gas between the surfaces is considered in Section 6.6. The assumption made initially is that both surfaces are black but we then progressively move to more realistic assumptions about them.

### 6.4.1 Radiation interchange between black surfaces

At first we assume that the grill consists of two surfaces only (Fig. 6.7). The upper surface is the heating surface; it is a black surface and the temperature is even all over it. The lower surface represents the food and is cooler than the upper one. It is also a black surface with an even temperature. These are very idealistic assumptions about the grill, but this situation represents the case where maximum radiant heat transfer is possible.

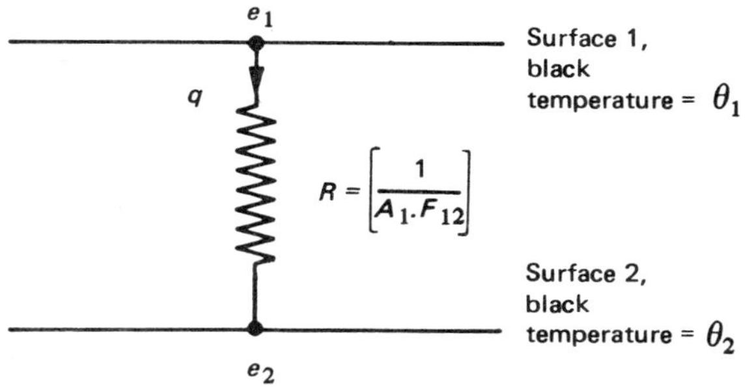


Fig. 6.7 – Parallel surfaces, both black.

Surface 1 radiates heat, as given by the Stefan-Boltzmann law, at a rate given by:

$$q_1 = A_1 \sigma \theta_1^4 = A_1 e_1$$

where $q_1$ = rate of radiant heat-transfer from surface 1 (W), $A_1$ = area of surface 1 ($m^2$), $\theta_1$ = temperature of surface 1 (K), $e_1$ = emissive power of surface 1 ($W/m^2$).

Surface 2 radiates heat at a rate given by:

$$q_2 = A_2 \sigma \theta_2^4 = A_2 e_2$$

where $q_2$ = rate of radiant heat transfer from surface 2 (W), $A_2$ = area of surface 2 ($m^2$), $\theta_2$ = temperature of surface 2 (K), $e_2$ = emissive power of surface 2 ($W/m^2$).

The net rate of interchange of radiant energy between the two surfaces is:

$$q_{12} = A_1 F_{12} \sigma (\theta_1{}^4 - \theta_2{}^4) = A_1 F_{12} (e_1 - e_2)$$

where $q_{12}$ refers to the net rate of flow from surface 1 to surface 2 ($q_{21}$ would refer to heat flowing from surface 2 to surface 1).

$F_{12}$ is the **view factor** and is defined as the fraction of heat leaving surface 1 that is interrupted by surface 2 (the subscript is 12, not 21). The view factor depends only on the geometry of the arrangement of the surfaces. View factors have been calculated for a number of geometries and sources of data for obtaining them are given in Appendix D.

The equation $q_{12} = A_1 F_{12} (e_1 - e_2)$ may be written as:

$$q_{12} = \frac{(e_1 - e_2)}{\left(\dfrac{1}{A_1 F_{12}}\right)}$$

which is analogous to the heat-flow equation and Ohm's law as introduced in earlier sections, that is

$$\text{flow} = \frac{\text{driving force}}{\text{resistance}} .$$

In radiation heat transfer, the driving force is the difference in emissive power $(e_1 - e_2)$ and the resistance to heat-transfer is $(1/A_1 F_{12})$. The electrical analogy is shown in Fig. 6.7. The geometrical situation is that of identical, parallel rectangular plates and appropriate view factors are given in Appendix D.1.5.

### 6.4.2 Radiation interchange between black surfaces separated by black side walls

A grill usually has sides forming a box-type enclosure. This situation is shown diagramatically in Fig. 6.8. We assume that the side walls, as well as the upper and lower surfaces, are black. Note that we are not complicating the situation by having a back wall to the grill.

In this case, radiant energy is transferred directly from surface 1 to surface 2 as in Section 6.4.1. It is also transferred indirectly via the side walls ($S$). This leads to the electrical network diagram shown in Fig. 6.8. There is no resistance between the side walls because both of them are at the same temperature, so that the driving force $(e_S - e_S)$ is zero and hence there is no heat-transfer between

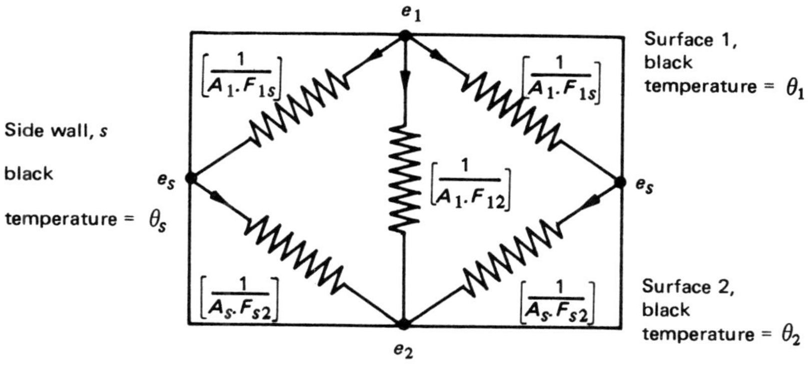


Fig. 6.8 – Parallel surfaces with side walls, all black.

them. The values of the resistances are shown in the figure and follow logically from the notation used, for example $(1/A_1F_{1S})$ is deduced because radiation is from surface 1 to surface $S$. The values of $F_{1S}$ and $F_{S2}$ may be obtained from Appendix D.1.7 because the surfaces involved are touching perpendicular planes. The total resistance of this network may be calculated by using the rules for adding electrical resistances together as shown in Fig. 6.9.

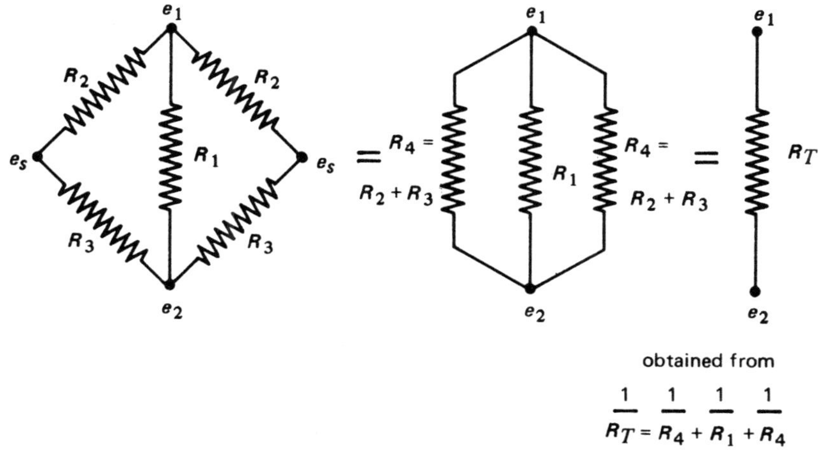


Fig. 6.9 – Calculation of total resistance of Fig. 6.8.

The net rate of heat-transfer from surface 1 to surface 2 is then given by:

$$q_{12} = \frac{(e_1 - e_2)}{R_T} .$$

### 6.4.3 Radiation interchange between grey surfaces separated by black side walls

In practice the upper and lower surfaces of the grill will be grey, not black as assumed so far. The effect of making the upper and lower surfaces grey is represented on the resistance diagram by adding an additional resistance for each grey surface as shown in Fig. 6.10 (compare this diagram with Figs. 6.8 and 6.9).

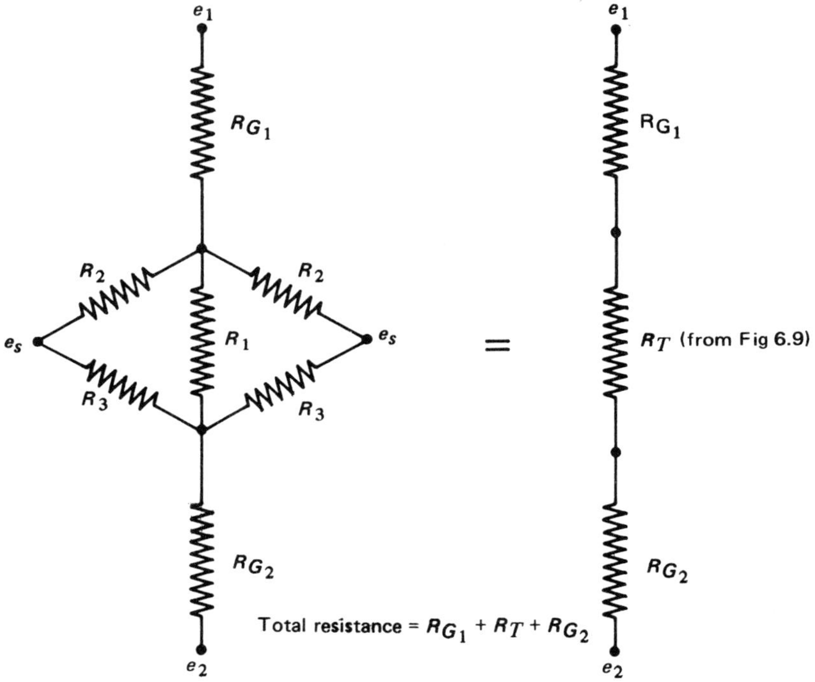


Fig. 6.10 – Parallel grey surfaces with black side walls.

The extra resistance due to to the surface being grey ($R_{G1}$ and $R_{G2}$) is $((1-\epsilon)/\epsilon A)$ where $\epsilon$ is the emissivity of the surface and $A$ is its area.

### 6.4.4 Radiation interchange between grey surfaces separated by grey side walls

Finally, we assume that all four surfaces (including the side walls) are grey. As in Section 6.4.3, the grey side walls are represented by additional resistances as shown in Fig. 6.11 (compare with Fig. 6.10).

The addition of these extra resistances ($R_{GS}$) for the grey side walls does not affect the flow of electricity (radiant heat) between $e_1$ and $e_2$ and therefore it does not matter that the side walls are grey or black as far as the heating of food on a grill is concerned.

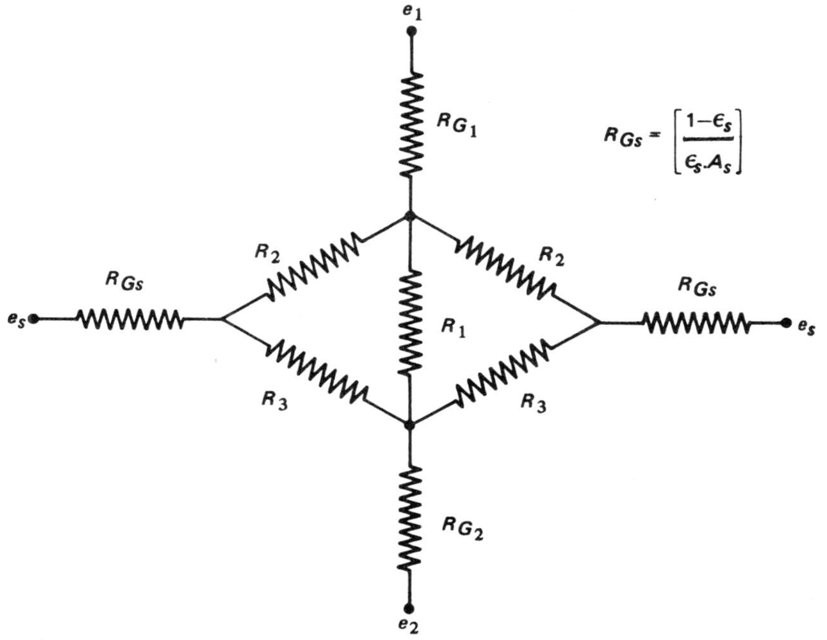


Fig. 6.11 – Parallel grey surfaces with grey side walls.

### 6.4.5 Radiation interchange between more than three surfaces

In the earlier sections the radiant heat transfer between three surfaces (1, 2 and $S$) was considered. Each arm of the network was of the form $((1-\epsilon)/\epsilon A)$ and this expression allows for the fact that the wall is grey rather than black. The radiation resistance between black surfaces is of the form $(1/AF)$. The same general principles apply for more than three surfaces. For example, if there are four black surfaces the diagram becomes that illustrated in Fig. 6.12.

If there are three surfaces there is radiation between surfaces 1 and 2, 1 and 3, and 2 and 3 (3 resistances). If there are four surfaces there is radiation between surfaces 1 and 2, 1 and 3, 1 and 4, 2 and 3, 2 and 4, and 3 and 4 (6 resistances). Thus as the number of surfaces increases, so the number of resistances increases, for example 5 surfaces give 15 resistances. If the surfaces are grey rather than black, an additional resistance at each junction has to be added to allow for the greyness of each surface.

In order to calculate the heat transfer between each pair of surfaces we need to know the temperature of each surface. This is because the rate of radiant heat-transfer between each pair of surfaces depends on the driving force, which in the case of radiation is the emissive power.

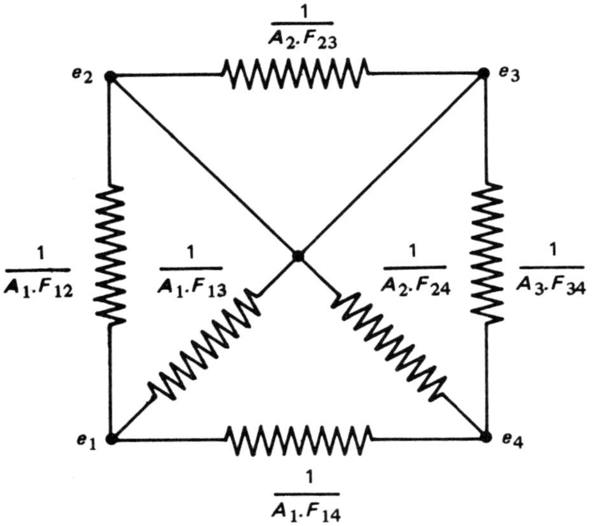


Fig. 6.12 – Radiant inter-change between four black surfaces.

## 6.5 RADIATION FROM GASES

Gases emit and absorb radiation in discrete energy bands: they are **selective** emitters. Two gas mixtures are used for heating food in catering: combustion products (composed of carbon dioxide, water vapour, oxygen and nitrogen) and air (composed of oxygen and nitrogen). Oxygen and nitrogen do not emit significant amounts of radiation and are also transparent to radiation. Carbon dioxide and water vapour however, do radiate and absorb significant amounts of radiation at certain frequencies. Therefore, in catering we can ignore air as an emitter and absorber of radiation but we cannot ignore combustion products in this respect. The determination of the absorption and emission of radiation by gases is difficult as it involves temperature, composition, density and geometry of the gas. However, several simplifications may be made which will allow estimations to be made.

Charts are available (Appendix D.3) for working out the emissivities of water vapour and carbon dioxide calculated on the basis of a hemispherical mass of gas at one bar pressure (radius of hemisphere = $L$). Other shapes may be treated by consideration of the mean beam length (see Appendix D).

## 6.6 GREY GAS IN AN ENCLOSURE

The heat transferred from a grey gas to a black surface is given by the equation:

$$q_{gs} = \sigma A(\epsilon_g \theta_g^4 - \alpha_s \theta_s^4)$$

where $q_{gs}$ = rate of heat transfer from gas to surface (W), $\epsilon_g$ = emissivity of the gas at $\theta_g$, $\alpha_s$ = absorptivity of the gas at $\theta_s$, so that $\alpha_s = \epsilon_s$, $\theta_s$ = temperature of surface (K), $\theta_g$ = temperature of gas (K), and $A$ = surface area ($m^2$).

The term $\alpha_s \theta_s^4$ allows for the absorption of radiant energy by the gas emitted from the black surface at $\theta_s$.

If the surface is grey then an allowance should be made for radiation reflected back from it. This complicates the calculation considerably. If the surface emissivity is $\geqslant 0.7$, the above equation may be used and a correction made for the surface being grey by multiplying the result obtained from the above equation by $((\epsilon_s - 1)/2)$ where as is the emissivity of the surface. The resistance network concept, may be used to represent a grey gas in an enclosure (Fig. 6.13).

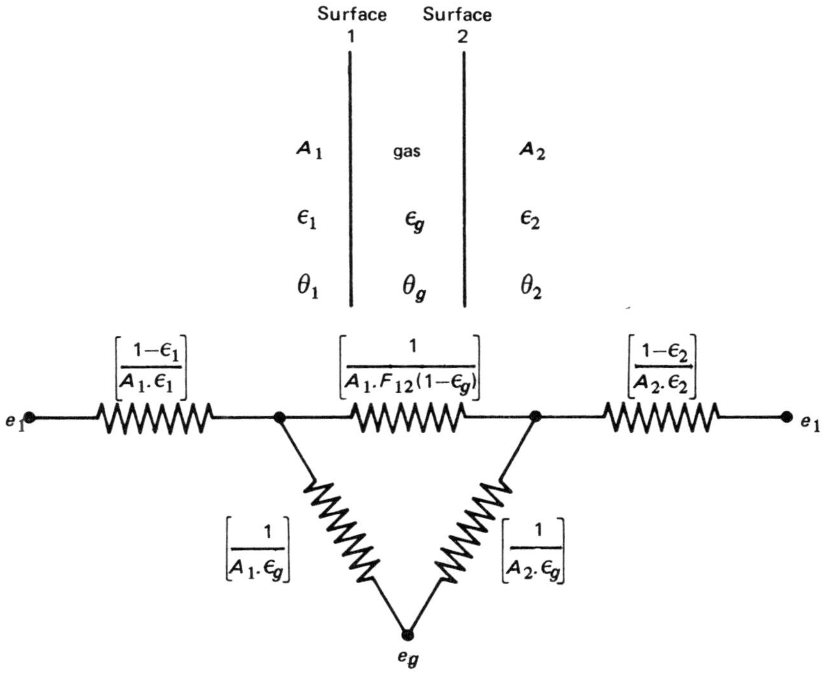


Fig. 6.13 – Grey gas in an enclosure.

The term $(1/(A_1 F_{12}(1 - \epsilon_g))$ is the resistance to the net exchange of heat between the two surfaces. The term $(1/A_1 \epsilon_g)$ is the resistance to radiation heat exchange between surface 1 and the gas. Lastly $(1/A_2 \epsilon_g)$ is the resistance to radiation heat exchange between the gas and surface 2. In the above example we had two parallel grey surfaces with no side walls. We may complicate this problem further by having two grey surfaces separated by a grey wall (as in Section 6.4.4) but this time filled with a grey gas. This situtation is illustrated in Fig. 6.14.

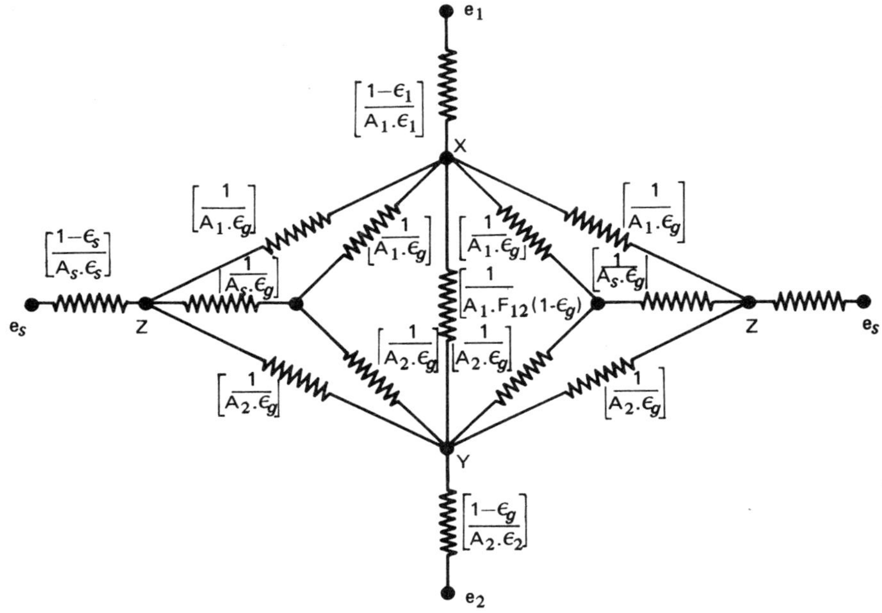


Fig. 6.14 – Grey parallel surfaces, grey side walls filled with grey gas.

$e_1$, $e_2$ and $e_S$ are grey surfaces.
$X$, $Y$ and $Z$ are the equivalent black surfaces to $e_1$, $e_2$ and $e_S$.

In Section 6.4.5 we considered radiation between more than three black surfaces. We could take that situation and modify it to allow for gas radiation also, although the network diagram would become exceedingly complex in this case.

## 6.7 APPLICATIONS OF RADIANT HEATING

### 6.7.1 Grillers

Three types of griller are available to the caterer. The **underfired grill** (Fig. 6.15) has its source of heat situated below the food. The radiant heat-source consists of pieces of volcanic lava or synthetic material which are heated by gas or electricity. The food is placed on bars above the heat-source. Fat droplets fall on the hot pieces of rock causing flares of flame (hence they are sometimes called flare grills). Some of these grills have a fan which produces a horizontal flow of air across the hot rock, which prevents excessive flaring. The flame, and smoke accompanying it, give a characteristic taste and appearance to the food. This type of grill was, originally, charcoal fired.

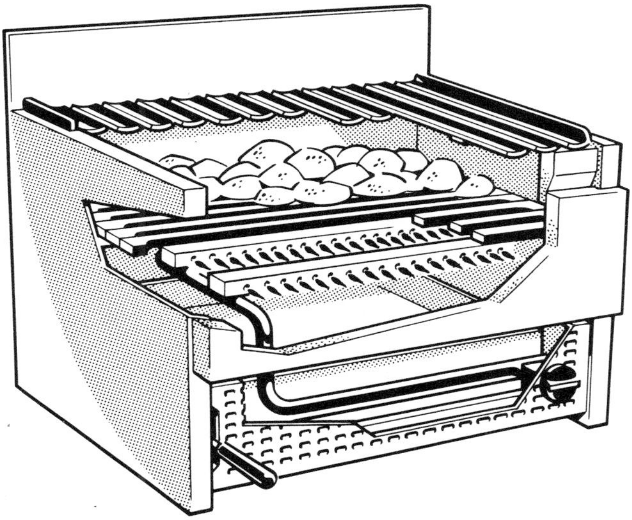

Fig. 6.15 – Countertop underfired grill (flare).

Good ventilation is required for this type of grill. It is limited in use to grilling whereas the next type to be described can be used for gratinating, browning, toasting and so on. The pieces of rock are cleaned by automatic exposure to the high temperatures used during grilling. Temperature control is achieved by controlling both the rate of heat input and the height of the bars above the pieces of rock.

The **overfired grill** (Fig. 6.16) has its heat-source situated above the food (salamander). The electrically-heated types may have two or three separate element circuits so as to permit several different heat outputs. Heat output is uneven because the elements are of finite size and this may cause uneven heating at low total outputs. Gas-heated types may have open flames heating a metal fret to generate radiant heat. The flames vary in length and hence eveness of heat output when turned up or down to vary the total heat output. Alternatively a porous plate burner is used (the plate radiating the heat). This gives an even heat output over its surface for the whole range of total heat outputs. The design of these burners is discussed in Chapter 11. The food is held on either an open metal grid or a solid metal plate, with the possibility of varying the element/burner to food distance so as to control the radiation intensity. The solid metal plate can be preheated before the food is placed on it and hence acts as a contact heat source at the start of cooking. Keniston (1977) has shown that automatic shut-off devices for grillers can result in substantial energy savings.

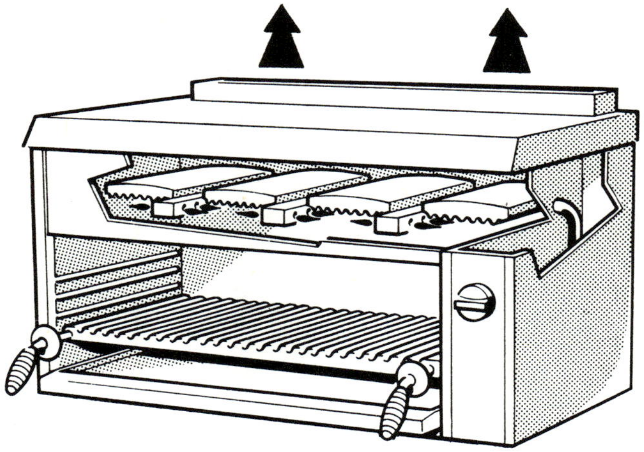

Fig. 6.16 – Overfired grill.

**Double-sided grills** have top and bottom electrically-heated quartz-tube elements (Chapter 11) to cook both sides simultaneously. Fischer (1977) has described a continuous, double-sided (grill) broiler (Fig. 6.17) capable of a wide range of heating times.

### 6.7.2 Toasters

Although toast may be prepared using an overfired grill, specialist equipment is widely used for preparing this particular product. The main requirement of the equipment is that it toasts the bread to an even brown colour all over its surface. Variations in design relate principally to the means of handling the bread/toast.

Batch-type pop-up toasters are top loading with automatic ejection of the toast. In continuous rotary toasters the bread moves vertically downwards between the heating elements. This method saves space compared with the third type, which is the continuous conveyer toaster. In this, the bread moves horizontally between the heating elements.

Marn (1963) investigated several gas-fired toaster designs and compared them with contemporary electric toasters. These studies were essentially empirical although the prototypes produced gave higher throughputs than contemporary

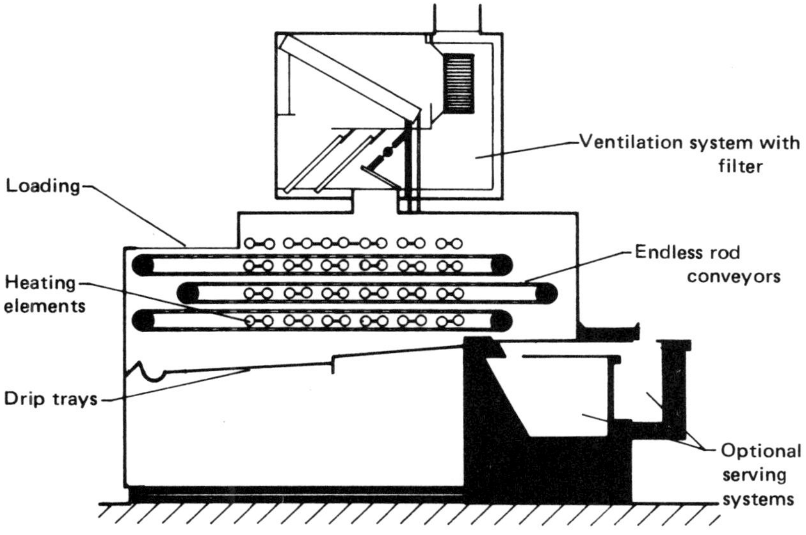


Fig. 6.17 – Continuous broiler.

equipment. Several types of radiant burner were tried and these were found to be more efficient and gave shorter heat-up periods than the metal-fret type of burner.

### 6.7.3 Infra-red ovens

Infra-red ovens utilize the quartz tube, which is heated by an electrical filament running down its centre, for the generation of infra-red energy. They are used in rapid-heating ovens and also in grills. The quartz tube radiates more heat than conventional electrical elements because it operates at a higher temperature and it has a higher emissivity. The quartz tube is fragile and may be protected by surrounding it with a larger, slotted or perforated light metal tube (see Chapter 11).

The quartz tube also emits radiation of shorter wavelengths than normal elements and this coincides with the ability of food to absorb radiation at these shorter wavelengths (2.5 – 5.0 $\mu$m). The radiation is absorbed within a short distance from the surface as discussed earlier. It is advantageous for the radiation to penetrate the food as much as possible so as to overcome unsteady-state heat-transfer problems. Ideally therefore the radiation should be at a slightly shorter wavelength (0.5 $\mu$m) than that at which the food mainly absorbs radiation, that is, it should be in the range 2.5 to 3.3 $\mu$m (medium-wave region; 1.3 to 2.5 $\mu$m is the short-wave region and 3.3 to 5.0 $\mu$m is the long-wave region). Microwaves penetrate the food even more, as will be seen in the next chapter.

A conventional metal-sheathed element is made of a chromium-nickel filament embedded in a ceramic and sheathed in stainless steel. Its surface temperature is below 700°C, and hence it radiates little energy in the infra-red band (see Fig. 6.5). Quartz tubes can withstand higher temperatures (quartz glass starts to bend only above 1200°C) and so it emits much more radiation in the infra-red region.

One design of infra-red oven, which is used specifically for cook-chill applications, employs quartz tubes at a lower temperature than normal to produce radiation in the long-wave region. The food is placed in a heavy ceramic plate and covered with a metal lid. The plate acts as a heat storage unit and the lid, by acting as a secondary emitter at a low temperature, also reduces the emissive power of the heating element. This combination produces rapid heating of the food with minimum cooking. Because of the heat stored in the ceramic plate, the food continues to heat for a time after it is taken from the oven.

Dewerth (1965) described a prototype small-size infra-red oven heated by a gas-fired radiant burner. The infra-red heat was directed onto the food by means of a teardrop-shaped cavity of mirror-finished aluminium. Its performance compared favourably to that of electric units.

Bengtsson and Dagerskog (1977) described a double-sided continuous IR frying tunnel. This work showed that, using short-wave quartz tubes (1 to 1.5 $\mu$m), the penetration depth was 1.25 mm compared with less than 0.5 mm for long-wave IR heating elements (penetration depth is defined as the depth under the surface where the incident power is reduced to about one-third). The higher penetration depth had a marked influence on the heating rate for food samples such as hamburgers, the effect diminishing with thickness. Pork chops (14 mm thick) and sliced beef (12.5 mm thick) were fried in times 31% and 26% quicker respectively by using short-wave heating instead of long-wave heating. Deeper penetration results in lower surface temperatures and hence the incident power level needs to be increased to reach the same degree of surface dehydration and browning, enabling further reductions in frying time to be achieved. Computer simulations showed that the main factors governing the food centre-temperature rise were incident power, absorption characteristics of the food surface and the sample thickness. Other factors were of secondary importance.

They considered that short-wave IR heating is considerably more versatile than double-sided contact frying with regard to both sample shape and size. It also reduces frying time by about 40%, for recipes which do not swell during heat treatment, but a higher energy input must be used (3 W/cm$^2$ compared with 2 W/cm$^2$).

## 6.8 NOMENCLATURE

$c$ speed of light (m/s)

$e$ emissive power (W/m$^2$)

$f$ frequency (1/s)
$q$ radiation heat of flux ($W/m^2$)
$x$ distance (m)
$A$ area ($m^2$)
$F$ view factor (–)
$S$ extinction coefficient (1/m)
$\alpha$ absorbtivity (–)
$\epsilon$ emissivity (–)
$\theta$ temperature (K)
$\lambda$ wavelength (m)
$\rho$ reflectivity (–)
$\sigma$ Stefan-Boltzmann constant ($W/m^2.K^4$)
$\tau$ transmissivity (–)

## 6.9 PROBLEMS

(1) Calculate the net rate of heat transfer to a loaf of bread in an oven with walls at a uniform temperature of 175°C. The emissivity of the bread surface is 0.85. Assume the surface area of the loaf is 0.05 $m^2$ and its surface temperature is 100°C. The loaf may be considered as a grey body completely surrounded by a grey surface.

(2) A gas-fired grill has the following dimensions:

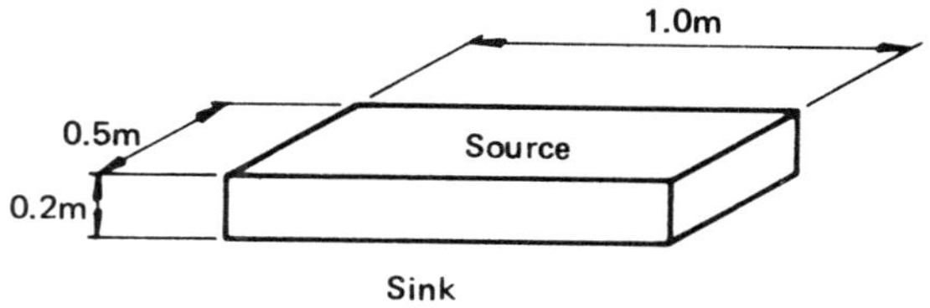


Assume throughout that the back and front are open to the kitchen. If the grill surface is at 1000 K and the food is at 300 K, find the rate of radiation heat-transfer under the following conditions.

(a) Assume the grill and the food are both black surfaces and that they can be regarded as segments of infinite parallel planes.
(b) Assume the grill and the food are both grey surfaces and that they can be regarded as segments of infinite parallel planes. In this case the grill and the food may be assumed to have emissivities of 0.9 and 0.8 respectively.
(c) Assume the grill and the food are both black surfaces but are separated by black side walls.

(d) As for (c) but with the grill and the food being grey surfaces with emissivities as given in (b).
(e) Comment on the results of the calculations.

(3) Investigate the effects of the following three parameters on the rate of radiation heat loss from a griddle surface of area 1.0 $m^2$.

(a) Temperature of griddle surface: 450 K (177°C) and 500 K (227°C).
(b) Emissivity of griddle surface: 1.0 (black) and 0.1 (bright stainless steel).
(c) Area covered by food (and hence not radiating heat to surroundings): 90%, 50% and 0%.

This situation may be taken as a small grey body surrounded by grey surfaces, with surroundings at 20°C.

Comment on the results.

## 6.10 REFERENCES

Bengtsson, N. *et al.* (1977), Paper in Glew, G. (ed.). *Catering equipment and systems design.* Applied Science Publishers.
Bengtsson, N. and Dagerskog, M. (1977), Paper in Glew, G. (ed.), loc. cit.
Deworth, D. W. (1965). AGA Research Report 1391.
Fischer, S. (1977), Paper in Glew, G. (ed.), loc. cit.
Keniston, L. (1977), *HCIMA Jnl.*, December, 11, 13–15, 17.
Marn, W. L. (1963), AGA Research Bulletin 95.

## 6.11 FURTHER READING

Gray, W. A. and Müller, R. (1974), *Engineering calculations in radiative heat transfer.* Pergamon Press.

CHAPTER 7

# Microwaves

## 7.1 INTRODUCTION

Heating food by convection has the limitation that after the heat arrives at the surface of the food, it then needs to be conducted from the surface into the interior. Radiant heating gives mainly surface heating with heat penetration limited to no more than about 1 mm below the surface. Conduction of heat from the surface or near to the surface into the centre of the food can be a relatively slow process. Microwave radiation, on the other hand, can penetrate some considerable distance into the food, thereby overcoming at least to some extent the limitation of the other more time-comsuming cooking and heating methods.

Microwave radiation, like thermal radiation and light, is part of the electromagnetic spectrum (Fig. 6.1). Microwaves however, have a higher frequency and longer wavelength than normal thermal radiation. In practice, two frequencies are used for microwaves:

| | |
|---|---|
| 915 MHz | wavelength 328 mm |
| 2450 MHz | wavelength 122 mm |

Catering equipment, in the form of small box-like ovens, usually makes use of the higher frequency.

Cooking by microwave radiation is based on the principle that all matter, including food is made up of electrically-charged atoms or molecules. Some materials are ionised and have an electric charge, but most atoms and molecules are electrically neutral. Despite this neutrality many molecules, called polar molecules, have their electrical charges positioned so that they are charged unevenly. The most important molecule to behave like this, as far as foods are concerned, is the water molecule. In this molecule, the negatively charged electrons are drawn towards the oxygen. This means that the oxygen has a small negative charge whilst the hydrogen atoms have a small positive charge. When polar molecules are subjected to microwave radiation, they try to line up with the electric field of the microwaves. They behave like tiny magnets in a magnetic field (Fig. 7.1). The electric field of the microwaves is alternating at, say, 2450 million times/second. Therefore the molecules switch backwards and forwards at the same rate (2450 million times/second).

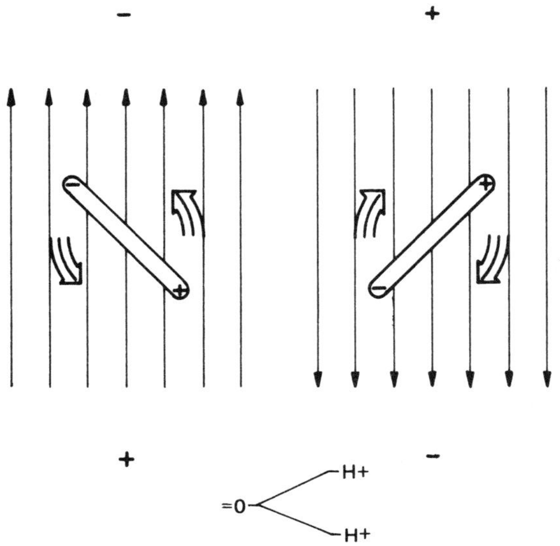


Fig. 7.1 – Action of polar molecules in a microwave field.

Water is a dipole (see Fig. 7.1) and it is this characteristic of water coupled with the quantity present in most foods that makes microwave cooking of food so fast. The rapidly moving molecules, trying to keep in line with the rapidly alternating field, are slowed down by a number of forces such as mechanical ones in the case of ice or solid fat and by viscous forces in the case of viscous liquids. The energy in the microwaves causes the movement of the molecules and therefore attempts to overcome these restraining forces. As a consequence of this some of the microwave energy is coverted into heat energy within the food itself. Ionised materials are also affected by the microwaves. A solution of, for example, salt in water contains the sodium and chlorine as free ions, the sodium having a positive charge and the chlorine a negative charge. The ions are accelerated by the electrical field, collide with other molecules, and thereby cause further (heat) energy to be released.

The heat is generated in the food as a result of the interaction between the food (or mainly the water in it) and the microwaves. Consequently, there is no 'heating-up time' with microwave ovens as there is with conventional ovens. With most ovens a small delay of 1 to 2 minutes is all that is required for the electrical components to warm up. Cooking or reheating times are short because, at least to a major extent, one overcomes the problem of the poor thermal conductivity of the food and the need to conduct heat from the surface of the food to the interior. However, the speed of heating demands more careful control of heating time if overcooking is to be prevented. A further disadvantage is that browning is not usually possible because the surface temperature is too low. This latter problem can be overcome by using another oven for browning, a browning element or browning dish in the microwave oven, or a forced-convection/microwave oven.

Microwave radiation is similar to thermal radiation in that it can be absorbed, transmitted or reflected. Materials which transmit microwaves include glass, paper, china and plastic. These make ideal container materials for the food in the microwave oven as they do not heat up or to a very limited extent only. Metals reflect microwaves and therefore do not heat up. The walls of microwave ovens are therefore made of metal and this metal reflects the microwaves until they are finally absorbed by the food. Because of these reflections, the food receives the microwave energy from all directions, and not from selected directions. Materials which are dielectrics and absorb microwaves are said to be 'lossy'. The greater the 'lossiness' of a material, the more microwave energy it will absorb.

## 7.2 THE POWER EQUATION

The power generated in a dielectric in a microwave field is given by the equation:

$$P = 55.61 \times 10^{-14} E^2 K'' \nu$$

where $P$ = power absorbed (W/cm$^3$), $E$ = voltage gradient (V/cm), $\nu$ = frequency (Hz), $K''$ = total dielectric loss factor (–) = $K''_{\text{dipole}} + K''_{\text{ionic}} = K' \tan \delta$, $K'$ = dielectric constant (–), $\tan \delta$ = loss tangent = $K''/K'$, $K''_{\text{dipole}}$ = loss factor due to dipole molecules and $K''_{\text{ionic}}$ = loss factor due to ionic molecules.

Examination of this equation shows that increases in all the parameters will increase the power absorbed. There are limitations to increasing each of them in practice. Any increase in voltage gradient ($E$) must be limited because if it reaches a sufficiently high value the air will break down and an 'arc' will result. An increase in the frequency of the microwaves ($\nu$) increases the power absorbed. The total dielectric loss factor ($K''$) is a property of the food. Goldblith (1975) has summarised the factors which affect this last parameter as:

(a) It is much less for ice than for water.
(b) It increases with increases in the water content and salt content of the food.
(c) It decreases as fat content increases.

(d) It is higher for raw meats than for cooked meat products.
(e) For meat products, it increases with decreasing frequency.
(f) It is affected by temperature: $K''_{\text{dipole}}$ decreases with an increase in temperature, while $K''_{\text{ionic}}$ increases with an increase in temperature.

In addition to the dielectric properties of the food, other physical properties can be important in relation to the temperatures achieved during heating. In particular, the specific heat and density are important. If a food has a low specific heat, it will rise to a higher temperature than a food with a high specific heat, for the same amount of absorbed microwave energy.

## 7.3 THE PENETRATION EQUATION

When a microwave hits the surface of the food a proportion of the radiation is reflected but the remaining radiation is absorbed by the food. This absorption causes the power flux to diminish from the surface of the food to the inside (the wave is attenuated) according to the equation:

$$P_x = P_o \exp(-2\alpha x)$$

where $P_o$ = power flux at surface (W/m$^2$), $P_x$ = power flux at a depth $x$ (m) from the surface (W/m$^2$) and $\alpha$ = attenuation coefficient (1/m).

The attenuation coefficient ($\alpha$) is related to the dielectric properties of the food by the equation:

$$\alpha = \frac{2\pi}{\lambda_o}\left[\frac{1}{2}K'\left(\sqrt{1+\tan^2\delta}-1\right)\right]^{\frac{1}{2}}$$

where $\lambda_o$ = wavelength in free space (m).

As the lossiness increases so does the value of the attenuation coefficient ($\alpha$) and this means that the wave will only penetrate a short distance into the food before it has been absorbed and converted into heat. This can be seen from Fig. 7.2 which shows the strength of microwave radiation as it penetrates into a food. Comparing two foods of different degrees of lossiness, we can see that the radiation penetrates much further into the low lossiness food. Bearing in mind that the heat produced depends upon the strength of the field, we can see that in the case of the low lossiness food the heat will be more even than in the case of the high lossiness food. In the case of the latter there will be intense heating within a distance of 10 to 20 mm below the surface. Any heating beyond this will rely largely on conduction from this heated layer. This situation can often lead to problems when two materials of different loss factors are heated simultaneously in a microwave oven. Ice has a relatively low loss factor and therefore heats very slowly in a microwave field. Water, on the other hand,

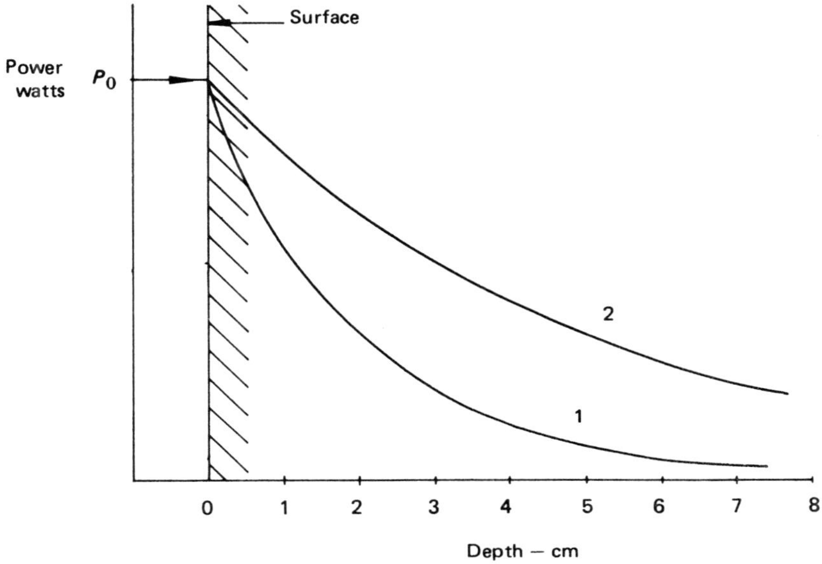


Fig. 7.2 – Effect of lossiness on microwave penetration.

has a high loss factor and heats very quickly. Because of this, when food is thawed in a microwave oven we often come across the problem known as **run-away** heating. Once a part of the frozen food thaws, that part will heat up very quickly and soon achieve 100°C whilst the remainder will remain frozen. This can only be overcome by allowing the food to rest during the heating period; this can be done either manually or automatically by using a pulsed microwave oven. The latter gives a cycle of heat/rest periods throughout the heating time.

The depth of penetration is often described by the half-power depth (HPD) or that thickness of food which reduces the power at the surface ($P_o$) to one half ($P_o/2$). It may be calculated from the following equation:

$$\text{HPD} = \frac{1}{\sqrt{K'} \tan \delta \ \nu} \ .$$

Thus, half-power depth is inversely proportional to frequency ($\nu$) whilst the power equation (Section 7.2) shows that the power absorbed ($P$) is directly proportional to frequency. A balance must therefore be struck between the power absorbed ($P$) on the food favoured by high frequencies ($\nu$) and the depth of penetration (HPD) favoured by low frequencies.

## 7.4 DESIGN AND CONSTRUCTION OF BATCH OVENS

This type of oven, as used in catering, utilises power levels of between 0.5 and 2.5 kW. It is used for the rapid heating of snacks, prepared food and vended foods and therefore needs to be versatile in terms of dish composition and dish geometry. This type of oven uses a standing-wave applicator, which is suitable for the large resonant cavity of the oven and the large and irregularly-shaped objects it needs to heat.

A stationary or standing wave is formed when two travelling waves are superposed on one another when travelling in opposite directions. In the resonant cavity of a microwave oven, standing waves occur because of the reflection of waves from the surfaces of the oven, the reflected waves being superimposed on the waves incident on the walls. The principle of a standing wave can be demonstrated using a string fastened between a tuning fork and a weight suspended over a pully. The weight keeps the string taught.

When the fork is vibrated, a travelling wave passes along the string to the right-hand end and is then reflected back. If the fork is kept vibrating the incident and reflected waves combine to form the standing wave shown in Fig. 7.3. The points marked $N$ are called nodes and here the string is at rest. The string vibrates between the nodes, the vibrations being at their greatest at the antinodes ($A$).

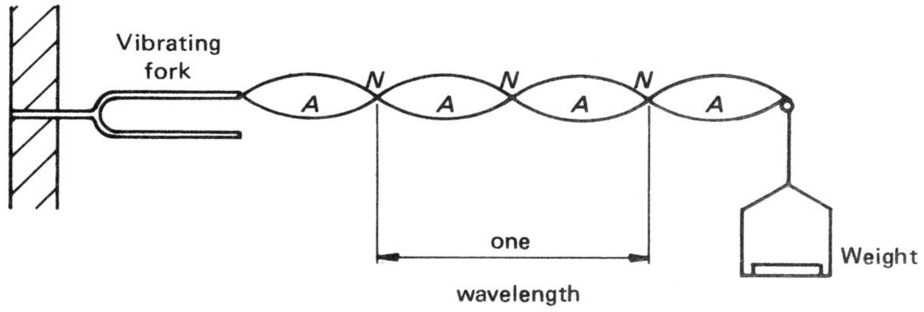


Fig. 7.3 – Principle of a standing wave.

In a microwave oven, maximum power is available at the antinodes and zero power at the nodes, that is, the power level in a resonant cavity varies from point to point in the oven. Therefore the disadvantage of the standing-wave applicator is that the microwave energy is not uniformly distributed throughout the oven cavity, so that the rate of heating differs at different points in the oven. In a sense this situation is analogous to the local convective-heat-transfer coefficient considered in Section 5.3.

A diagram of an oven is shown in Fig. 7.4. The mains supply is converted into a high voltage to feed the magnetron, and this generates the microwaves which are passed down a waveguide into the oven cavity (see Chapter 11 for more details). At a frequency of 2450 MHz, magnetrons are available at power levels up to 5 kW; if more power than this is required a number of magnetrons must be used simultaneously. A cooling fan is needed to keep the magnetron cool. The oven door must be designed to ensure leakage of microwaves is kept below the safe level. This is achieved by using a door seal and a safety switch which cuts off the microwave supply when the door is opened. Often the door of an oven incorporates a perforated plate which allows the operator to see inside the oven. Some standards require the perforated screen to be covered by glass but a glass door in itself does not prevent the leakage of microwaves as glass is transparent to them. The purpose of the mode stirrer will be explained later. The magnetron is approximately 60% efficient in converting electrical energy into electromagnetic radiation, and for loads of 500 g or over, 70% of the electromagnetic radiation is absorbed by the food in the oven and is converted into heat.

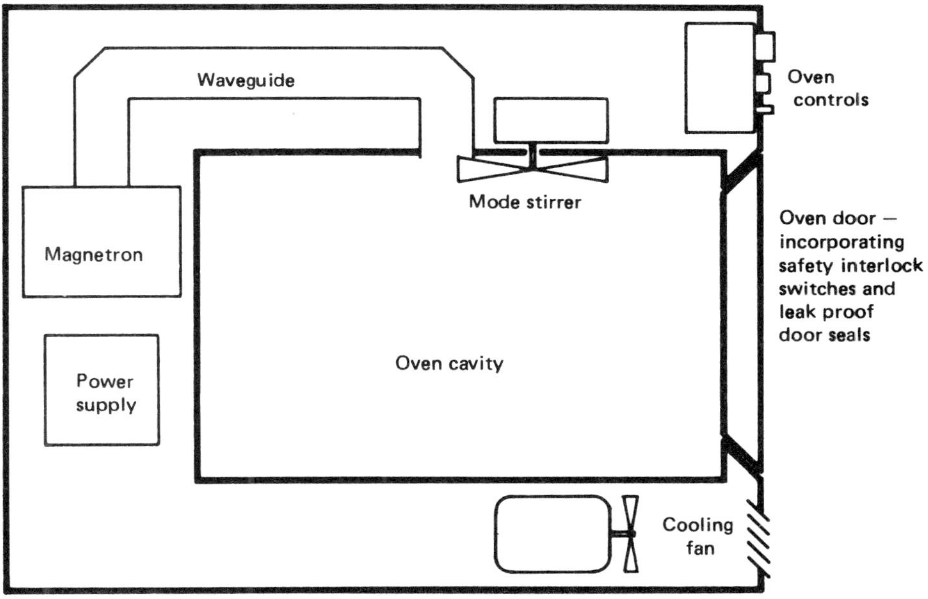


Fig. 7.4 – Microwave oven with mode stirrer.

The oven cavity, of cubical or rectangular shape, is constructed of stainless steel or aluminium. When microwave radiation is incident at a metal surface, the incident wave and reflected wave produce a stationary interference pattern as described above. The dimensions of the cavity are designed to support this standing-wave pattern throughout the whole of the cavity. Thus there will be a pattern of high and low energy density, with maxima separated by a distance of $\lambda/2$ (equivalent to 6 cm for a frequency of 2450 MHz and 17 cm for 896 MHz).

With loads over 500 g the efficiency does not increase further and hence the time to heat a load of 1000 g is twice that for 500 g. Stand-by consumption of energy, which is required to keep the electrical components warm, is about 10% of that when food is actually being heated.

A cavity can support more than one mode of resonance, that is more than one standing-wave. As the size of the cavity increases, relative to the wavelength of the radiation, so does the number of possible modes of resonance. The object of microwave-oven design is to maximise the number of modes present in the cavity. This is called the multi-mode cavity principle. The effect of propagating a large number of modes simultaneously is that each will produce its own pattern of high and low energy regions and the overall effect will be to even out the energy distribution pattern.

For a 100 litre capacity cavity operating at a frequency of 2450 MHz there are 18 theoretically possible modes. Since microwave generators do not produce a single wavelength of radiation, but operate in a narrow bandwidth, this increases the number of possible modes. Also, the method by which the microwave energy is transferred from the waveguide to the cavity can affect the number and types of modes present.

Other devices have been used to influence the energy distribution in the resonant cavity. For example, mode stirrers (Fig. 7.4) are incorporated in batch microwave ovens. A mode stirrer is a rotating fan-like device, which is designed to operate in the path of the energy-coupling port. It is made of metal and when it is in line with the entry port, spurious reflections are caused and the resonant modes are disturbed. Its action is to change continuously the mode patterns set up within the cavity. Because of the presence of mode stirrers, and also because of the varying load conditions, the magnetron has to be designed to withstand a high level of reflected waves. Another method of improving the energy distribution is by the use of a turntable for holding the food (Fig. 7.5). This accepts the non-uniform energy distribution and attempts to move all portions of the food through high and low energy regions.

The energy distribution in a commercial microwave oven, even with multimoding and mode stirrers, is not uniform. The load in the cavity also has a profound effect on the energy distribution pattern. The amount of energy absorbed by the load, and the amount of energy reflected from it, affects the matching of the generator (magnetron) and load and hence the modes of resonance. For this reason a microwave oven can only be assessed for energy uniformity in relation to a particular load. This is difficult in catering applications where the shape and size of the load vary so much. It should be remembered too that the attenuation of the microwaves in the food (Section 7.3) will also cause non-uniformity of energy distribution, and hence temperature, within the food. We would expect that a graph of temperature against depth would

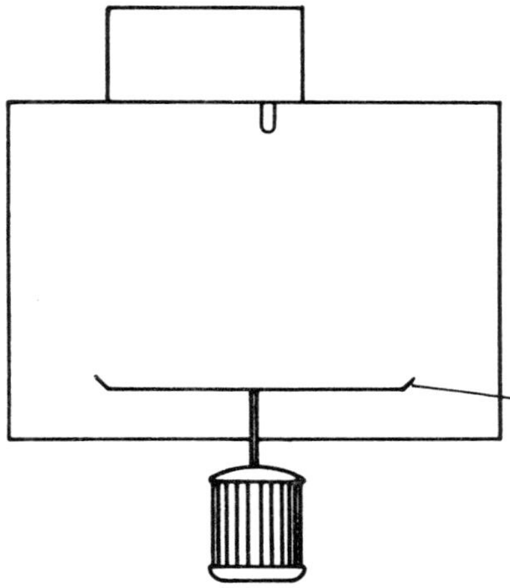


Fig. 7.5 – Oven with turntable.

have a similar shape to the graph of power against depth. However, the lossiness of food varies with temperature; with most food the lossiness decreases as its temperature rises. This will lead to an evening out of the heating because the radiation will be able to penetrate further into the food once the surface layers have heated up. With foods containing a lot of salts and ions however the lossiness increases with temperature.

A further complicating factor is also connected with the attenuation effect. If no other modes of heat transfer are present, then maximum energy generation will take place at the suface, with attenuation occurring below the surface. The walls and the air in the oven cavity however are not heated by the microwaves and therefore there will be a convective and radiative heat loss from the surface of the food. In addition to this, heat will be lost from the surface by the evaporation of water. The effect of these losses will be to reduce the temperature of the surface of the food to below that of food immediately beneath the surface.

The unevenness of the temperature in the food will also cause some heating and cooling of the food to take place by unsteady-state conduction between adjoining parts of the food at different temperatures.

## 7.5 USE OF MICROWAVE OVENS

Bengtsson and Ohlsson (1975) reviewed the use of microwave ovens for rapid heating and concluded that they should permit the very rapid heating of ready meals and meal components with less weight loss and better retention of sensory and nutritional quality that conventional heating methods, provided good control of the time-temperature development can be maintained. A small error in heating time may have considerable effects. They gave additional complicating factors as the selective heating occurring at the corners and edges of the food and uneven energy distribution caused by oven design and the food itself. Control problems are much less in tunnel microwave ovens than in small batch ovens. They believe that developments in control methods should enhance further the value of this form of heating. (Recently, a microwave oven fitted with a temperature probe to measure and control the generation of heat has been marketed.) The use of microwave ovens for food reheating could eliminate the need for the warm-holding of food, which causes large deteriorations in sensory and nutritional quality. The value of a microwave oven for this purpose would be enhanced considerably if correct heating of the food from the frozen state could be achieved routinely by this method.

Although the vast majority of microwave ovens are of the batch type as described in Section 7.4, there are examples in the catering industry of the use of conveyorised tunnel ovens, see for example Klaassen (1977) and Doyon (1977). A tunnel oven consists of a number of microwave resonant cavities joined together with a conveyor belt passing through them. The number of cavities required depends on two factors, one being the throughput required and the other the initial and final temperatures of the food. If the food is at −20°C it requires 500 to 700 kJ/kg to heat it to 100°C, whereas if the food is at 5°C it only requires 300 to 400 kJ/kg for the same final temperature. Because of this, if frozen foods are used then more microwave cavities and a longer conveyor belt are needed than is the case if the foods are chilled. Even with this additional power it is still difficult to heat frozen foods quickly because of the problem of run-away heating.

## 7.6 SAFETY OF MICROWAVE OVENS

There has been much concern over the safety aspects of using microwave ovens. Just as food (biological tissue) is heated in a microwave oven, so human tissue can be heated if it comes into contact with microwaves leaking from an oven. This heating of body tissue can result in burns, which may be repairable (for example, hands) but may also be unrepairable (for example, eye burns).

Standards have been drawn up based on two parameters. The first is the incident power density at the location of personnel using an oven. The main problem in sealing an oven is the door and therefore a safety interlock in incorporated into it. This shuts off the microwave generator when the oven door is opened. The second parameter used is the duration of exposure to the radiation. A typical standard specifies a maximum emission level of 5 mW/cm$^2$ at any point 50 mm or more from the external surface of the appliance, when the oven is operating at its maximum power setting and has a load of 275 ml of drinking water initially at 20°C and containing 1% NaCl, placed at the centre of the load-carrying surface provided by the manufacturer.

Microwave ovens should be checked regularly to make sure that there is no accumulation of dirt on the hinges and oven door seals. Also ovens should be checked for microwave leakage with a leakage monitor at regular service intervals.

## 7.7 CALCULATION OF HEATING EFFECTS BY MICROWAVES

Microwave heating provides a good example of the interactions which take place between the various types of heat transfer in practical cooking situations. We have looked at conduction, convection and radiation heating in isolation. In practice, when a food is heated in a microwave oven all of these processes are taking place simultaneously.

As an aid to the understanding of both microwave energy and its relationship to other heat-transfer mechanisms, we are going to look at a technique for calculating the effect of unsteady-state heat transfer. This uses a numerical method to solve the differential equations which describe heat transfer. The method used is to divide the material up into a number of thin slabs; the heating time is divided up in a similar manner into time intervals. The method is very similar to the Schmidt method of calculation described in Section 4.7. The accuracy of the calculation depends upon the use of very small values for the thickness of the slab and the length of the time interval. Because of this, a long time is required for manual calculation and it is normal to use a computer for its solution. We are going to look at the steps involved in this calculation for the simplest case which is that of one-dimensional heat transfer in a slab of material with microwave radiation incident on one surface only.

Although a theoretical derivation of the equations is complex, a simple understanding may be obtained by looking at the heat balance for one of these slabs. This elaborates the derivation of the Schmidt method of Section 4.7. Both an internal slab and the surface slab must be examined. An internal slab is illustrated in Fig. 7.6.

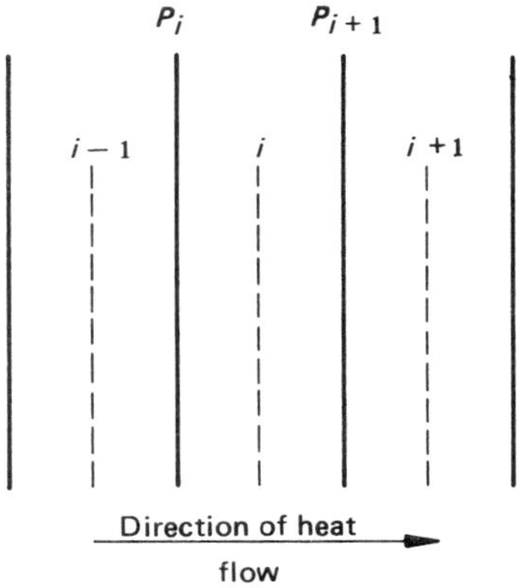


Fig. 7.6 – Microwave heating in an internal slab.

The heat balance for one slab over one time increment is:

$$\begin{bmatrix}\text{Gain in heat}\\ \text{by conduction}\\ \text{through surface}\\ \text{of slab}\end{bmatrix} - \begin{bmatrix}\text{Loss in heat by}\\ \text{conduction through}\\ \text{other surface}\\ \text{of slab}\end{bmatrix} + \begin{bmatrix}\text{Internal heat}\\ \text{generation by}\\ \text{absorption of}\\ \text{microwave energy}\end{bmatrix}$$

$$= \begin{bmatrix}\text{Increase in enthalpy of the slab}\end{bmatrix}.$$

This may be expressed in mathematical terms as:

$$\underbrace{\frac{\lambda A}{\Delta x}(\theta_{i-1}^{n}-\theta_i^{n})}_{\substack{\text{Rate of heat}\\ \text{gain (W)}}} - \underbrace{\frac{\lambda A}{\Delta x}(\theta_i^{n}-\theta_{i+1}^{n})}_{\substack{\text{Rate of heat}\\ \text{loss (W)}}} + \underbrace{q_i A}_{\substack{\text{Rate}\\ \text{of}\\ \text{heat}\\ \text{generation (W)}}} = \underbrace{\frac{A\Delta x\, c\rho}{\Delta t}(\theta_i^{n+1}-\theta_i^{n})}_{\substack{\text{Rate of increase}\\ \text{in enthalpy (W)}}}$$

where $\lambda$ = thermal conductivity of food (W/m.K), $\Delta x$ = thickness of slab (m), $A$ = area through which heat is flowing ($m^2$), $c$ = specific heat of food (J/kg.K), $\rho$ = density of food ($kg/m^3$) and $\Delta t$ = time interval (s).

$q_i$ is the rate of heat absorption per unit area of slab ($W/m^2$) from the microwave radiation. If the microwave radiation at the surface of the $i$th slab is $P_i$ and that at the rear face (which is the surface of the $i+1$ slab) is $P_{i+1}$, the relationship between the two (Section 7.3) is:

$$P_{i+1} = P_i \exp(-2\alpha\Delta x)$$

and

$$q_i = P_i - P_{i+1}\,.$$

In the heat balance we may delete $A$ and multiply through by $\Delta x/\lambda$. This gives:

$$\theta_{i-1}^{n} - \theta_i^{n} - \theta_i^{n} + \theta_{i-1}^{n} + q_i\frac{\Delta x}{\lambda} = \frac{\Delta x^2 c\rho}{\Delta t\lambda}(\theta_i^{n+1}-\theta_i^{n})\,.$$

The term $\Delta t\lambda/\Delta x^2 c\rho$ is the Fourier Number, $Fo$ (Section 4.9), and therefore we may re-write the equation as:

$$\theta_i^{n+1} - \theta_i^{n} = Fo\left(\theta_{i-1}^{n} + \theta_{i+1}^{n} - 2\theta_i^{n} - q_i\frac{\Delta x}{\lambda}\right).$$

If the Fourier number is fixed at $\frac{1}{2}$ by the choice of suitable values of $\Delta x$ and $\Delta t$, the equation simplifies to:

$$\theta_i^{n+1} = \frac{1}{2}\left(\theta_{i-1}^{n} + \theta_{i+1}^{n} + q_i\frac{\Delta x}{\lambda}\right). \qquad (7.1)$$

This equation can be expressed in words as: **The temperature in an incremental slab at the end of one time increment is equal to one half of the sum of the temperatures of the two adjacent slabs at the end of the previous time increment together with the heat generated internally.** In the case of conduction without microwave heating, the expression involving $q_i$ is removed and the equation becomes identical to that used in the Schmidt method, where the temperature at a point is the average of the temperatures at the two adjacent points at the previous time interval (Section 4.7).

It should be noted that for a stable solution of the one-dimensional case, the Fourier number must be set equal to or less than $\frac{1}{2}$ (Kreith, 1976).

The surface slab is illustrated in Fig. 7.7. In this case the heat balance is carried out in exactly the same way except that the heat gain through the left-hand side of the slab is by convection and not by conduction.

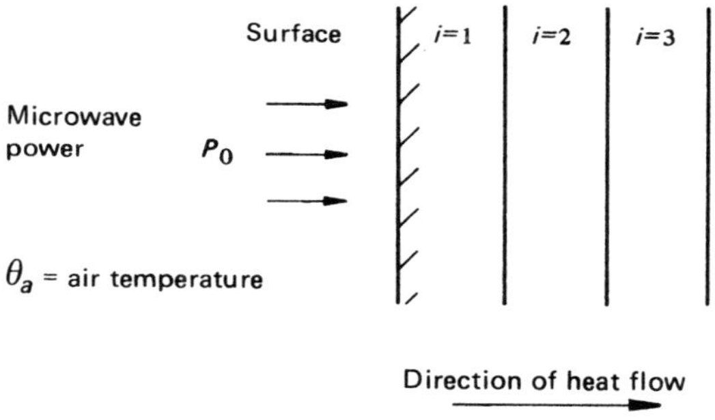


Fig. 7.7 – Microwave heating in surface slab.

The heat balance for the surface slab over one time increment is:

$$hA(\theta_a - \theta_1^n) - \frac{\lambda A}{\Delta x}(\theta_1^n - \theta_2^n) + q_1 A = \frac{A \Delta x c \rho}{\Delta t}(\theta_1^{n+1} - \theta_1^n)\,.$$

| Rate of heat gain by convection (W) | Rate of heat loss (W) | Rate of heat generation (W) | Rate of increase in enthalpy (W) |
|---|---|---|---|

For the case of the surface slab, the microwave radiation at the left-hand surface is equal to the microwave power flux at the surface, $P_o$. Hence $q_1 = P_o - P_2$, and $P_2 = P_o \exp(-2\alpha\Delta x)$.

The heat balance equation for the surface slab can be simplified in the same way as that for the internal slab, to give:

$$\theta_1^{n+1} = \frac{1}{2}\left[\frac{h\Delta x}{\lambda}(\theta_a - \theta_1^n) + q_1\frac{\Delta x}{\lambda} + \theta_1^n + \theta_2^n\right] \qquad (7.2)$$

An illustration of the method of calculation will now be given. A slab of food, 40 mm thick and originally at 0°C, is heated in a microwave oven, with the microwave energy incident on the upper surface only, for three minutes. The microwave power flux is 100 kW/m$^2$ and the air temperature in the oven is 20°C. The physical properties of the food are:

density 1000 kg/m$^3$ ($\rho$),
specific heat 4200 J/kg.K ($c$),
thermal conductivity 1.07 W/m.K ($\lambda$),
surface convection coefficient 5 W/m$^2$.K ($h$),
microwave attenuation coefficient 35 m$^{-1}$ ($\alpha$).

If the food is divided into eight slabs this makes $\Delta x = 0.005$ m. Since *Fo* is set at $\frac{1}{2}$, the value of $\Delta t$ is fixed. Substituting the data in

$$Fo = \frac{\Delta t \lambda}{\Delta x^2 c \rho} = 0.5,$$

gives $\Delta t = 49.1$ s. This means that we must carry out the calculation for three time intervals to give us (approximately) the three minutes of microwave application. The first stage in the calculation is to assign a column to each of the eight slabs, with the upper surface being represented by the left-hand column and the lower surface by the right-hand column. Initially each of these slabs is at a temperature of 0°C. The next step in the calculation is to calculate $q_i$ for each of the slabs for the first time increment.

Upper surface slab:

$$\begin{aligned} P_o &= 100\,000 \text{ W/m}^2 \\ P_2 &= P_o \exp(-2\alpha\Delta x) \\ &= 100\,000 \exp(-0.35) \\ &= 70\,468 \text{ W/m}^2, \\ q_1 &= P_o - P_2 \\ &= (100\,000 - 70\,468) \text{ W/m}^2 \\ &= 29\,532 \text{ W/m}^2. \end{aligned}$$

Second slab:

$$\begin{aligned} P_2 &= 70\,468 \text{ (from above)}, \\ P_3 &= 70\,468 \exp(-0.35) \\ &= 49\,657 \text{ W/m}^2, \\ q_2 &= P_2 - P_3 \\ &= (70\,468 - 49\,657) \text{ W/m}^2 \\ &= 20\,811 \text{ W/m}^2. \end{aligned}$$

This calculation is then repreated for the remaining six slabs to give the values of $P_i$ in Table 7.1. Then, the temperature change in the surface slab at the end of the first time interval is calculated using equation (7.2):

upper surface slab:

$$\theta_1^1 = \frac{1}{2}\left[\frac{5 \times 0.005}{1.07}(20-0) + \frac{29\,532 \times 0.005}{1.07} + 0 + 0\right] = 69.2°\text{C}\,.$$

Similarly, Equation (7.1) is used for the internal slabs. The results are presented in Table 7.1.

In general:

Temperature in first (upper surface) slab at the end of one time interval (Equation (7.2)) $= \frac{1}{2}$ [ $\frac{h\Delta x}{\lambda}$ (Air temperature − temperature of same slab at end of previous period.) + $\frac{\Delta x}{\lambda}$ (Heat generated in slab during this time interval, $q_i$.) + Temperature of same slab at end of previous time interval. + Temperature of adjacent slab to the right at end of previous time interval. ]

Temperature in an internal slab at the end of one time interval (Equation (7.1)) $= \frac{1}{2}$ [ Temperature of adjacent slab to the left at end of previous time interval. + Temperature of adjacent slab to the right at end of previous time interval. + $\frac{\Delta x}{\lambda}$ (Heat generated in slab during this interval, $q_i$.) ]

Lower surface slab: as for upper surface except that the adjacent slab is the one to the left; there is of course no adjacent slab on the right.

**Table 7.1 – Stages in calculation of microwave heating.**

| Slab number | 1 upper surface | 2 | 3 | 4 | 5 | 6 | 7 | 8 lower surface |
|---|---|---|---|---|---|---|---|---|
| Initial temp. | 0 | 0 | 0 | 0 | 0 | 0 | 0 | 0 |
| $P_i$ ($P_o$ = 100 000) | 70468 | 49657 | 34992 | 24659 | 17376 | 12245 | 8629 | 6080 |
| $q_i$ | 29532 | 20811 | 14665 | 10333 | 7283 | 5131 | 3616 | 2536 |
| $\theta_i^1$(end of 1st time period) | 69.2 | 48.6 | 34.3 | 24.1 | 17.0 | 12.0 | 8.4 | 6.2 |
| $P_i$ ($P_o$ = 100 000) | 70468 | 49657 | 34992 | 24659 | 17376 | 12245 | 8629 | 6080 |
| $q_i$ | 29532 | 20811 | 14665 | 10333 | 7283 | 5131 | 3616 | 2536 |
| $\theta_i^2$(end of 2nd time period) | 127.3 | 100.4 | 70.6 | 49.8 | 35.0 | 24.7 | 17.5 | 13.4 |
| $P_i$ ($P_o$ = 100 000) | 70468 | 49657 | 34992 | 24659 | 17376 | 12245 | 8629 | 6080 |
| $q_i$ | 29532 | 20811 | 14665 | 10333 | 7283 | 5131 | 3616 | 2536 |
| $\theta_i^3$(end of 3rd time period) | 181.6 | 147.5 | 109.4 | 76.9 | 54.3 | 38.3 | 27.5 | 21.5 |

The calculation as shown in Table 7.1 cannot give accurate results for a number of reasons. First, the accuracy depends on choosing small values for $\Delta x$ and $\Delta t$; the values chosen were not small. Then, in practice, foodstuffs containing high percentages of water will not rise above 100°C until drying of the surface has taken place. Our simple model can allow for this by replacing any temperatures over 100°C by this figure. Lastly physical properties are temperature-dependent. In particular q is a very siginificantly affected by temperature changes, as the following data show.

| Temperature (°C) | $\alpha$ (water) |
|---|---|
| 0 | 62.4 |
| 20 | 35.2 |
| 40 | 20.9 |
| 60 | 14.2 |
| 80 | 9.5 |
| 100 | 7.3 |

In order to achieve more accurate results, it is necessary to make the calculations much more laborious by using much smaller values of $\Delta x$ and $\Delta t$. This is best done by using a computer. A full discussion of computer simulations, including microwave heating, is included in Chapter 9.

## 7.8 NOMENCLATURE

$c$ specific heat (J/kg.K)
$i$ number of slab (–)
$n$ number of time interval (–)
$q_i$ rate of heat absorbtion ($W/m^2$)
$x$ depth below surface (m)
$\Delta x$ thickness of slab (m)
$\Delta t$ time interval (s)
$A$ area ($m^2$)
$E$ voltage gradient (V/cm)
$Fo$ Fourier number (–)
HPD half power depth (m)
$K'$ dielectric constant (–)
$K''$ dielectric loss factor (–)
$P$ power absorbed ($W/cm^3$)
$P_o, P_x$, etc. power flux ($W/m^2$)

$\alpha$ attenuation coefficient (1/m)
tan $\delta$ loss tangent (–)
$\theta$ temperature (°C or K)
$\lambda$ thermal conductivity (W/m.K)
$\nu$ frequency (Hz)
$\rho$ density (kg/m$^3$)

## 7.9 PROBLEMS

(1) Repeat the calculation in Section 7.7 but calculate the values of $P_i$, and hence $q_i$, using a value of $\alpha$ appropriate to the temperature in each slab at the end of the previous time interval.

(2) Discuss the effect of allowing for the valuation in q with temperature.

(3) What will be the effect of leaving the food for ten minutes after the three minutes of microwave heating?

## 7.10 REFERENCES

Bengtsson, N. and Ohlsson, T. (1975), Paper to 6th European Symposium on Food: engineering and food quality, IUFoST, Cambridge.

Doyon, P. (1977), Paper in Glew, G. (ed.), *Catering equipment and systems design*. Applied Science Publishers.

Goldblith, S. A. (1975), *Royal Society Discussion on Food Technology in the 1980's*. The Royal Society, 55-68.

Klaassen, A. (1977), Paper in Glew, G. (ed.), loc. cit.

Kreith, F. (1976), *Principles of heat transfer*, 3rd ed. Harper & Row.

## 7.11 FURTHER READING

Decareau, R. V. (ed.) (1975), *Microwave ovens and frozen foods make cents.* Transactions of the International Microwave Power Institute, Vol. 5.

Kirk, D. and Holmes, A. W. (1975), The heating of foodstuffs in a microwave oven. *J. Fd. Technol.*, **10**, 375-384.

Okress, E. C. (ed.) (1968), *Microwave power engineering*, Vol. 2, Applications, Chapter 5. Academic Press.

Osepchuk, J. M. (1978), *A review of microwave oven safety. Jnl. Microwave Pwr.*, **13**(1), 13-26.

Tinga, W. R. (ed.) (1976), *Cooking with microwaves.* Transactions of the International Microwave Power Institute, Vol. 6.

Van Zante, H. J. (1973), *The microwave oven.* Houghton Mifflin.

CHAPTER 8

# Freezing and Thawing

## 8.1 INTRODUCTION

The freezing and thawing of foods are problems in unsteady-state heat transfer with the additional complication of a change of state, that is a liquid (mainly water in the food) becoming a solid with the removal of latent heat and a solid becoming a liquid with the addition of latent heat. The latent heat involved is large compared with the sensible heat, as shown for water in Fig. 8.1. The advent of cook-freeze catering units and the increasing use of frozen convenience foods in conventional catering operations involve an understanding of these two basic processes. It is necessary to know how long it will take to freeze or thaw a particular item. An appreciation of freezing and thawing equipment and chilled- and frozen-food holding equipment (refrigerators, cold stores and chilled-food cabinets) is also required.

## 8.2 FREEZING AND DEFROSTING CURVES

Freezing and defrosting curves describe the temperature-time relationship in a food whilst it is being frozen or thawed. Pure substances, such as water, freeze and thaw at one particular temperature but foods freeze and thaw over a range of temperatures. During freezing, latent heat is removed from the food, whilst during thawing latent heat is added to the food. This removal/addition of latent heat is in addition to the removal/addition of sensible heat which alters the temperature of a substance or food without freezing/thawing taking place. Typical freezing and defrosting curves are shown in Fig. 8.2.

Freezing takes place in three stages. In the **precooling period** (*A*), the temperature falls to the freezing point because sensible heat is removed. During the **phase change period** (*B*), for a pure substance, the liquid becomes solid at constant temperature with the removal of latent heat. For a food, the liquid in the food (principally water) becomes solid, but in this case the process takes place over a range of temperatures. Even at comparatively low temperatures (below commercial food-storage temperatures) some unfrozen water may still be present. The temperature range over which, say, 90% of the water content freezes, varies from food to food. Finally, there is usually a **tempering period** (*C*). For a pure substance the temperature begins to fall again, whilst for a food it falls more rapidly than in period *B*. Sensible heat is removed during this stage. With foods the temperature of the changeover from stage *B* to stage *C* is not clearly defined for the reasons given above.

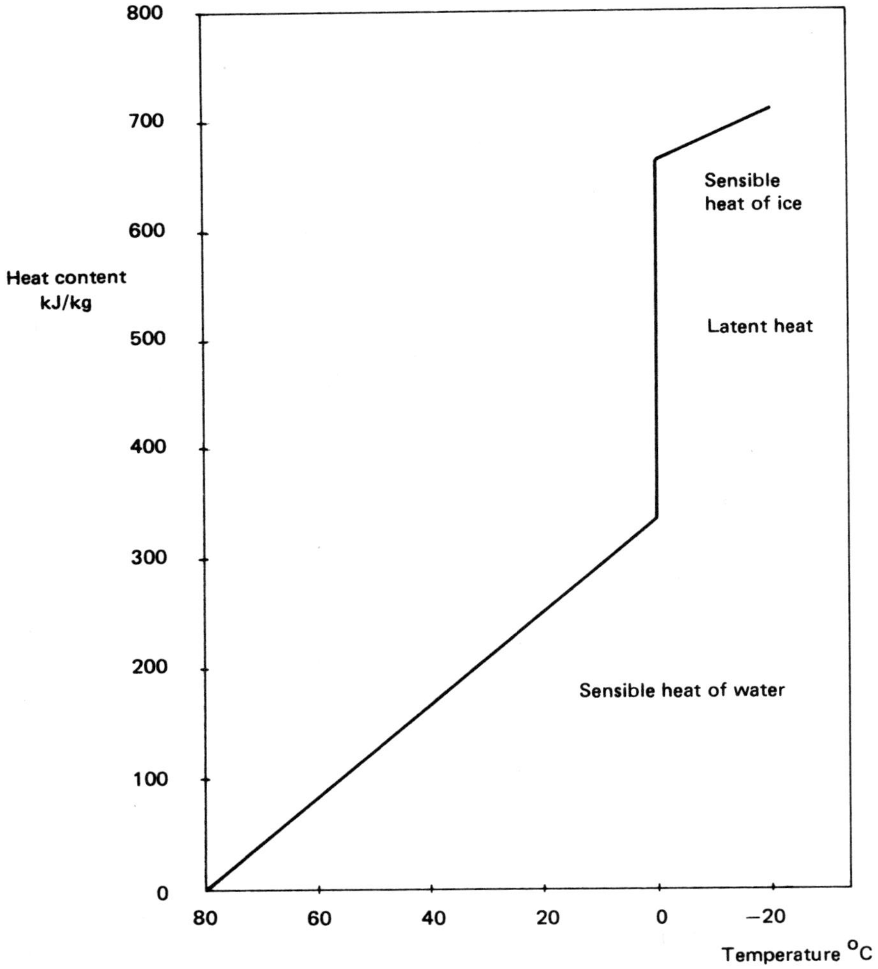


Fig. 8.1 – Heat to be removed (kJ/kg) from water at 80°C to bring it to the temperature shown on the other axis.

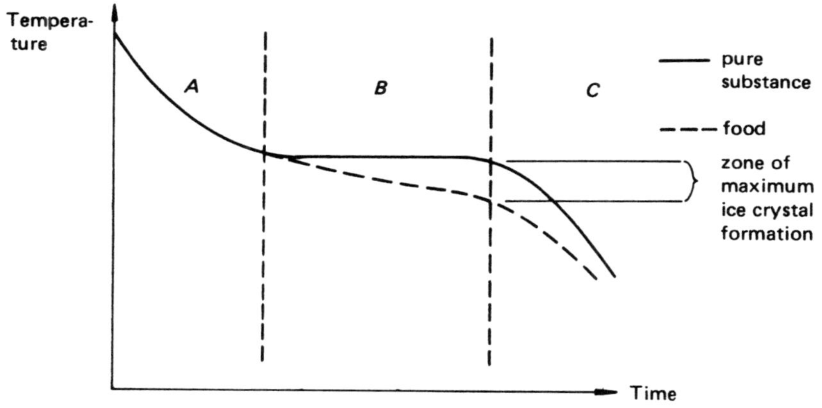


Fig. 8.2 – Typical freezing/defrosting curves.

Consequently it is not possible to define a clear endpoint to the freezing process of foods. There are two approaches to a practical definition of freezing time, one relevant to the quality of the product and the other to the throughput of the freezing plant. The **freezing time** of a body may be defined as the time taken for its thermal centre to fall through the zone of maximum ice crystal formation. Alternatively the **effective freezing time** may be defined as the time required to lower the temperature of the product from its initial average value to a given value at the thermal centre.

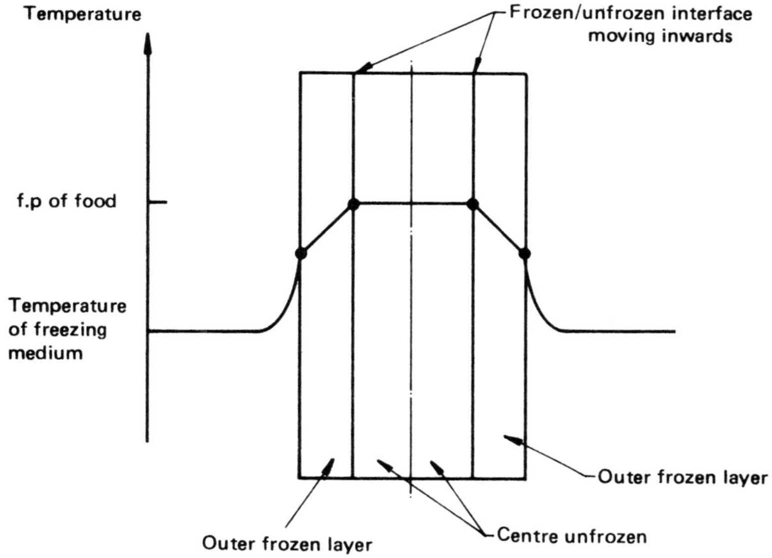


Fig. 8.3 – Temperature distribution during freezing.

The thawing process is the reverse of the freezing process except that, under similar conditions, thawing is a much slower process than freezing because the thermal diffusivity of ice is approximately nine times that of water. In order to freeze food, the food is placed in a medium which is at a temperature below the freezing point of the food. The food starts freezing on the outside and the frozen/unfrozen interface moves inwards progressively as time passes (Fig. 8.3). It is generally considered that, from the point of view of food quality, as fast a rate of freezing as possible should be used, coupled with as low a final temperature as possible. These objectives may be achieved by using a cooling medium with as low a temperature as possible (this increases the driving force removing heat from the food) and minimising the resistance to heat transfer in the freezing equipment.

The effect of freezing rate on the temperature-time curve is illustrated in Fig. 8.4.

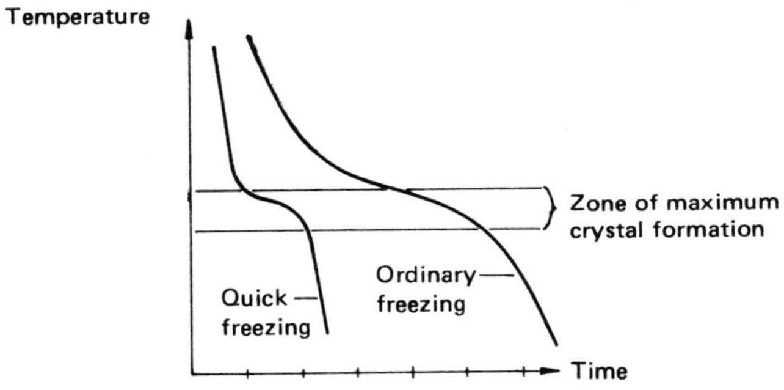


Fig. 8.4 – Effect of freezing rate in temperature-time curve.

## 8.3 CALCULATION OF FREEZING AND THAWING TIMES

The equation most commonly quoted in the literature for calculating the freezing/thawing time of a food is that derived by Plank:

$$t = \frac{L\rho}{\Delta\theta}\left(P\frac{a}{h} + R\frac{a^2}{\lambda}\right)$$

where $t$ = the freezing (or thawing) time (s), $\rho$ = the density of food (kg/m$^3$), $L$ = the latent heat of food (J/kg), $\Delta\theta$ = the freezing temperature of food – temperature of freezing medium (°C), $a$ = the smallest dimension of the food (m), $h$ = the overall heat-transfer coefficient to food surface (W/m$^2$.K), $\lambda$ = the thermal conductivity of the frozen food (or unfrozen food in case of thawing) (W/m.K) and $P$ and $R$ = dimensionless constants dependent on geometry of food. For an infinite cylinder $P = 1/4$, $R = 1/16$; for an infinite slab $P = 1/2$, $R = 1/8$; for a sphere $P = 1/6, R = 1/24$ and for a cube $P = 1/6, R = 1/24$. For brick-shaped solids, $P$ and $R$ are obtained from Fig. 8.5, where $\beta_1$ and $\beta_2$ are the ratios of the two longest sides to the shortest.

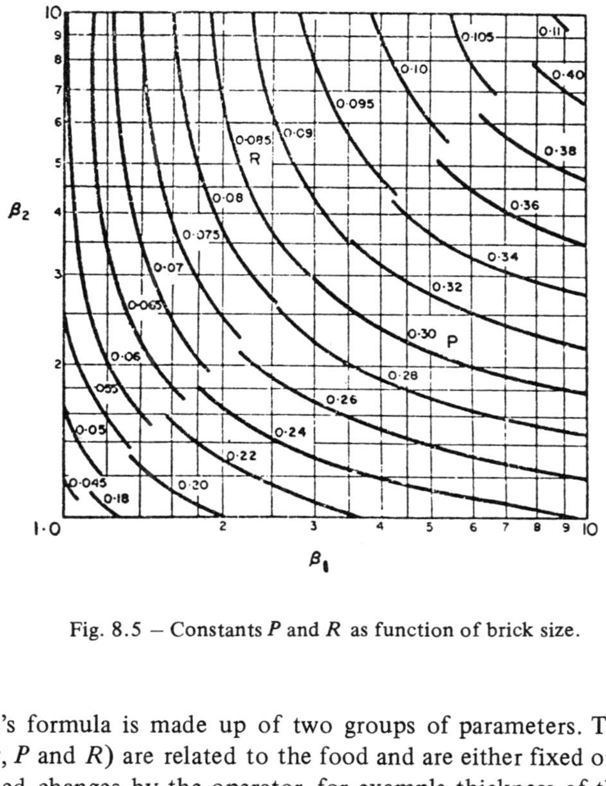


Fig. 8.5 – Constants $P$ and $R$ as function of brick size.

Plank's formula is made up of two groups of parameters. The first group ($L$, $\rho$, $\lambda$, $a$, $P$ and $R$) are related to the food and are either fixed or only capable of restricted changes by the operator, for example thickness of the food pack. The other group ($\Delta\theta$ and $h$) are dependent on the heat-transfer equipment. It is these two parameters which may be influenced by equipment design.

The assumptions made by Plank in deriving his formula were:

(a) freezing occurs at a definite temperature (as for a pure substance);
(b) freezing starts with all the food at the freezing temperature; the pre-cooling and tempering periods are excluded;
(c) heat transfer from the freezing front in the food to the cooling medium is in steady-state, that is freezing takes place relatively slowly;
(d) food density is the same in both the frozen and unfrozen states;
(e) thermal conductivity is independent of temperature;
(f) the freezing medium is at a constant temperature.

In practice, these assumptions are rarely correct and therefore one would not expect Plank's equation to give very accurate predictions of freezing/thawing times. If the food is packaged then the value of the overall heat-transfer coefficient ($h$) in Plank's equation will not be simply the convective heat-transfer coefficient. The value of this overall coefficient may be calculated more generally from the equation:

$$\frac{1}{h} = \frac{1}{h_c} + \frac{x_p}{\lambda_p} + \frac{1}{h_r}$$

where $h$ = the overall heat-transfer coefficient, as used in Plank's equation ($W/m^2.K$), $h_c$ = the convective heat-transfer coefficient to the surface ($W/m^2.K$), $x_p$ = the thickness of packaging material covering the food (m), $\lambda_p$ = the thermal conductivity of packaging material (W/m.K) and $h_r$ = the radiation heat-transfer coefficient to the surface ($W/m^2.K$).

It should be evident from the above equation that the presence of a package reduces the effective heat-transfer rate between the coolant fluid and the food.

Earle (1965) tested Plank's formula over a range of conditions in an air-blast freezer. He found that the theoretical predictions of freezing time needed multiplying by 1.42 (range over 18 tests: 1.35 to 1.54) to obtain agreement with the actual freezing times. Part of this factor (15–20% of the freezing time) was required because the initial temperature of the meat was 10–13°C. The remainder of the increase in time required was attributed to either data inaccuracies or Plank's assumptions. In a subsequent paper, Earle and Fleming (1967) presented a similar comparison for lamb and mutton freezing. The multiplying factor in this paper (calculated by the present authors) is in the range 0.72 to 1.21 (average 0.99). However, in this case the latent heat term in the Plank equation was replaced by the enthalpy change. This enthalpy change is the heat content due to the initial temperature being above freezing point plus the latent heat of freezing. By using enthalpy change instead of latent heat only, one is effectively including the pre-cooling heat removal in the equation. Vanichseni *et al.* (1972) tested Plank's equation for predicting thawing times of meat. They found it necessary to modify Plank's equation (to account for the differences between theoretical and experimental values) as follows:

(a) they replaced the latent heat term by the enthalpy change between the initial and final centre temperatures;
(b) they used a multiplying factor of 1.2.

Brennan *et al.* (1976) and Walker (1970) have made similar suggestions to Vanichseni's regarding the replacement of the latent heat term by an enthalpy term, although there is no suggestion that an additional correction factor is required. Brennan *et al.* however pointed out that if the food is cooled considerably below its freezing point, then the simple substitution of enthalpy change for latent heat is likely to underestimate the duration of the process. It should also be noted that Walker was calculating thawing times and generally limiting the temperature rise of the food to just above the thawing temperature. In these circumstances he found a good correlation between practical and theoretical results.

It would seem that if the pre-cooling and tempering stages are significant parts of the freezing/thawing process, then Vanichseni's method is likely to give the best agreement between calculated and experimental results. If there is no significant tempering period (during freezing) or heating period (above thawing point in thawing) then enthalpy change may be used instead of the latent heat term, without the need for a correction factor.

Recently, Cleland and Earle (1976) have developed a more complex form of Plank's equation for slabs which they showed was in error by less than ±3%. Plank's equation was put in the following dimensionless form:

$$Fo = P\left(\frac{1}{Bi\,Ste}\right) + R\left(\frac{1}{Ste}\right)$$

where $Fo$ = Fourier number, $\alpha t/a^2$ (Section 4.8), $\alpha$ = thermal diffusivity of a food ($m^2/s$), $Ste$ = Stefan number, $c_s\Delta\theta/\Delta H$, $c_s$ = specific heat of frozen food ($J/m^3.K$), $\Delta H$ = enthalpy difference between initial freezing temperature and final centre temperature ($J/m^3$), $Bi$ = Biot number, $ha/\lambda$ (Section 4.8), $P = 0.5072 + 0.2018\,\text{Pk} + Ste\,(0.3224\,\text{Pk} + 0.0105/Bi + 0.0681)$, $R = 0.1684 + Ste\,(0.2740\,\text{Pk} - 0.0135)$, $Pk$ = Plank number $= c_L\Delta\theta'/\Delta H$, $c_L$ = specific heat of unfrozen food ($J/m^3.K$) and $\Delta\theta'$ = initial temperature of food minus freezing temperature of food (°C).

This method was verified experimentally over the following range of variables:

initial temperatures up to 40°C;
freezing medium temperature −15°C to −45°C;
slab thickness up to 0.12 m;
heat-transfer coefficient 10 to 500 $W/m^2.K$.

Plank's equation shows that a decrease in the temperature of the coolant medium will increase $\Delta\theta$ and hence reduce the freezing time. An increase in the convective heat-transfer coefficient ($h$) will also reduce the freezing time. This may be achieved by increasing the air velocity. In turn, this will increase the power requirements for blowing the air over the food. Power requirements are proportional to (air speed)$^{2.6-3}$ whilst the heat-transfer coefficient is proportional to (air speed)$^{0.6-0.9}$ according to Ciobanu *et al.* (1976). This means that cost-wise there is a limit to maximum air speeds which are economical. The same authors quote Plank as calculating that the minimum overall energy consumption for a 50 mm thick product was obtained at an air velocity of 6 to 8 m/s. Thicker products are frozen most economically at lower air velocities than this. These are the two principal parameters which may be controlled. It should also be noted that an increase in pack thickness ($a$) increases freezing time, and therefore the food pack should be as thin as possible consistent with production and customer constraints. If two packs are placed on top of each other and are touching then this effectively doubles the pack thickness. It is essential that there is a free flow of air around each food pack in the freezer.

## 8.4 REFRIGERATION CYCLE

It is convenient in this chapter to look at the means of producing cold air for freezing or chilling foodstuffs, that is to look at what are called refrigeration cycles. In catering, cold air is produced either to hold food and drinks chilled but not frozen, to hold food frozen or to actually freeze food in a blast freezer. Air is not however always the heat-transfer medium. In an ice-maker, water is frozen in contact with the evaporator surface of the refrigeration cycle. Whatever the purpose, however, the principle is the same.

The principle of a refrigeration cycle is for the heat to be received by a working fluid (refrigerant) at a low temperature (in a cold room, freezer or cellar) and for the heat to be rejected at a higher temperature (into a kitchen usually). This take-up of heat by the refrigerant needs to be accompanied by the doing of work, usually by means of electrically-driven compression of the working fluid. We may represent a refrigerator and the corresponding refrigeration cycle as shown in Figure 8.6.

There are four stages in the refrigeration cycles. First the refrigerant (working fluid) in the vapour phase is compressed (in a compressor). This is represented by stage 1 to 2 in Fig. 8.6. Work is done on the refrigerant ($W_{12}$) and it increases in temperature as well as pressure. Secondly refrigerant leaving the compressor is passed through a condenser (stage 2 to 3). This is a tube system which is cooled by the air in the kitchen since the refrigerant leaves the compressor above kitchen temperature. It is thereby cooled and condenses to a liquid with heat ($Q_{23}$) being expelled to the kitchen. Then the liquid refrigerant passes through an expansion device (stage 3 to 4) where its pressure is reduced and it, the refrigerant, does work ($W_{34}$) on the expansion device. Finally liquid refrigerant passes through the evaporator (stage 4 to 1) which is situated in the freezer, cold room or whatever. It evaporates and in so doing abstracts heat ($Q_{41}$) from the refrigerator cabinet, cold room and so on. This heat has originally been conducted into the cabinet from the warmer kitchen. The cycle then repeats itself.

The cycle is represented in Fig. 8.6 by means of a $T$-$S$ or reversed Carnot cycle diagram. $T$ is the absolute temperature of the refrigerant and $S$ is its entropy. Entropy is defined as total heat content ($H$) over absolute temperature ($T$) and is a useful concept in considering the refrigeration cycle because stages 1 to 2 (compression) and 3 to 4 (expansion) are carried out at constant entropy.

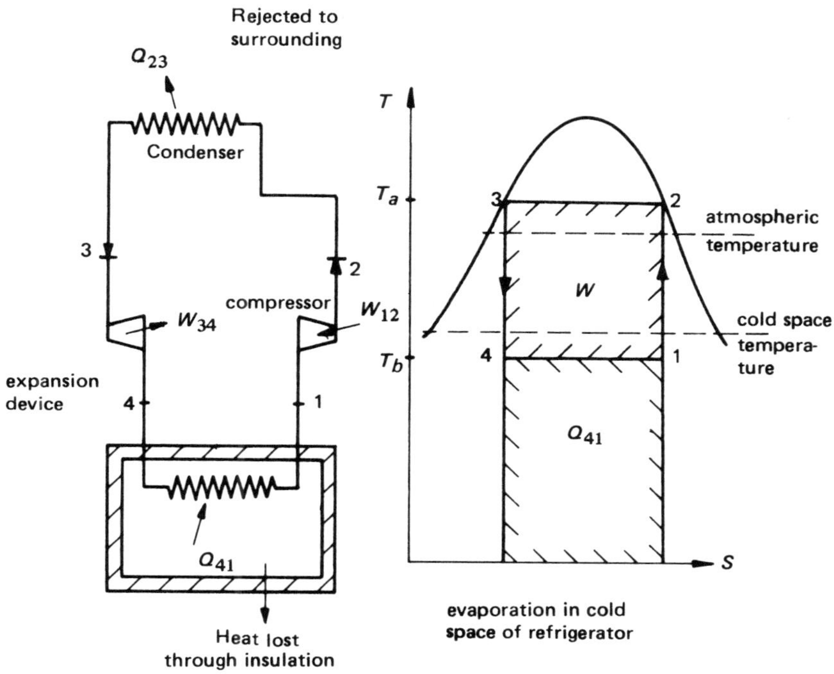


Fig. 8.6 – Ideal refrigeration cycle.

The aim of the refrigerator is to abstract as much heat ($Q_{41}$) as possible in the evaporator for a net expenditure of work ($W_{12} - W_{34} = W$). These two areas are shown in Fig. 8.6. The performance of the refrigeration cycle is expressed as: coefficient of performance $= Q_{41}/W$ where $Q_{41}$ and $W$ are represented by the areas shown in Fig. 8.6. The coefficient of performance may also be defined as:

$$\frac{T_b}{T_a - T_b}.$$

$T_a$ is the temperature in the condenser, which must be at least 10°C above the kitchen temperature. If it were not, a much larger surface area would be required to transfer the heat by convection from the condenser to the kitchen. If the kitchen is warm or if the refrigeration equipment is large there may be insufficient temperature difference between the kitchen temperature and the refrigerant temperature. Also, the refrigerator equipment may contribute to hot working conditions. In these cases, the condenser coil may be situated outside the building. Similarly $T_b$ must be at least 10°C lower in temperature than the cold room for the same reason. These necessary temperature differences limit the coefficient of performance attainable.

The effect of cold room temperature on the coefficient of performance may be illustrated as follows. If a kitchen temperature is 20°C and the cold room is held at −20°C, then:

$$T_a = 20° + 10°\text{C} = 30°\text{C} = 303\ \text{K},$$
$$T_b = -20 - 10°\text{C} = -30°\text{C} = 243\ \text{K}.$$

Therefore

$$\text{the coefficient of performance} = \frac{243}{303 - 243} = \frac{243}{60} = 4.05\ .$$

If, however, the food only needs to be held at 5°C:

$$T_a = 303\ \text{K},$$
$$T_b = 5 - 20°\text{C} = -15°\text{C} = 258\ \text{K}.$$

Therefore

$$\text{coefficient of performance} = \frac{258}{303 - 258} = \frac{258}{45} = 5.73\ .$$

Thus, holding the food at 5°C instead of −20°C increases the coefficient of performance from 4.05 to 5.73.

In practice, an actual refrigeration cycle is not as ideal as the reversed Carnot cycle. The expansion device used in practice is a simple throttle valve (because it is cheap) and therefore the expansion is not carried out at constant entropy. The effect of this is to reduce the coefficient of performance of the cycle. The compression stage, similarly, is not at constant entropy because of difficulties in controlling the process. Again this reduces the coefficient of performance. Nevertheless, the practical cycle is much cheaper to achieve than an (ideal) reversed Carnot cycle. The coefficient of performance defined above gives the maximum value attainable. Real cycles deviate from this and the designers' aim will be to get as near to ideal as possible while taking full account of the economics of the situation.

A practical cycle will be as shown in Fig. 8.7. It is seen from the diagram that the area which represents the net expenditure of work ($W$) is greater than in the case of the real cycle and hence the coefficient of performance is smaller.

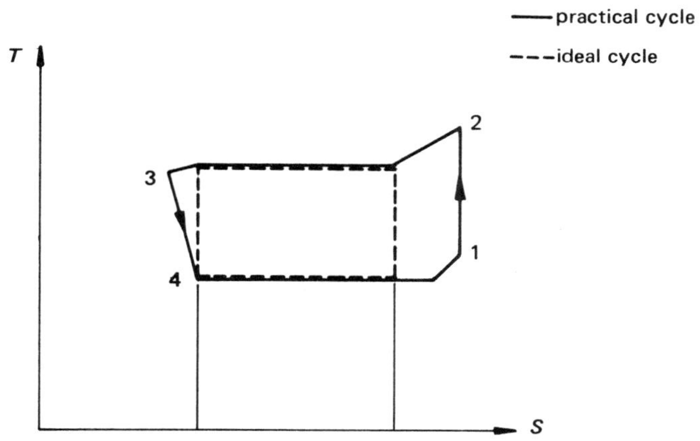


Fig. 8.7 – Practical refrigeration cycle.

## 8.5 FREEZING AND CHILLING EQUIPMENT

Maximum freezing times are often set by authorities for reducing the temperature of the food from, say, 70°C to −18°C after cooking. Similar standards may also be set for chilling food to temperatures below 5°C. It is likely that these standards will be more stringent in the future. Such standards have a great influence on the techniques used for freezing/chilling food in the catering industry.

There are three methods of freezing food available for catering operations. Other methods of freezing are not feasible, either because of the size of the packs to be frozen or because of the variability of pack size likely when freezing prepared meals or meal components.

### 8.5.1 Air-blast freezers

This is the most-used method of freezing food in catering. The air-blast freezer uses a high velocity stream of cold air circulating over the food. The food warms up the air because the air extracts heat from the food, and the warmed air is recirculated through a refrigerated heat exchanger to reduce its temperature again. The air temperature must not be higher than −32°C for efficient freezing and should, preferably, be −40°C. The air velocity must not be less than 5 m/s and should be, preferably, 6.6 m/s. Heat-transfer coefficients are increased considerably by increases in air velocity (see Chapter 5) as shown by the following typical date:

| | |
|---|---|
| still air | 5.7 $W/m^2.K$ |
| air velocity 2.5 m/s | 17.0 $W/m^2.K$ |
| air velocity 5.0 m/s | 28.4 $W/m^2.K$. |

As pointed out in Section 8.3, there is an economic limit to increasing the air velocity to achieve higher heat-transfer rates. Furthermore, the packaging of food reduces the effective heat-transfer rate, although packaging is essential from other points of view (see Chapter 10). Nevertheless, freezing times of say 75 to 90 minutes are achieved in this type of equipment compared with about 8 hours in still-air units (the domestic type). Air-blast freezers may be quite small in size with batch operation. Large units are semi-continuous, arranged as a push-through tunnel, with trolleys of unfrozen food pushed in on one side and trolleys of frozen food removed into a cold store on the other side. Even larger units operate as continuous tunnels and are arranged so that the air flow is in the opposite direction to the food flow (counter-flow operation). Blast freezers may also be used to chill food rather than freeze it.

### 8.5.2 $CO_2$ freezing

In this type of freezer liquid carbon dioxide ($CO_2$) is piped from a storage tank and is injected into the freezer through specially-designed nozzles. The liquid carbon dioxide changes, in the freezer, to a mixture of gas and solid carbon dioxide snow, both at a temperature of about −65°C to −70°C. Fans cause intensive circulation of the cold gas around the food to provide even and rapid freezing. These units may also be used for chilling food and crust freezing (their operating temperature range is +10°C to −70°C). Faster freezing is obtained with $CO_2$ freezers than with air-blast freezers because of the lower temperatures of the coolant (higher $\Delta\theta$) and also higher convective heat-transfer rates. No values of the latter have been published in the literature but evaporation (boiling) of the solid (or liquid) carbon dioxide on the food surface will enhance the value of the heat-transfer coefficient ($h$) obtained by rapid circulation of the $CO_2$ gas in the unit (see Section 5.5.1).

The freezer is thermostatically controlled to ensure economical use of refrigerant while timers are incorporated into the unit for the same purpose. The liquid $CO_2$ is fed automatically from storage to the freezer. The $CO_2$ is held in storage at a pressure of 20 bar and at a temperature of −18°C. The doorframe of the freezer is heated to prevent the build-up of ice near the door hinges, ensuring that the door opens freely. A typical freezing time would be 45 minutes (cf. 75 to 90 minutes with an air-blast freezer). Against this gain must be set the capital cost of the system (the freezer cabinet is cheaper than an air-blast cabinet including its refrigeration unit, but a storage tank for the $CO_2$ must be provided) and the cost of the carbon dioxide.

A batch-type unit is available for catering applications although tunnels are available if the throughput is high enough.

### 8.5.3 Liquid $N_2$ freezing (cryogenic freezing)

This is similar to $CO_2$ freezing except that liquid nitrogen is used. The temperature of liquid nitrogen is −196°C, giving an even higher temperature difference

($\Delta\theta$). The liquid nitrogen boils on coming into contact with the food, thereby enhancing the heat-transfer coefficient ($h$). Typical freezing times are up to 25 minutes, which is faster than with liquid $CO_2$. Here again the freezing tunnel costs less than an equivalent, mechanically-refrigerated blast freezer but storage of liquid $N_2$ is expensive.

Chilling equipment is similar to freezing equipment and is used to reduce the temperature of the food to below 3 to 5°C but keeping it above 0°C. Blast-air chillers, liquid $CO_2$ and liquid $N_2$ units are all used. The main problem is to maximise the chilling rate without at the same time freezing the crust of the food. For cook-chill systems, particularly when the food is to be stored for several days, the food should be chilled to and stored at a temperature below 3°C. The shelf-life of a chilled food is influenced by the rate of cooling after cooking and some form of rapid chilling device is essential for such a system.

## 8.6 THAWING EQUIPMENT

The increasing use of frozen foods in catering, and the slowness with which they thaw, has led to consideration of the possibilities of thawing foods rapidly prior to reconstitution. Jason (1974) has suggested three potential methods: electrical resistance heating (thawing), vacuum thawing and radio-frequency (r.f.) power source, with the food in sealed plastic bags.

Starke (1977) has described the development of a forced-convection cabinet for such a purpose. The cabinet (Fig. 8.8) is not unlike a forced-convection oven.

Thaw times of $2\frac{1}{2}$ hours were achieved for thawing from the cold store temperature to +5°C with maximum air temperatures of 15°C. Lidding the packs was found to lengthen thaw times by 60 to 70%. Dorney and Glew (1977) considered that only two methods of thawing in catering situations have potential for development:

(a) air thawing: (i) by natural convection;
(ii) by forced convection;

(b) water thawing: (i) by immersion;
(ii) by spray.

Tests were carried out on thawing by natural and forced air convection and partial water immersion (the containers were not fully water tight). It was concluded that the following methods were both controlled (that is, avoided excessive temperatures in the product) and sufficiently fast: partial immersion in water at 10°C and forced air convection at 10°C. The latter was selected for further study as water immersion may cause physical damage to the food. The tests were additional to those described by Starke (ibid.). The use of a modified version of Plank's equation for predicting the results of the tests was considered earlier in the chapter.

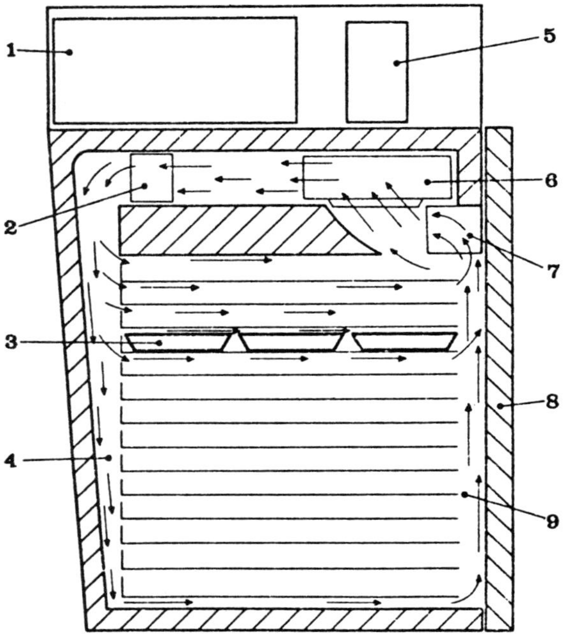


Fig. 8.8 – Rapid thaw cabinet – individual pack version. (1) condensing unit, (2) heating elements, (3) individual pack, (4) tapered delivery duct, (5) fan motor, (6) fan, (7) cooling coil, (8) door, (9) air return duct.

## 8.7 NOMENCLATURE

- $a$ thickness of a food pack (m)
- $c$ specific heat (J/m$^3$.K)
- $h$ heat-transfer coefficient (W/m$^2$.K)
- $t$ freezing/thawing time (s)
- $x$ thickness (m)
- $Bi$ Biot number (–)
- $Fo$ Fourier number (–)
- $H$ heat content (enthalpy) (J)
- $\Delta H$ enthalpy change (J/m$^3$)
- $L$ latent heat (J/kg)
- $P$ shape constant (–)
- $Pk$ Plank number (–)
- $Q$ heat (J)
- $R$ shape constant (–)
- $Ste$ Stefan number (–)
- $T$ temperature (K)
- $W$ work (J)

$\alpha$ thermal diffusivity ($m^2/s$)
$\beta$ length ratio (–)
$\theta$ temperature (°C or K)
$\lambda$ thermal conductivity (W/m.K)
$\rho$ density ($kg/m^3$)

## 8.8 PROBLEMS

**(1)** (a) Frozen beef steak (not packaged) is placed on a preparation table for defrosting in a room at 20°C. The side in contact with the table is effectively insulated so that thawing is taking place on one direction only. Calculate the thawing time of the steak.

Data: thickness of steak = 8 mm;
latent heat of steak ($L$) = 251 kJ/kg;
heat-transfer coefficient ($h$) = 7.4 $W/m^2.K$;
freezing point of steak = –2°C;
thermal conductivity of steak ($\lambda$) = 0.48 W/m.K;
density of steak ($\rho$) = 1040 $kg/m^3$.

Size of each steak may be approximated to 200 mm × 125 mm × 8 mm.

(b) Recalculate the thawing time for steaks of thickness 16 mm and 24 mm.

(c) Comment on the results.

**(2)** (a) Estimate the freezing times of a food pack in various types of freezer using the following data.

Data: thickness of pack (a) = 50 mm;
other pack dimensions = 100 mm × 75 mm;
cardboard wrapping thickness ($x_p$) = 1 mm;
thermal conductivity of cardboard ($\lambda_p$) = 0.06 W/m.K;
thermal conductivity of frozen food ($\lambda$) = 1.53 W/m.K;
latent heat of food ($L$) = 256 kJ/kg;
density of food ($\rho$) = 1088 $kg/m^3$;
food freezes at –2°C.

Equipment data:

| | $h_c$($W/m^2.K$) | Coolant temperature (°C) |
|---|---|---|
| Still air freezer | 5.7 | –30 |
| Air blast freezer (2.5 m/s) | 17.0 | –30 |
| (5.0 m/s) | 28.4 | –30 |

(b) Comment on the results.

## 8.9 REFERENCES

Brennan, J. G. *et al.* (1976), *Food engineering operations* (2nd ed.). Applied Science Publishers.

Ciobanu, A. *et al.* (1976), *Cooling technology in the food industry*, 104 and 146. Abacus Press.

Cleland, A. C. and Earle, R. L. (1976), *Bulletin de l'institut du froid*, annexe 1976-1, 369-375.

Dorney, D. C. and Glew, G. (1977), *J. Fd. Technol.*, **12**, 523-534.

Earle, R. L. (1965), paper to 1st International Conference on Food Science and Technology, 1962. Gordon and Breach.

Earle, R. L. and Fleming, A. K. (1967), *Food Technology*, **21**, 79-84.

Jason, A. C. (1974), *IFST Proceedings*, 7(3), 146-157.

Starke, C. (1977), paper in Glew, G. (ed.), *Catering equipment and systems design*. Applied Science Publishers.

Vanichseni, S. *et al.* (1972), *J. Fd. Technol.*, **7**, 259-270.

Walker, R. B. (1970), M.Phil. thesis, University of Leeds; cited by Dorney, D. C. and Glew, G. (1977), *J. Fd. Technol.*, **12**, 523-534.

## 8.10 FURTHER READING

Astrom, S. (1972), *Refrigeration and air conditioning*, June, 76, 79, 80 and 82.

Dorney, D. C. and Glew, G. (1974), *Hospitals, JAHA*, **48**(13), 128, 130, 131 and 134.

Dorney, D. C. and Glew, G. (1974), *Refrigeration and air conditioning*, **77**(918), 26, 28, 29, 31, 33, 35 and 36.

CHAPTER 9

# Multi-mode Heat-Transfer

## 9.1 INTRODUCTION

In the earlier chapters of this book we have considered equipment in which there was, essentially, only one mode of heat transfer taking place to the food surface. In this chapter we consider equipment where heat transfer to the food surface takes place by more than one method. This type of equipment represents both the oldest and most traditional methods of cooking food (ranges, general purpose and roasting ovens) and also more modern types of equipment where two or more methods of heat transfer have been deliberately combined together. Predicting rates of heat transfer in such equipment is more complicated than for a single mode of heat transfer and therefore this chapter is the most appropriate for a discussion of computer simulations of heat-transfer processes.

## 9.2 TRADITIONAL COOKING EQUIPMENT

**General purpose ovens**, as their name implies, are used to roast or bake a complete range of products of many different sizes. Electrically-heated versions have partially-shielded elements. The gas-heated version may be either internally heated with the flames inside the cooking space (Fig. 9.1) or semi-externally heated (Fig. 9.2) with the burners placed under a metal plate (sole). In the latter case the hot gases enter around the edges of the sole or through vents in the inner walls of the oven.

Internally-heated ovens are hotter at the top than at the bottom, enabling different foods to be cooked at the same time. Roasting ovens are usually of this type. Semi-externally heated ovens tend to give even heating and hence the same food can be cooked on all shelves simultaneously. Both types may be combined with a boiling table to form a cooking range.

**Pastry ovens** are designed specifically for cooking pastry goods. The oven compartment is shallow compared with general purpose ovens, giving close control over temperature range throughout the oven. This close control of temperature range ensures uniformity of appearance amongst the products.

Fig. 9.1 – Open top range (direct heat).

Fig. 9.2 – General purpose oven.

Such control is necessary over the range of temperature settings required by different products.

A continuous roasting oven (Fig. 9.3) has been developed. The food is carried on trays and the conveyor system is arranged so that the food travels principally in a vertical direction, thus reducing the floor space required.

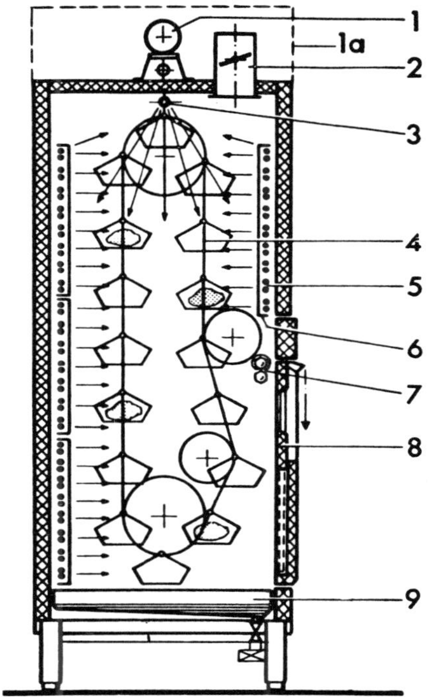


Fig. 9.3 – Continuous roasting oven.

In all types, heat transfer to the food is by convection and radiation, and also by conduction along the shelves from the oven walls. In an electric oven, air circulation will be by natural convection (with cool air near to warm walls) whilst in a gas-fired oven, natural convection will be supplemented by forced convection brought about by the gas/air mixture issuing from the burner nozzles. In addition, the carbon dioxide and water vapour in the combustion products of a gas-fired oven will radiate heat to the food surface, together with radiation from the walls.

The American Gas Association (1952) reported measurements in a gas-fired domestic oven which showed that 45.3% of the heat absorbed by a cake batter was convected heat and 54.7% radiated heat (thermostat set at 191°C). However, different parts of the cake received different proportions of convected and radiated heat:

| | | |
|---|---|---|
| cake top | 23.1% convected | 76.9% radiated, |
| cake sides | 68.6% convected | 31.4% radiated, |
| cake bottom | 61.4% convected | 38.6% radiated. |

An oven with shiny walls gave essentially the same proportion of convected and radiated heat as an oven with black walls. The relative amounts of convected and radiated heat depended to a great extent on the surface emmisivities of the food and its container.

Dunning (1953) quoted 23% as the radiation contribution to heat transfer in a gas oven at 217°C, whilst for two different electric ovens at 232°C the radiation contribution was 42%. He also quoted a 20% radiation contribution for baking scones in a gas oven with a centre-oven temperature of 221°C. It is evident from these data that the proportions of convected and radiated heat can vary widely in practice without adversely affecting the product.

## 9.3 NEWER COOKING EQUIPMENT

The number of possible deliberate combinations of heat-transfer mechanisms for heating food is large, as the following list of available types of equipment indicates.

(i) **Convection plus steam**: a variety of forced-convection ovens are available with the capability of either (a) injecting a fine spray of water into the air stream or (b) having a tray of water in the base. Forced convection steamers are also now available.

(ii) **Infra-red radiation plus convection**: this entails some penetration of heat below the surface; the combination of infra-red radiation plus refrigerated air permits the maximum penetration of the infra-red radiation whilst the refrigerated air reduces surface scorching and moisture evaporation from the surface.

(iii) **Microwaves plus convection**: the microwave radiation penetrates into the food whilst hot air covection heating achieves the desired surface heating effects. Smith (1978) described an oven of this type but with the warm air in the form of impinging jets rather than the equivalent of a parallel flow over the food as is used in existing equipment of this type. The use of impinging jets instead of a parallel flow enhanced considerably the convective heat-transfer rate, as would be expected from the discussion of impinging jets in Section 5.6.2.1.

(iv) **Microwaves plus radiation**: this has a similar effect to (iii) but more pronounced surface effects are possible.

(v) **Steam plus radiation**: this is a steamer with heated walls and gives convection heating under condensing conditions in a wet atmosphere together with the surface effects possible with radiant heating.

(vi) **Steam plus microwaves**: this combination is not available as commercial equipment.

There remains the possibility that more than two modes of heat transfer operating together in a pre-determined way may give better results than one or two modes only. Capital cost however is likely to increase as more modes of heat-transfer are used in the same piece of equipment. Consequently, increasing complexity may be self-defeating in terms of cooking/food-heating costs, even if it produces a marginally better product. The effects of using several modes together may best be tested, at first, by building a computer model to give an overall impression of the likely temperature effects in the food (see next section).

## 9.4 A GENERAL COMPUTER MODEL

### 9.4.1 Rationale of computer models

So far in this book the quantitative approach to equipment has been emphasised. After discussing concepts, we have nearly always turned to numerical examples to determine the magnitude of any changes we might expect to take place as a result of varying the design parameters of the equipment. Almost invariably however, situations have been dealt with when only one mode of heat-transfer was involved. In reality at least two modes are always involved: one mode in the food itself and another mode for getting the heat to the surface of the food (radiation or convection or microwaves or combinations of these). The complicating effects of freezing and thawing on the unsteady-state conduction of heat in the food have also been considered. The simplest case of two modes of heat transfer taking place simultaneously was dealt with in the case of the convection oven. In this case however, the technique used only enabled us to deal with centre- and surface-temperatures. It was not possible to calculate the temperatures between the food surface and the food centre. Furthermore, the methods presented so far are only capable of dealing with relatively simple shapes. Many foods in catering are not of simple shape.

All these considerations mean that for most predictive studies in equipment we need to resort to a computer. This enables us to handle the large numbers of calculations involved in a very short time, a task which would probably take several man-years using simple calculating machines. But why do we use computer simulations, as they are called, of catering equipment? The answer lies in the fact that experimental measurements in equipment are very costly to carry out in terms of manpower and physical facilities. Very many experiments would be required to investigate thoroughly even one piece of equipment. The computer enables us to simulate all these experiments and to narrow down appreciably the range of conditions to be tested experimentally. The predictions give us a feel for what is happening. The predictions do not however, replace the need for all experimental work. In the end, confirmation of the computer-model predictions by experiment must take place to confirm the validity of the computer model and to check parameters such as sensory properties which cannot be predicted by computer models of this type.

In this chapter, we discuss first the approach to the development of a general computer model of catering equipment. The aim is to show the basis on which it is developed rather than the detail of the program writing. Particular attention will be paid to the assumptions made, as these will affect the accuracy of any predictions obtained. We also stress the importance of the need for data to use with the model. Subsequently we review the literature on computer modelling and the results obtained to date.

The development of the computer model described below takes place in four stages:

(a) setting-up of three simultaneous heat balances, using the heat-transfer equations for convection, radiation, steady-state and unsteady-state conduction (Chapters 2, 4, 5 and 6);
(b) providing data for the heat balance equations;
(c) deriving a finite-difference equation for calculating the temperature distribution in the food, involving the solution of the unsteady-state conduction equation (Chapter 4) with the inclusion of internal heat generation by microwave heating (Chapter 7);
(d) establishing a computation procedure to allow the previous three stages to be drawn together.

The model excludes consideration of a change of state as in freezing and thawing (Chapter 8) and neglects mass-transfer effects (Chapter 10). The simulation is based partly on a computer model developed by Lucas *et al.* (1969) for predicting heat transfer in gas-fired processes, although they have not applied it to catering equipment. It is also based partly on a computer model developed by Kirk and Holmes (1975) for heating with microwaves.

### 9.4.2 Simultaneous heat balances

In this section heat-transfer to the surface of the food is considered and in Section 9.4.4. heat transfer from the food surface inwards. The first assumption made, so as to simplify the explanation, is that the food is cylindrical in shape and that it is placed in a cylindrical oven. This is one of the simplest geometries possible from the point of view of a computer model. The oven, illustrated in Fig. 9.4, is gas-fired because a model of a gas-fired oven may be simplified to that of an electric oven but not vice versa.

There are three heat balances which need to be satisfied simultaneously at any moment in this situation. They are all rate equations (units W or J/s), that is each term is a rate of heat transfer.

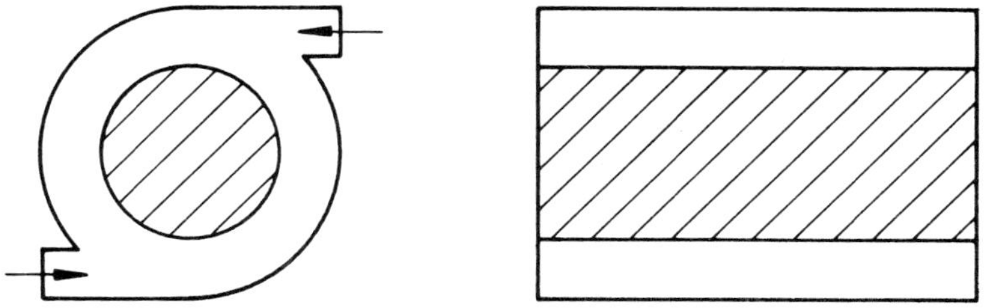

Fig. 9.4 – Cylindrical oven.

The heat balance for the combustion products is:

the rate of heat transfer to the food = the rate of heat loss from the combustion products — the rate of loss through the walls of the oven.

$$\text{The rate of heat transfer to the food} = V_f \frac{\partial I_f}{\partial \tau}$$

where $V_f$ = volume of food ($m^3$), $I_f$ = heat content of food ($J/m^3$) and $\tau$ = time (s).

$$\text{The rate of heat loss from the combustion products} = Q(I_{in} - I_{out})$$

where $Q$ = flow of combustion products into the oven ($m^3/s$), $I_{in}$ = heat content of the combustion products as they enter the oven ($J/m^3$) and $I_{out}$ = heat content of combustion products as they leave the oven ($J/m^3$).

The rate of heat loss through the oven walls is considered later.

The heat balance for the food is:

the rate of heat transfer to the food = the rate of heat transfer by radiation from the walls to the food + the rate of heat transfer by radiation from the combustion products to the food + the rate of heat-transfer by convection from the combustion products to the food.

$$\text{The rate of heat transfer to the food} = V_f \frac{\partial I_f}{\partial \tau}$$

$$\begin{array}{c}\text{The rate of heat transfer by radiation}\\ \text{from the walls to the food}\end{array} = A_f \,\bar{\mathbf{F}}_{fw}\, \sigma (T_w^4 - T_f^4)$$

where $A_f$ = surface area of food (m$^2$), $\bar{F}_{fw}$ = radiation interchange factor between the food and the oven walls (—) and involves view factors ($F$) and emissivities ($\epsilon$), $T_w$ = oven wall temperature (K) and $T_f$ = food surface temperature (K).

$$\begin{array}{c}\text{The rate of heat transfer by radiation from}\\ \text{the combustion products to the food}\end{array} = A_f \bar{F}_{fg} \sigma (T_g^4 - T_f^4)$$

where $\bar{F}_{fg}$ = radiation interchange factor between the food and the combustion products (—) and $T_g$ = temperature of combustion products (K).

$$\begin{array}{c}\text{The rate of heat transfer by convection from}\\ \text{the combustion products to the food}\end{array} = A_f h_{gf} (T_g - T_f)$$

where $h_{gf}$ = convective heat transfer coefficient between the combustion products and the food surface (W/m$^2$.K).

The heat balance for the oven walls is:

> the rate of heat loss through the oven walls + the rate of heat-transfer by radiation from the oven walls to the food = the rate of heat-transfer by radiation from the combustion products to the oven walls + the rate of heat-transfer by convection from the combustion products to the oven walls.

The rate of heat loss through the oven walls is considered later.

$$\begin{array}{c}\text{The rate of heat transfer by radiation}\\ \text{from the oven walls to the food}\end{array} = A_f \bar{F}_{fw} \sigma (T_w^4 - T_f^4) .$$

$$\begin{array}{c}\text{The rate of heat transfer by radiation from}\\ \text{the combustion products to the oven walls}\end{array} = A_w \bar{F}_{wg} \sigma (T_g^4 - T_w^4)$$

where $A_w$ = area of oven walls (m$^2$) and $\bar{F}_{wg}$ = radiation interchange factor between the oven walls and the combustion products.

$$\begin{array}{c}\text{The rate of heat transfer by convection from}\\ \text{the combustion products to the oven walls}\end{array} = A_w h_{gw} (T_g - T_w)$$

where $h_{gw}$ = convective-heat-transfer coefficient between the combustion products and the oven walls.

The following assumptions have been made so far:

(a) The oven is 'well-stirred', that is the temperature of the combustion products in the oven throughout their volume is constant and is equal to the temperature at which the combustion products leave the oven.

This assumption is necessary for the use of one value of $T_g$ in the equation. If this assumption is not made then the oven must be split up into zones and the computation carried out for each zone with each zone having its own constant temperature.

(b) The convective heat-transfer coefficients are uniform. Local variations occur because of local changes in gas velocity and turbulence. Uniform coefficients are therefore likely in ovens of simple shape. The error of this assumption increases as the contribution of total heat transfer by convection increases.

(c) The radiation factors are uniform. This implies that the emissivities of the combustion products, the food surface and the wall surface are each uniform. Strictly speaking this implies that no combustion takes place in the oven itself.

(d) The surface temperatures of the food and the oven wall are uniform. This assumption follows from the previous three assumptions because local temperature variations can only be caused by local variations in combustion products temperature and heat-transfer coefficients.

### 9.4.3 Data for the heat-balance equations

The terms $V_f$, $I_f$ and $\tau$ are considered in the next section. The term $Q(I_{in} - I_{out})$, for methane, may be re-written as:

$$Q_g(R_a + 1.0)\left(\frac{C}{R_a + 1.0} - I_{out}\right)$$

where $Q_g$ = volume flow rate of methane ($m^3/s$), $R_a$ = air/gas ratio, that is a measure of the proportion of excess air used (–) and $C$ = net calorific value of methane ($J/m^3$). The areas $A_w$ and $A_f$ can be measured.

In this arrangement of oven, food and combustion products there is radiation interchange between a non-grey gas entirely enclosed by two grey surfaces (the oven walls and the food surface). The load cannot see itself and therefore the following relationships hold:

$$\bar{F}_{fg} = \frac{\epsilon_g \epsilon_f}{m}\left[(1 - \epsilon_w)\frac{A_f}{A_w} + \frac{1}{n}\right],$$

$$\bar{F}_{wg} = \frac{\epsilon_g \epsilon_w}{m}\left[(1 - \epsilon_f)\frac{A_f}{A_w} + \frac{1}{n}\right],$$

$$\bar{F}_{fw} = \epsilon_f \epsilon_w\left[\frac{x}{y} + \frac{1 - x}{m_{n=1}}\right]$$

where
$$l = \frac{\epsilon_g{}^2}{2\epsilon_g - \epsilon_{2g}},$$

$$m = [1-\epsilon_w]\,[1-(1-\epsilon_f)n]\,\frac{A_f}{A_w} + \frac{1}{n} - (1-\epsilon_w)\,,$$

$$n = 1 - \frac{\epsilon_g}{l},$$

where $\epsilon_g$ = emissivity of the combustion products (–), $\epsilon_f$ = emissivity of the food surface (–), $\epsilon_w$ = emissivity of the oven walls (–), and $\epsilon_{2g}$ = gas emissivity evaluated for twice the beam length. The radiation interchange factors change with temperature because the emissivities change with temperature. This may be taken into account in the method of solution. The effect of temperature on the emissivity of combustion products is available, but very little information is available for the emissivities of the solid materials (oven walls and food). The convective-heat-transfer coefficients were determined experimentally by Lucas *et al.* (1969) for this geometric situation for both the food and the oven walls. The physical properties of the combustion products are available for these calculations.

If the heat stored in the walls of the oven does not vary with time, then the wall losses may be calculated as a steady-state conduction loss requiring a knowledge of oven dimensions, thermal conductivity of the wall materials, and the temperatures on the inside ($T_w$) and the outside of the walls. The outside wall temperature may be set at some appropriate value. The temperature of the food surface, $T_f$, is set at an initial value which is the temperature of the food at the start of the heating period.

It has now been shown that three simultaneous equations (heat balances) may be set up with the appropriate data. These equations can now be solved to give values to the three unknown terms, that is $T_w$, $T_g$ and $\partial I_f/\partial\tau$.

### 9.4.4 Calculation of the temperature distribution in the food

The calculations give a value of the term $V_f\,\partial I_f/\partial\tau$ which is a heat-transfer rate to the surface of the food. This may now be used to calculate the changes of temperature within the food itself. The food is cylindrical in shape. The heat transfer within such a shape with internal heat generation (applicable if microwaves are used) is given by the following differential equation:

$$\frac{\partial T}{\partial\tau} = \alpha\left[\frac{\partial^2 T}{\partial r^2} + \frac{1}{r}\frac{\partial T}{\partial r} + \frac{1}{r^2}\frac{\partial^2 T}{\partial\theta^2} + \frac{\partial^2 T}{\partial z^2} + \frac{q}{\lambda}\right]$$

where $T$ = temperature (K) in the cylinder at a point defined by specifying $z$, $r$ and $\theta$ (Fig. 9.5), $\tau$ = time (s), $q$ = rate of internal heat generation by microwave energy (W), $\lambda$ = thermal conductivity of the food (W/m.K), and $\alpha$ = thermal diffusivity of food ($m^2/s$).

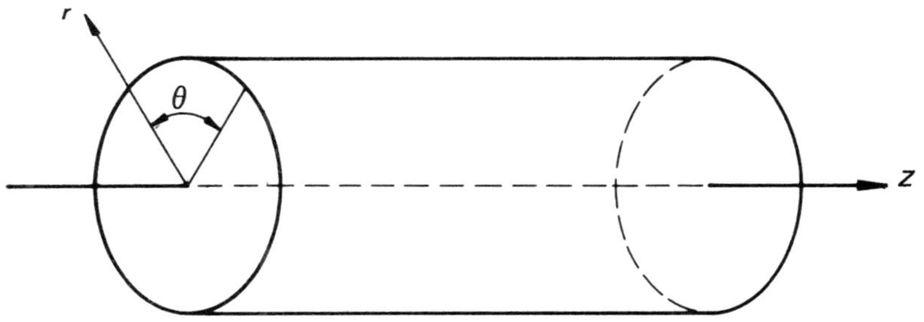


where $Z$ = distance along cylinder axis

$r$ = distance from axis of cylinder

$\theta$ = angle around the axis from the $r$-axis

Fig. 9.5 – Co-ordinate system for the food, where $z$ = distance along cylinder axis, $r$ = distance from axis of cylinder, $\theta$ = angle around the axis from the $r$-axis.

If it is assumed that the heat arrives uniformly on the curved surface only, heat-transfer in the $z$-direction can be neglected. Since the heat flux is uniform over the surface, there will be no temperature variation in the $\theta$ direction and both

$$\frac{\partial^2 T}{\partial z^2} = 0 \quad \text{and} \quad \frac{\partial^2 T}{\partial \theta^2} = 0\,.$$

The differential equation then simplifies to:

$$\frac{\partial T}{\partial \tau} = \alpha \left[ \frac{\partial^2 T}{\partial r^2} + \frac{1}{r}\frac{\partial T}{\partial r} + \frac{q}{\lambda} \right].$$

This equation involves temperature variations in the $r$ direction only. There are many ways of solving differential equations such as this. If the equation were integrated by pure mathematics this would be described as an **analytical solution**. In practice this is difficult, particularly where $q$ is temperature dependent. An alternative approach is to replace the terms in the above differential equation by finite-difference increments. This then leads to a **numerical solution**. There are many different ways of carrying out the numerical solution, the differences in methods being characterised mainly by the simplicity of the method, the stability and the accuracy of the solution.

In Chapter 7 one method of carrying out a finite difference procedure was considered. Because the calculation is carried out manually it is of necessity simple and its accuracy is limited by using only a few increments. Stability can also be a problem. A more complicated method of solution will now be described which gives a more accurate solution. The calculations are complex, so the approach will be restricted to six cylinder segments (Fig. 9.6).

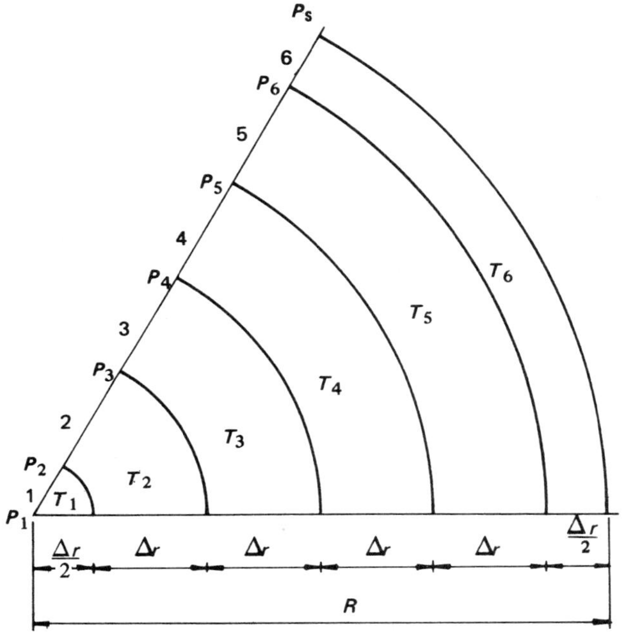


Fig. 9.6 – Food divided into six segments.

The total radius of the food cylinder = $R$. The sections are numbered (from the middle) 1, 2, 3, 4, 5 and 6 with sections 1 and 6 of thickness $\Delta r/2$ whilst the other sections have a thickness $r$, that is $R = 5\Delta r$. The temperature in each section is assumed to be uniform throughout that section and is $T_1 \ldots T_6$ for sections 1 . . . 6 respectively.

The absorbtion of microwave power may be considered in a similar manner to that described in Section 7.7. The microwave radiation falling on unit length of the surface of the cylinder is $P_s$. The microwave power leaving section 6 and entering section 5 is $P_6$, and so on. The microwave energy absorbed by section 6 is that falling on the surface minus that leaving section 6, that is

$$(P_s - P_6) \quad \text{where } P_6 = P_s \exp\left(-2\alpha' \frac{\Delta r}{2}\right)$$

$$= P_s \exp(-\alpha' \Delta r)$$

where $\alpha'$ = attenuation coefficient (1/m). For the 5th section, the energy absorbed is that entering minus the energy leaving, that is

$$(P_6 - P_5) \quad \text{where } P_5 = P_6 \exp(-2\alpha'\Delta r).$$

and so on until for the 1st section the energy absorbed is:

$$(P_2 - P_1) \quad \text{where } P_1 = P_2 \exp(-\alpha'\Delta r)\,.$$

There is an error in this method, which is likely to increase as the diameter of the cylinder decreases, because the effects of any residual energy on the cylinder axis ($P_1$) is ignored. There is experimental evidence to suggest that for small cylinders and spheres the mechanism of microwave penetration changes and it appears that resonance occurs within the food. This results in an antinode of resonance on the axis of the cylinder giving rise to a 'focussing' effect.

It is now possible to write the differential equation in finite-difference form (for ring $r$ where $r = 2$ to 5):

$$\frac{(T_r' - T_r)}{\Delta\tau} = \frac{\alpha}{2}\left[\frac{(T_{r-1} - 2T_r + T_{r+1}) + (T_{r-1}' - 2T_r' + T_{r+1}')}{\Delta r^2} + \frac{1}{(r-1)\Delta r}\left(\frac{(T_{r+1} - T_{r-1}) + (T_{r+1}' - T_{r-1}')}{2\Delta r}\right) + \frac{(P_{r+1} - P_r)}{\lambda}\right]. \quad \text{(A)}$$

$T_r'$ refers to the temperature in segment $r$ at the end of the heating period ($\Delta\tau$) under consideration and $T_r$ refers to the temperature in the same segment at the beginning of the same heating period. $T_{r+1}$ referes to the adjacent outer segment to $r$ and $T_{r-1}$ refers to the adjacent inner segment to $r$. The outer segment ($r = 6$) is considered in terms of a heat balance which includes heat transfer by convection and radiation to its outer surface.

The heat balance for this segment may be stated as:

> the rate of increase in the heat content of the outer segment over a time increment, $\Delta\tau$ = the rate of arrival of heat on the outer food surface by convection and radiation + the rate of heat absorbtion from the microwaves − the rate of heat loss by conduction through the segment,

that is,

$$\frac{V_s\lambda_s}{\alpha_s}\left(\frac{T_f' - T_f}{\Delta\tau}\right) = \frac{V_f\,\partial I_f}{\partial\tau} + (P_s - P_6) - \frac{\lambda_{s,6}\,A_{s,6}}{2\partial r}\,(T_f - T_{f-1} + T_f' - T_{f-1}') \quad \text{(B)}$$

where $V_s$ = volume of the outer segment and $A_s$ = mean surface area of the outer segment, $T'$ = the temperature of the segment at the end of the heating period

$(\Delta\tau)$, $T$ = the temperature at the start of the same heating period, $\lambda_{s,6}$ and $\alpha_{s,6}$ are food properties at the temperature of the outer segment and $f$ and $f-1$ refer to food surface and inner surface of the outer segment respectively ($f = s$ and $f-1 = 6$).

A similar heat balance for the inner segment ($r = 1$) may be stated as:

> the rate of increase in the heat content of the inner segment over a time interval $\Delta\tau$ = the rate of heat absorption from the microwaves + the rate of heat conduction through the segment,

that is,

$$\frac{V_1 \lambda_{1,2}}{\alpha_{1,2}} \left(\frac{T_1' - T_1}{\Delta\tau}\right) = (P_2 - P_1) + \frac{\lambda_{1,2} A_{1,2}}{2\Delta r}(T_2 - T_1 + T_2' - T_1') \qquad \text{(C)}$$

where $V_1$ = volume of the inner segment and $A_{1,2}$ = mean surface area of the inner segment, $\lambda_{1,2}$ and $\alpha_{1,2}$ are food properties at the temperature of the inner segment, 1 and 2 refer to the outer and inner surfaces of the segment, $T'$ = the temperature of the segment at the end of a heating period $(\Delta\tau)$, and $T$ = the temperature at the start of the same heating period.

These equations for the outer, inner and in-between segments have to be re-arranged to give the required temperature [$T_r'$ in (A), $T_f'$ in (B) and $T_1'$ in (C)] in terms of the other variables. The temperatures ($T$) at the start of each time interval $(\Delta\tau)$ are known, but those at the end of each time interval ($T'$) need to be calculated. Re-arranging equation (A) gives:

$$T_r' = \frac{A}{1+2A}\left[T_r\left(\frac{1}{A} - 2\right) + T_{r+1}(1+B) + T_{r-1}(1-B) + T_{r+1}'(1+B) + T_{r-1}'(1-B) + C(P_{r+1} - P_r)\right]$$

where $A = \alpha\Delta\tau/2\Delta r^2$, $B = 1/2(r-1)$, and $C = \Delta r^2/\lambda$.

This may be simplified further to:

$$T_r' = \frac{A}{1+2A}\left[D + T_{r+1}'(1+B) + T_{r-1}'(1-B)\right] \qquad \text{(X)}$$

where $D = [T_r(1/A - 2) + T_{r+1}(1+B) + T_{r-1}(1-B) + C(P_{r+1} - P_r)]$. This equation applies to all the segments except the inner and outer ones, that is for $r = 2, 3, 4$ and 5.

Re-arranging equations (B) and (C) gives two further equations of a similar type, equations (Y) and (Z). These equations ((X), (Y) and (Z)) enable us to calculate $T_f'$, $T_5'$, $T_4'$, $T_3'$, $T_2'$ and $T_1'$.

### 9.4.5 Computational procedure

The computational procedure to calculate the temperature progression in the food with time consists of the following steps:

(1) The temperatures $T_f$, $T_5$, $T_4$, $T_3$, $T_2$ and $T_1$ are set at the start of the computational period, for example all at 20°C if the food is at room temperature.

(2) The three simultaneous equations set up from the three heat balances of Section 9.4.2 are solved to give a value of $V_f\,\partial I_f/\partial\tau$ for the first time period, using the data of Section 9.4.3.

(3) This value of $V_f\,\partial I_f/\partial\tau$, together with the values of microwave absorption in each segment and the initial set temperatures, are used to solve equations (X), (Y) and (Z) of Section 9.4.4. This gives the temperature profile in the food at the end of the first time period. This is done by an iterative process until a convergent solution is obtained, that is until a new set of temperatures is obtained which is negligibly different† from the last set. These temperatures then represent a convergent solution of the temperature distribution at the end of the first time period.

(4) The new surface temperature (calculated in step (3)) is then used to solve the three simultaneous equations again (as in step (2)) to give a new value of $V_f\,\partial I_f/\partial\tau$ for the second time period.

(5) The iteration of step (3) is then repeated, using values of the physical properties relevant to the new temperature for each segment calculated previously in step (3). This gives the temperature profile after two time intervals.

(6) Steps (4) and (5) are repeated until some previously defined heating time is reached.

The computer can be programmed to print out the temperature profile after each time interval or as often as is desired. The time the computer takes to carry out the computation will depend upon a number of factors, the most important one being the number of segments used, the length of the time interval set and the temperature difference specified for a convergent solution. The larger the first two factors and the smaller the third, the more accurate the solution obtained, but the longer the computing time will be. Because computing time is very expensive, the accuracy required of the computer solution must be carefully considered.

A term allowing for the thawing of frozen food has not been included in this discussion but this can be done by adding a latent heat term to equations (A), (B) and (C). Similarly a term could also be included which would allow for the penetration of infra-red radiation into the food.

† Note: We define in the programme what is negligibly different but the smaller the difference, the longer the computation time.

Often we need to know not just the changes in temperature in the food but how much cooking will have taken place as well. Cooking is a time-temperature phenomenon and can be incorporated in a computer model as a number, generated by the summation of a temperature-dependent function for each time increment. This will identify differences in cooking effects at different points in the food, indicating zones of under- and over-cooking.

## 9.5 REVIEW OF COMPUTER-MODELLING LITERATURE

In Section 9.4.1 it was stated that experimental measurements are required to confirm the accuracy of computer predictions of the temperature progression in food when it is processed in catering equipment. If it can be shown that computer predictions are in close agreement with experimental measurements, then the value of using computer-models of catering equipment will be clear. The mathematical modelling of heat transfer is well established in other industries but in catering it was only introduced in the early 1970's and only began to gain momentum in the middle of the decade. A brief review of published work is presented here to demonstrate the value of such an approach. It should also be borne in mind that experimental validation of the simpler calculation procedures outlined in earlier chapters serves the same purpose.

Ohlsson and Bengtsson (1971) developed a one-dimensional model for the heating by microwaves (and allowing for unsteady-state conduction) of an infinite parallel slab of food of given thickness. The differential equation was solved using a finite-difference technique. Temperature-dependent dielectric properties were used but a constant thermal diffusivity was assumed. A surface convective-heat-transfer coefficient was estimated as was power density from calorimetric measurements and from measured temperature profiles. The field was assumed to be a plane wave propagating perpendicular to the surface. Good agreement was obtained between predicted and measured temperature profiles for salted ham, simulated meat and beef, in view of the assumptions made. The model was not valid for temperatures approaching 100°C because heat losses due to evaporation were not taken into account. It was evident that heat conduction played an important role in microwave heating of foods.

Charm *et al.* (1972) used a finite-difference method for calculating freezing and thawing times for cylinders and slabs of food. (It was suggested in the previous section that freezing/thawing could be incorporated into a more general model.) Fig. 9.7 gives two comparisons between their calculated and experimental data.

Kirk and Holmes (1975) used a finite-difference method to calculate the temperature rise in an infinte slab (one-dimensional problem) heated by microwaves. The assumptions made included fixed values of thermal conductivity, the use of surface heat-transfer coefficients which did not correspond to actual

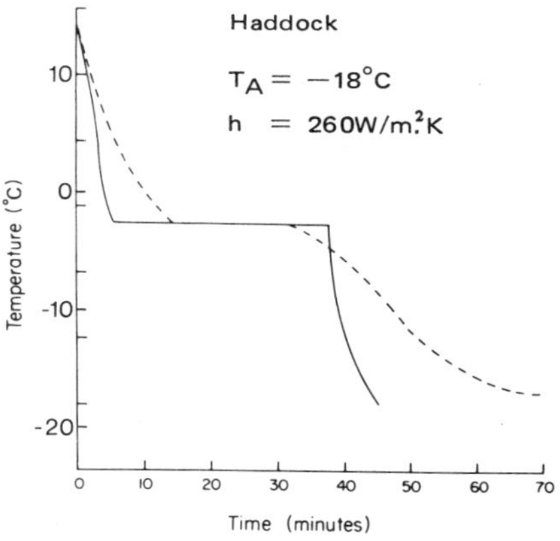


Fig. 9.7a – Time-temperature relationship for freezing in a slab (Haddock). Solid line is calculated, dotted line is experimental data.

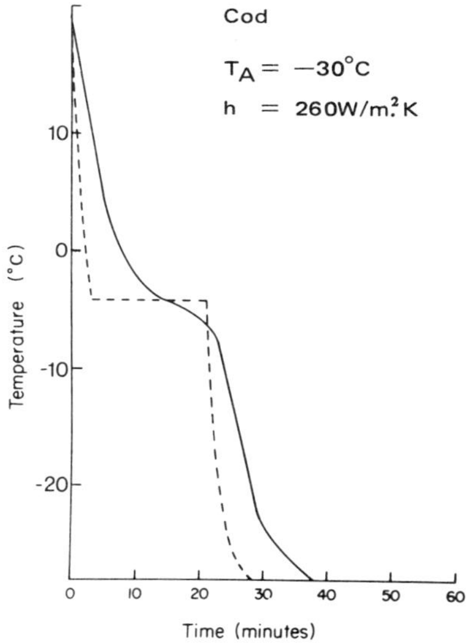


Fig. 9.7b – Time-temperature relationship for freezing in a slab (Cod). Dotted line is calculated, solid line is experimental data.

surface temperatures and simplifications of the structure of the solid. Experimental measurements were made using agar, with the sides shielded from the microwaves by metal. Good agreement was obtained between the theoretical and experimental results, as illustrated in Fig. 9.8.

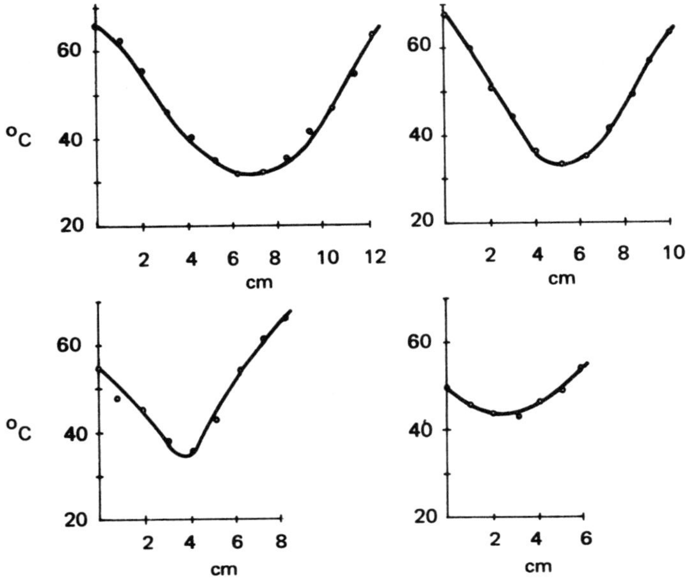


Fig. 9.8a – Temperature profiles for slabs with radiation on two surfaces, at an initial temperature of 20°C. ○ = mean of experimental values, (——) = calculated values.

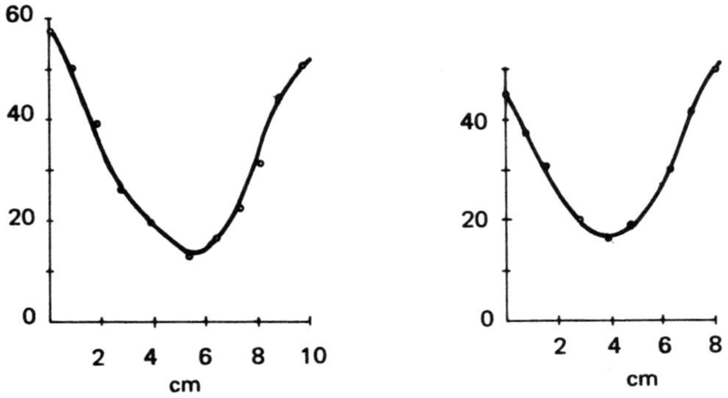


Fig. 9.8b – Temperature profiles for slabs with radiation on two surfaces, at an initial temperature of 5°C. ○ = mean of experimental values, (——) = calculated values.

Nykvist and Decareau (1976) constructed a cylindrical oven for heating meat roasts by microwaves (two frequencies), steam (three pressures) and radiant heat (six oven-temperature settings). The energy sources could be applied singularly or in any manual or automatic combination. The oven and meat were modelled mathematically as a finite cylinder (allowing for temperature variations along the length of the cylinder as well as from the surface to the axis) using a finite-difference procedure. It was assumed that the infra-red and microwave energy were uniformly distributed over the surface, that dielectric and thermal properties did not change with moisture content during cooking (but did change with temperature) and that the evaporative loss was uniform over the surface. Heat-transfer coefficients, heat exchange due to evaporation and drippings and radiant heat-exchange were included as boundary conditions. The results reported in the paper deal only with temperature variations from the axis to the surface (and not along the length of the cylinder). Comparisons between predicted and experimental temperature profiles were given for conventional heating, microwave heating and a combination of microwave and conventional heating. Agreement was excellent for conventional heating and good for the combination but less satisfactory for microwave heating alone. The authors commented that end-heating effects are significant although they did not discuss them further in the paper.

Bailey *et al.* (1974) developed a computer program to calculate thawing times. It was based on a finite-difference procedure and allowed for changes in thermal conductivity and thermal diffusivity with temperature. The latent heat effect was also allowed for continuously during the calculation procedure. Very close agreement between calculated and experimental results was obtained.

Ohlsson (1976) has simulated the time-temperature distribution in meat patties using a combination of double-sided contact frying (belt grill) for browning and microwave heating. Good agreement was made between predicted and measured values of centre and average temperatures. The calculations showed that for thicknesses of 10 mm or less, there was no advantage in using microwaves in addition to the grill. The calculations were made for an infinite slab and were mainly valid for the central part of a patty. In practice, edges and corners were heated more rapidly but this was not taken into account in the calculations.

Bengtsson *et al.* (1976) have simulated the roasting of beef using a three-dimensional finite-difference method for temperature and moisture content. Good agreement between predicted and experimental profiles (Fig. 9.9) was obtained except that the predicted temperature rise above 60°C was slightly too rapid. Moisture profiles were calculated from the temperature profiles. Agreement between experimental and calculated moisture profiles (Fig. 9.10) was fairly good initially (after cooking to 40°C centre-temperature) but calculated moisture retention was somewhat less than that found experimentally for cooking to a centre temperature of 70°C.

Cleland and Earle (1977) have compared the advantages and disadvantages

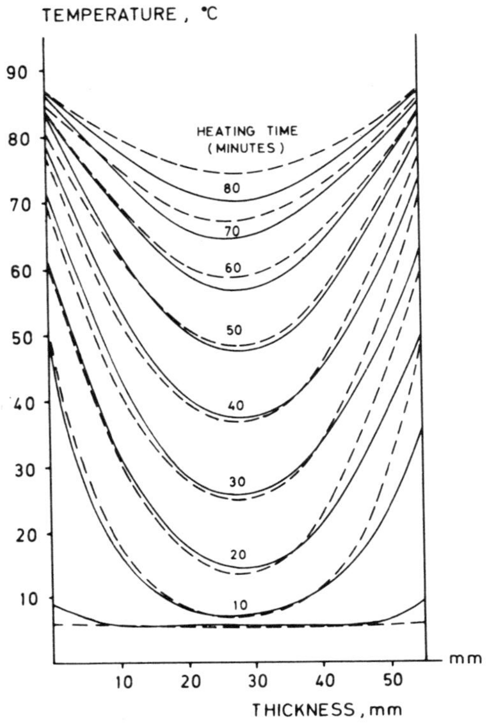


Fig. 9.9 – Comparison between experimental (solid lines) and calculated (dashed lines) temperature profiles at 175°C oven temperature.

of the numerical (finite-difference method) and analytical (for example their modification of Planck's equation: see Chapter 8) approaches to the calculation of freezing times. In the case of the freezing of slabs, they found that the analytical approach was at least as accurate as the much more complex numerical approach. It should be noted however that the freezing of a slab is a relatively simple problem compared with the many problems we meet in catering situations.

Bengtsson *et al.* (1977) have reviewed the use of computer simulation at the Food Research Institute, Sweden. Examples were given of the use of computer simulation in the study of the cooking of potatoes, microwave heating, frying and thawing/freezing. They demonstrated that it is possible to obtain good agreement between predicted and experimental results. They view the value of computer simulation as a screening device for the separate or combined effects of large numbers of variables. This reduces the number of experiments required to optimise processes or equipment with a consequent saving of time

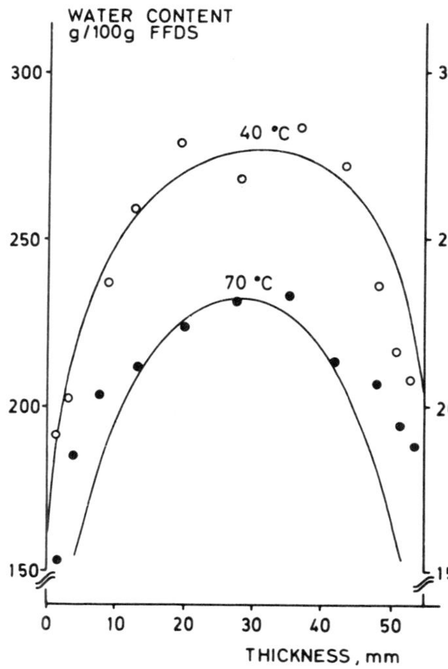


Fig. 9.10 – Calculated water profiles (solid lines) for end center temperature of 40°C and 70°C oven temperature in comparison with experimental values (points).

and money. They also remind the reader that computer simulations are based on simplifying assumptions. Therefore, experiment and simulation must always be combined for successful development work.

Kirk (1977) has outlined the development of a general computer model for food heating and has given examples of its use. He sees such a model as an interface between equipment, food and the catering system. Singh and Segerlind (1974) and De Baerdemaeker *et al.* (1977) have described a finite-element method for calculating the temperature distribution in foods of irregular shape, in contrast to the regular shapes considered throughout this book. Simulations reported in this paper, relevant to catering, were:

(a) temperature progression during the cooking of a chicken leg in a liquid at 100°C, using different thermal properties for the meat and the bone;
(b) temperature progression during the pan frying of a beef steak, allowing for the thermal conductivity of the beef varying with direction and for turning the beef on the griddle plate.

However, no comparisons between computer and measured temperature distributions were made.

Rebellato *et al.* (1978) also reported the use of the finite-element method, in their case for freezing processes. Predictions were made for freezing a lamb carcass and a side of beef. Experimental temperature measurements were not reported for these situations although the authors claim agreement between measured and computed values of the order of 1 to 2% for one-dimensional freezing of several biological materials.

The finite-element method, is, potentially, an important advance in computer simulations for catering equipment because food used in catering is not always of simple shape nor is it often homogeneous, for example a pie with a crust and meat/fruit filling. As yet, however, the accuracy of this prediction method has not been determined experimentally.

## 9.6 NOMENCLATURE

$h$ convective-heat-transfer coefficient ($W/m^2.K$)
$q$ rate of heat generation by microwave energy (W)
$r$ cylindrical co-ordinate (m)
$\Delta r$ thickness of cylindrical segment (m)
$z$ cylindrical co-ordinate (m)
$A$ area ($m^2$)
$C$ net calorific value ($J/m^3$)
$F$ view factor (–)
$\bar{\mathbf{F}}$ radiation interchange factor (–)
$I$ heat content ($J/m^3$)
$P$ microwave energy flux ($W/m^2$)
$Q$ flow rate ($m^3/s$)
$R$ radius of cylinder (m)
$R_a$ air/gas ratio (–)
$T$ temperature (K)
$V$ volume ($m^3$)
$\alpha$ thermal diffusivity ($m^2/s$)
$\alpha'$ attenuation coefficient (1/m)
$\epsilon$ emissivity (–)
$\theta$ cylindrical co-ordinate (–)
$\lambda$ thermal conductivity (W/m.K)
$\sigma$ Stefan-Boltzmann constant ($W/m^2.K^4$)
$\tau$ time (s)

## 9.7 REFERENCES

American Gas Association (1952), Research Report No. 1182.

Bailey, C. *et al.* (1974), *J. Sci. Fd. Agric.*, **25**, 81-97.

Bengtsson, N. *et al.* (1976), *Journal Food Science*, **41**, 1047-1053.

Bengtsson, N. *et al.* (1977), paper in Glew, G. (ed.), *Catering equipment and systems design.* Applied Science Publishers.

Charm, S. E. *et al.* (1972), *ASHRAE Jnl.*, **14**(11), 39-45.

Cleland, A. C. and Earle, R. L. (1977), *Journal of Food Science*, **42**(5), 1390-1395.

De Baerdemaeker, J. *et al.* (1977), *Journal of Food Process Engineering*, **1**, 37-50.

Dunning, E. W. B. (1953), Gas Council Research Communication GC 12.

Kirk, D. and Holmes, A. W. (1975), *J. Fd. Technol.*, **10**, 375-384.

Kirk, D. (1977), paper in Glew, G. (ed.), loc. cit.

Lucas, D. M. *et al.* (1969), Gas Council Research Communication GC 151.

Nykvist, W. E. and Decareau, R. V., *Jnl. Microwave Power*, **11** (1), 3-24.

Ohlsson, T. and Bengtsson, N. (1977), *Microwave energy applications newsletter*, **4**(6), 3-8.

Ohlsson, T. (1976), paper presented at the 22nd European meeting of meat research workers, Malmö, Sweden.

Rebellato, L. *et al.* (1978), *Journal of Food Science*, **43**, 239-243, 250.

Singh, R. P. and Segerlind, L. J. (1974), ASAE Paper No. 74-6015.

Smith, D. P. (1978), *Activities Report*, **30**, 97-104.

CHAPTER 10

# Mass Transfer

## 10.1 INTRODUCTION

So far we have discussed the movement or transfer of fluids and the transfer of heat by conduction, convection and radiation. In this Chapter we consider the transfer of materials, or mass transfer. While fluid flow is caused by a pressure difference and the flow of heat by a temperature difference (as shown earlier), mass-transfer is caused by a concentration difference. Whenever there is a difference in concentration, there will be a transfer of mass (material) by the process of diffusion, with the diffusing substance moving from a place of high concentration to one of low concentration. This should be contrasted to the movement of, say, a fluid when it is pumped along a pipe (see Chapter 3).

Mass transfer is analogous to heat transfer in several ways. It can take place under steady-state conditions, where concentrations are constant with time, or under unsteady-state conditions, where concentrations are changing with time. Convective mass transfer takes place in fluids; it may be laminar or turbulent. It can also be more complicated than heat transfer because we may have more than one material being transferred at the same time. In catering we can identify the following mass-transfer processes.

**Drying processes** are important in several situations including the drying of food during cooking, during storage and during chilling and freezing operations. In these situations, a wet solid (food) is in contact with dry or partially-dry air/combustion products and there is a flow of moisture from the food to the gaseous medium. Sometimes drying is desirable, as in crust formation. In other cases it is undesirable, for example drying of the surface of frozen meat during freezing and/or storage, known as 'freezer burn'.

**Leaching processes** may be deliberately carried out as in coffee percolation. In this operation coffee flavours are extracted from coffee granules when water at 93–99°C is passed over them. Similarly, during tea making, hot water is poured on a tea bag. The area around the bag is very dark initially, indicating the high content of 'tea extract' there. Eventually the dark colour diffuses through the remainder of the water. If we stir the water with a teaspoon the process takes place more quickly, under forced-convective mass-transfer conditions (cf. normal natural-convection mass-transfer conditions when it is not stirred).

Losses of flavour and nutrients from food during cooking are mass-transfer processes which we usually wish to avoid. On the other hand, we deliberately simmer bones in water to produce a stock (when soluble compounds in the bones are extracted with the water). In these situations, a solvent (usually water) extracts certain constituents from a solid (food).

**Absorption processes** often involve the immersion of a solid food in a liquid with some of the liquid being absorbed into the solid food. One example of this is the rehydration of dry and semi-dry foods, such as dry soup mixes and dehydrated potatoes. Oil absorption by foods occurs during frying. Both processes are desirable but too much oil absorbtion is costly and leads to lower customer satisfaction.

**Migration processes** also involve mass transfer. Two examples are the transfer of volatile materials from the crust into the centre of the food, for example during bread-making, and the rendering of fat from fatty tissues in meat, sausages and so on (migration of liquids out of the food).

## 10.2 SIMPLE MASS TRANSFER

### 10.2.1 Diffusion in fluids

The concept of a boundary layer has been explained for both velocity and temperature in earlier chapters. There is a corresponding boundary layer for mass transfer. Consider air flowing alongside a meat suface as shown in Fig. 10.1.

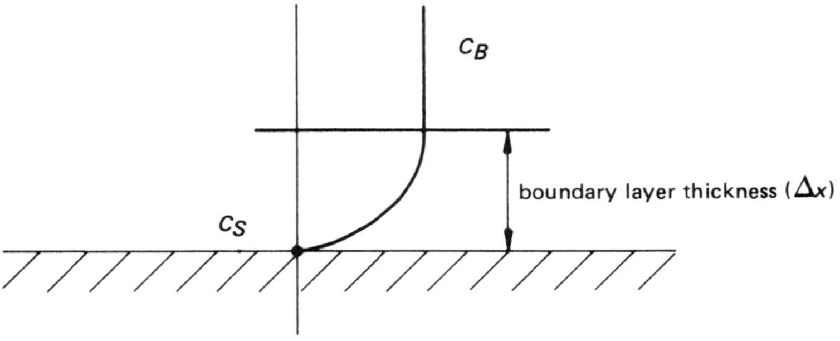


meat containing a large proportion of moisture

$C_S$ = concentration of water vapour in saturated air at the surface temperature

$C_B$ = concentration of water vapour in the air in the main air stream

Fig. 10.1 – Mass transfer caused by air flow along a meat surface.

In this situation mass transfer takes place under steady-state conditions and only one component (water vapour) is diffusing. This steady-state situation assumes that there is always sufficient moisture in the meat to prevent the surface drying out. In other words, as moisture leaves the surface it is replaced by other moisture from the interior of the meat. An analogous process takes place during leaching; in this case water replaces the air in the above example. A mass-transfer coefficient may be defined by the equation:

$$m = kA(C_S - C_B) = \frac{(C_S - C_B)}{\left(\dfrac{1}{kA}\right)}$$

where $m$ = mass flow rate (of water vapour) (kg/s), $C_S$ = concentration at the surface (kg/m$^3$), $C_B$ = concentration in the bulk fluid (kg/m$^3$), $A$ = area over which mass transfer is taking place (m$^2$), and $k$ = mass-transfer coefficient (m/s). The units of the mass-transfer coefficient ($k$) depend on how the concentration is expressed. The mass-transfer coefficient, $k = D/\Delta x$, where $D$ = diffusion coefficient for water vapour in air (m$^2$/s), and $\Delta x$ = boundary-layer thickness (m).

The mass-transfer equation is of the form:

$$\text{flow of material} = \frac{\text{driving force}}{\text{resistance}}.$$

The driving force in this case is a concentration difference.

The mass-transfer resistance = $(1/kA)$ is analogous to a heat-transfer resistance. The value of $k$ may be obtained (for mass-transfer in fluids) from correlations involving the following groups:

$$\text{laminar flow} \quad Sh = f(Gr, Sc)$$

$$\text{turbulent flow} \quad Sh = f(Re, Sc)$$

where $f$ = 'some function of' and

$$\text{the Grashof number, } Gr = \frac{gl^3 \Delta\rho}{\rho}\left(\frac{\rho}{\mu}\right)^2.$$

In this case it is a concentration gradient ($\Delta\rho$) which causes the natural convection and hence this Grashof number contains the term $\Delta\rho$ instead of the coefficient of expansion term as in natural-convection (Section 5.4.2).

$$\text{The Sherwood number, } Sh = \frac{kl}{D}$$

and

$$\text{the Schmidt number, } Sc = \frac{\mu}{\rho D} = \frac{\nu}{D}$$

where $k$ = mass-transfer coefficient (m/s), $l$ = characteristic length (m), $D$ = diffusion coefficient ($m^2/s$), $\mu$ = dynamic viscosity (kg/m.s), $\rho$ = denisty ($kg/m^3$), and $\nu = \mu/\rho$ = kinematic viscosity ($m^2/s$).

The Sherwood number is analogous to the Nusselt number and the Schmidt number to the Prandtl number when comparing mass transfer to convective heat transfer. The Schmidt number is dependent on fluid properties only (as is the Prandtl number). The diffusion coefficient depends on both the material being transferred and the fluid in which it is being transferred.

In mass-transfer correlations, a number of other dimensionless groups are also used:

$$\text{Péclet number, } Pe = Re\,Sc\ ,$$

$$\text{Stanton number, } St = Sh/Re\,Sc$$

and

$$j_D = St\,Sc^{\frac{2}{3}}.$$

These are given here for reference pruposes. Some mass-transfer correlations are given in Appendix E and some relevant data in Appendix I.

### 10.2.2 Diffusion in solids

Diffusion in solids may take place by several mechanisms and it is often not possible to say with certainty which is occurring inside food materials. Treybal (1968) distinguishes between two types of solid diffusion process and these will be described briefly.

**Structure-insensitive diffusion** occurs when the solute is dissolved to form a homogeneous solid solution. If the diffusion takes place under steady-state conditions, then the equation which applies is:

$$m = \frac{(C_S - C_B)}{\left(\dfrac{\Delta x}{DA}\right)}\ .$$

This is similar to the equation already given for steady-state mass-transfer in fluids and is analogous to the steady-state heat conduction equation. In this case $\Delta x$ would be the thickness of the slab of material (or an equivalent may be used in the case of cylinders or spheres). If the process is under unsteady-state conditions, the Dalgleish charts may be used, where the Biot number is replaced by the Sherwood ($Sh$) number, that is ($kr_m/D$). The Fourier number is replaced by the dimensionless group ($Dt/r_m{}^2$) and the reduced temperature is replaced by a reduced concentration ($C_B - C/C_B - C_O$) where $r_m$ = distance from mid-plane to surface, $C_B$ = concentration in bulk fluid, $C_O$ = original concentration in solid and $C$ = concentration in solid at position $r$ and time $t$.

The second type of solid diffusion process is **structure-sensitive diffusion**. This refers to diffusion in porous or granular solids, which permit the flow of a fluid (liquid or gas) through the interstices or capillaries in the solid. Two situations relevant to catering may be described.

In **diffusion through a uniformly porous solid** a substance which is uniformly porous is assumed, that is the nature and percentage of voids are constant in all directions and in every part of the solid. Imagine the pores and interstices are filled with a solution of salt in water. If the solid is immersed in water, there will be a continuous liquid path leading from the middle of the solid to the surrounding water via its pores and interstices. In time the salt will diffuse out of the brick into the water. In this case the equation for structure-insensitive diffusion would apply, although the exact length of the path and the containing walls of the pores and interstices will lead to the use of an 'effective diffusivity'. This would be applicable when the internal diffusion of moisture controls the rate of drying (see the next section).

The second situation is the **flow of gases through porous solids**. If there is a concentration difference (measured as a pressure difference) for a gas in a porous solid, the gas will flow through the solid. This may be treated as a fluid flow problem, using Poiseuille's equation, because the velocity will be small and the flow laminar. In using this equation entrance and exit losses would be ignored and this would be regarded as satisfactory for this purpose. The length of the capillaries will, of course, be larger than the dimensions of the solid and measurements would need to be made to estimate these.

## 10.3 DRYING

Drying is used in the food industry to reduce the moisture content of foods so that they can be stored. In catering our main concern is often to prevent over-drying during food preparation and storage, although in some preparation processes, for example bread making and cake making, a degree of drying is required.

When food is heated in an oven the air/combustion products supply the necessary sensible and latent heat of evaporation to the moisture in the food. They also carry away the moisture from the food surface. Thus heat and mass transfer take place simultaneously. During the freezing and chilling of food too there is a movement of dry air over food causing mass transfer (loss of moisture) to take place. Moreover, when food is stored in a freezer a temperature gradient often exists, with the result that there will be convection currents of air moving over the food surface causing a loss of moisture. Defrost cycles will give rise to temperature variations, which will also occur when freshly-frozen food is loaded and doors opened. The warmer air takes up moisture from the food and then deposits it again (as ice) on coming into contact with colder surfaces. This transfer of moisture is much reduced if the food is wrapped in a suitable packaging material. The same effect can occur within the package if there is sufficient air in it. This will cause 'frost' to be deposited on the interior walls of the pack. Both these processes cause the food to be partially dehydrated with a consequent loss of quality.

Food not protected by packaging may suffer sufficient dehydration of the surface layers to cause discolouration due to the food pigments being concentrated and oxidised on the surface. The surface may also turn greyish-white if sufficient water has been removed from it to leave behind cavities. Where this phenomenon occurs in meat it is called 'freezer burn'. If the freezer burn is not too severe the meat may revert back to its original state on cooking, but severe freezer burn may leave permanent effects.

Before discussing the drying process in more detail, we need to define the different types of moisture present in food. These terms are illustrated in Fig. 10.2.

The **equilibrium moisture** ($X^*$) is the moisture content of a substance when at equilibrium with a given atmosphere containing moisture. The **bound moisture** is the moisture contained in a substance which exerts an equilibrium vapour

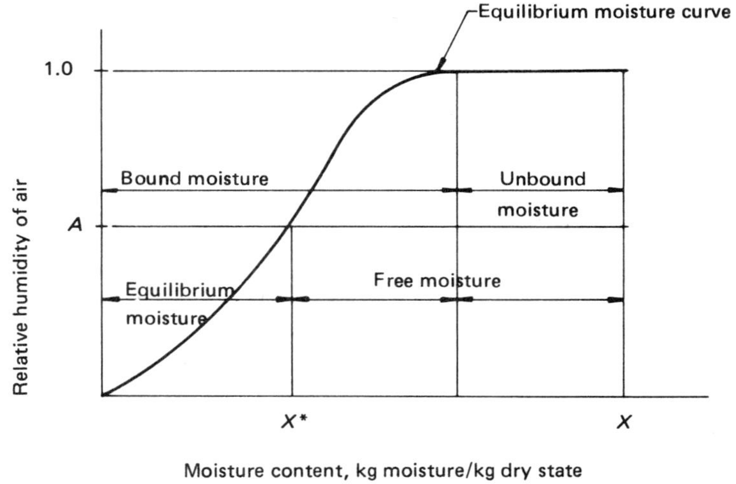


Fig. 10.2 – Moisture contents of food.

pressure less than that of pure water at the same temperature. **Unbound moisture** is the moisture contained in a substance which exerts an equilibrium vapour pressure equal to that of pure water at the same temperature. **Free moisture** is the moisture contained in a substance in excess of the equilibrium moisture. Only the free moisture can be evaporated from a food. For a given food the bound moisture content depends on the chemical nature of the food. The equilibrium moisture and the free moisture depend not only on the chemical nature of the food but on the humidity of the air in contact with the food as well.

The curves in Fig. 10.3 represent the drying of a wet solid under **constant drying conditions**, that is exposure to air at constant temperature, humidity and velocity. The letters on the two curves correspond with each other, the lower curve being plotted from the upper curve.

Initially the surface of the solid may be covered with a thin film of water. The surface (in the case of oven heating) will be colder than the equilibrium temperature and therefore the evaporation rate rises in the initial adjustment period (from A to B) until the temperature of the surface reaches this equilibrium temperature. If the surface is warmer (as in freezing or chilling) than the equilibrium temperature, the converse happens ($A'$ to B). This period is usually so short that it is ignored. During the **constant rate period** (B to C), the rate of drying is constant and the rate of mass-transfer is given by the equation given previously, that is

$$m = kA\,(C_S - C_B)$$

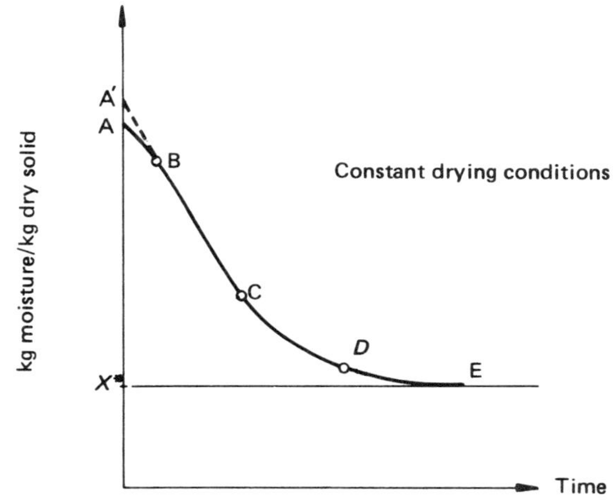


Fig. 10.3 (a) – Drying curves for a solid (constant drying conditions).

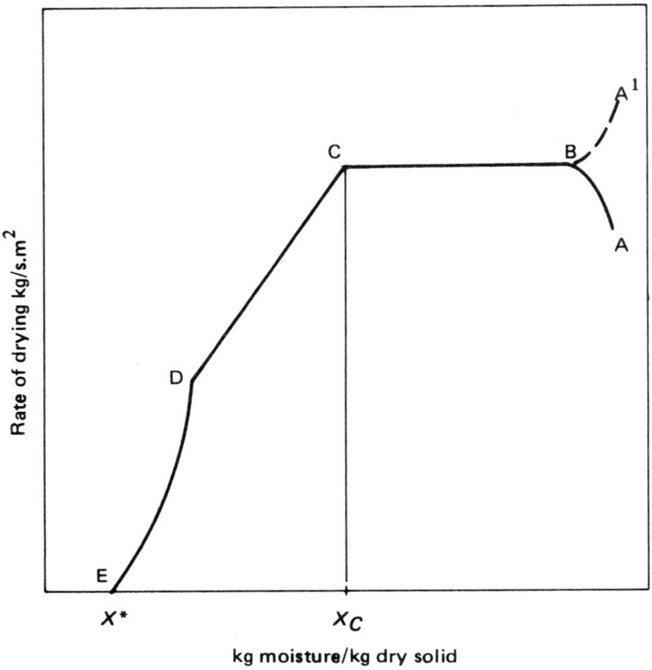


Fig. 10.3(b) – Drying curves for a solid (constant drying conditions).

where $C_B$ = concentration of water vapour in the drying air (or combustion products), and $C_S$ = concentration of water vapour in the air at the surface temperature.

It is assumed that the air at the surface is saturated with moisture and therefore $C_S$ depends only on the surface temperature. Although water is evaporating from the surface, it is replaced by the migration of unbound water from inside the solid. In this stage the rate of evaporation depends on the rate of heat transfer to the surface. Therefore if the air flow is speeded up as in a convection oven or freezer/chiller cabinet, the rate of drying is speeded up too.

The rate of convective heat transfer is given by the equation (Section 5.2.1):

$$q = hA(\theta_B - \theta_S)$$

where $q$ = rate of heat transfer (W), $h$ = convective-heat-transfer coefficient ($W/m^2.K$), $A$ = heat-transfer area ($m^2$), $\theta_B$ = air temperature (°C), and $\theta_S$ = surface temperature (°C), and since equilibrium exists:

$$mL = q$$

where $L$ = latent heat of evaporation (J/kg) of water at the surface temperature, $\theta_S$ or

$$m = \frac{hA}{L}(\theta_B - \theta_S) .$$

The factors controlling the rate of drying during this period are, according to this equation:

(a) the surface area over which drying is taking place ($A$);
(b) the temperature difference ($\theta_B - \theta_S$);
(c) the heat-transfer coeffficient.

Lowering the humidity of the drying air ($C_B$) also increases the rate of drying but this is allowed for by corresponding changes in surface temperature and hence $C_S$. Increases in oven temperature, air velocity and surface area all increase the rate of drying in this period. In some equipment, heat radiation to the food surface also occurs and, in this case an allowance for it must be incorporated into the '$h$' valve.

In the first part of the **falling rate period** (C to D) the surface begins to dry out and dry spots begin to appear on the surface. These occupy increasingly larger proportions of the surface. Hence the effective drying area decreases and the rate of drying falls because the drying rate is calculated for the whole surface area. The rate at which moisture arrives at the surface from the interior of the solid is less than the rate of evaporation, hence the appearance of dry spots on the surface. The moisture content at $C$ is called the **critical moisture content** ($X_c$). Also, the surface temperature begins to rise and continues to do so as drying proceeds, making it possible for chemical reactions to take place.

In the second part of the **falling rate period** (D to E) all the surface moisture has disappeared and the plane of evaporation moves from the surface into the solid itself, causing a further fall in drying rate. The surface temperature of the solid continues to rise. At point E, the solid is in equilibrium with the drying air and drying ceases.

In the falling rate period, the external factors which were important in the constant rate period become less so. The internal movement of moisture in the solid, caused by the concentration gradients in the food, now controls the rate of drying. The mechanism(s) causing this movement of moisture within the solid are important but are not thoroughly understood. The mechanisms proposed for this internal movement of moisture include liquid diffusion, as in structure insensitive diffusion in solids, and capillary movement, with the unbound moisture flowing through the solid in the same way as oil flows through a wick by means of a mechanism involving surface tension.

The above description referes to a simple system. Foodstuffs are not simple systems because they are made up of a complex of components some of which hold bound (rather than free) water. The water also is not pure but contains dissolved solids. The structure of food is cellular, unlike that of a porous solid.

Nevertheless, the above picture of drying is an approximation to what one would expect to happen when foodstuffs are heated in catering situations. If the surface dries out and its temperature increases, then irreversible changes may occur at the surface of the food, for example browning or burning and flavour loss. However, bread with a brown crust results from a deliberate moisture loss at the surface. It has been shown that water migrates within the food as drying is taking place. The soluble compounds in the water will be transferred at the same time, although some may be held back because the cell walls act as semi-permeable membranes. As a result some soluble compounds will concentrate at the surface when the water carrying them leaves them there as it evaporates. However, there is then a concentration gradient of these compounds, higher at the surface and lower inside the food. Therefore there will be mass transfer of these compounds back into the food.

The authors have made some calculations to illustrate the effect of several parameters on the moisture loss from a food surface heated under different conditions. The food pack was assumed to be 0.5 m long × 0.2 m wide with moisture loss taking place from one surface only. It was assumed that the other surface is prevented from drying by protecting it with a container. The results of these calculations are recorded in Table 10.1.

These results show that the most significant effect on the drying rate of the food is an increase in surface temperature from 50°C to 80°C for all the conditions studied. Air speed is significant when the surface temperature is 80°C but not when it is 50°C. Air temperature is less important. Direct gas heating gives a significantly lower drying rate than electric heating; this is because of the higher humidity of the combustion products. The modern trend in convection oven design is to make provision for the continuous injection of a fine spray into the air stream. This saturates the air and reduces the rate of moisture loss. Another type of oven uses air which is almost completely saturated with water vapour and operates at temperatures below 100°C. The authors calculate that for this type of oven operating at 90°C with air of 95% relative humidity, an air velocity of 10 m/s and a food surface temperature of 80°C (the worst conditions for drying, as indicated in Table 10.1) there will be a weight gain of 0.3 g/s. In reality, of course, the weight gain is limited in extent because the food would eventually become saturated with water. Again, the limitations of the model are seen. In this case, is assumes water can be absorbed by the food without limit, which is not true. The advantages of this type of oven or a forced-convection oven with moisture injection, from a mass-transfer viewpoint, are thus clearly demonstrated.

In conclusion, we should point out that these calculations were made for unwrapped food. If the food is wrapped properly then clearly most of this moisture loss can be avoided. Correct wrapping of food, particularly for storage and freezing/chilling operations, involving much longer times than cooking, is just as essential.

**Table 10.1 – Effect of several parameters on moisture loss in foods.**

| Heating medium | Velocity of heating medium (m/s) | Heating medium temperature (°C) | Surface temperature of food (°C) | Rate of weight loss (g/s) |
|---|---|---|---|---|
| Dry air (electric heating) | 5 | 150 | 50 | 0.06 |
| | 5 | 150 | 80 | 0.43 |
| | 5 | 200 | 50 | 0.06 |
| | 5 | 200 | 80 | 0.41 |
| | 10 | 150 | 50 | 0.09 |
| | 10 | 150 | 80 | 0.60 |
| | 10 | 200 | 50 | 0.08 |
| | 10 | 200 | 80 | 0.58 |
| Combustion products (direct gas heating) | 5 | 150 | 50 | −0.01 |
| | 5 | 150 | 80 | 0.35 |
| | 5 | 200 | 50 | −0.01 |
| | 5 | 200 | 80 | 0.34 |
| | 10 | 150 | 50 | −0.01 |
| | 10 | 150 | 80 | 0.50 |
| | 10 | 200 | 50 | −0.02 |
| | 10 | 200 | 80 | 0.49 |

Note: A minus sign indicates a weight *gain* by the food.

The only attempt to investigate moisture gradients in food during cooking (as opposed to total moisture loss) is that of Bengtsson *et al.* (1976) which was referred to in the previous chapter (Fig. 9.10). They obtained reasonable agreement between predicted and experimental profiles by assuming the moisture content was a function of the temperature of the food.

## 10.4 LEACHING PROCESSES

Leaching processes, such as coffee and tea making, are governed by the steady-state diffusion equation (Section 10.2.1):

$$m = kA\,(C_S - C_B)\ .$$

If one wishes to attain a certain concentration of extract in the product, as in coffee making, then $C_B$ is fixed: $C_S$ will however decrease as the soluble extractives are leached from the ground coffee. In simmering bones to produce a stock, $C_B$ will increase during this batch process. The principal variable in the process is the surface area of the solid ($A$) from which leaching is taking place. The effect of particle size on surface area is considerable as will be discussed in Section 12.2. This is why coffee beans are ground for coffee making and why 'instant' tea and coffee dissolve very rapidly in water.

Tea is traditionally made in a teapot. The tea is contained in bags only for the convenience of disposing of the spent tea leaves. There is a wider variety of equipment available for making coffee but perhaps the simplest form is the percolator (Fig. 10.4). In this the grains of coffee are contained in an infuser.

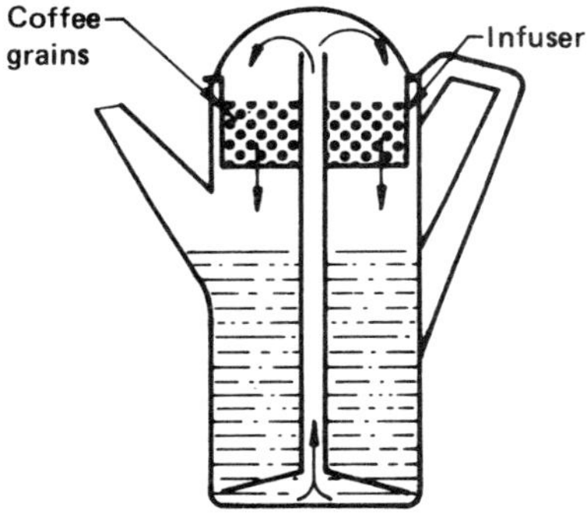


Fig. 10.4 – Coffee percolator.

As the water heats up, steam is produced which forces the hot water up a central pipe. On leaving the pipe it falls over the coffee grains, extracting the coffee and then it re-mixes with the bulk water. Eventually, when the concentration of coffee extract in the water ($C_B$) is high enough, percolation stops. A wide variety of sizes of this type of equipment is made. Another common type of equipment is the funnel and flask type, which has a separate water boiler. The coffee grains are held on a filter paper in a funnel and hot water is caused to flow over the layer of coffee grains in a 'once-through' manner to be collected in a flask below, which may be kept hot by an independent heater. This type may be a specialised unit or may be built into more complex tea/coffee sets (Section 5.6.2.3). 'Expresso' coffee machines are pressure boilers working at 140–280 $kN/m^2$ pressure. They are designed for making individual cups of coffee by forcing the water through the coffee grains under pressure. The coffee grains are held on a metal strainer.

The development of an automatic gas-fired coffee maker was described by Schaab and Dewerth (1966).

## 10.5 ABSORBTION PROCESSES

Carlheim-Gyllensköld (1970) reported the results of tests on the penetration of salt into gelatine and several food stuffs. Unfortunately his work makes no attempt to correlate his experimental results with theoretical predictions.

## 10.6 MIGRATION PROCESSES

The only work reported on migration processes in food is that for fat content profiles in patties reported by Bengtsson *et al.* (1976) who concluded that the profiles seem to be affected more by raw material variations than by heating conditions.

## 10.7 NOMENCLATURE

$h$ convective-heat-transfer coefficient ($W/m^2.K$)
$j_D$ a combination of dimensionless groups ($St\,Sc^{\frac{2}{3}}$)
$k$ mass-transfer coefficient (m/s)
$l$ length (m)
$m$ rate of mass-transfer (kg/s)
$q$ rate of heat-transfer (W)
$r$ distance (m)
$\Delta x$ thickness (m)
$A$ area ($m^2$)
$C$ concentration ($kg/m^3$)
$D$ diffusion coefficient ($m^2/s$)
$Gr$ Grashof number (–)
$L$ latent heat (J/kg)
$Sc$ Schmidt number (–)
$Sh$ Sherwood number (–)
$X$ moisture content (kg moisture/kg dry food)
$\theta$ temperature (°C or K)
$\mu$ absolute/dynamic viscosity (kg/m.s)
$\nu$ kinematic viscosity ($= \mu/\rho$) ($m^2/s$)
$\rho$ density ($kg/m^3$)

## 10.8 PROBLEMS

(1) Unwrapped food is being heated in a convection oven. The food is in the form of a flat plate (thin pack) 0.5 m long × 0.2 m wide, with one surface in direct contact with the heating air/combustion products.

Calculate the mass-transfer coefficients for the water-air system under the four conditions obtained by combining the following variables:

air velocity 5 m/s and 10 m/s
air temperature 150°C and 200°C .

(2) Calculate the transfer of moisture for the four conditions in question (1) under the following conditions.

(a) Surface temperature of food = 20°C, 50°C and 80°C.
(b) Heating with (i) air which has a relative humidity of 50% at 20°C and (ii) combustion products of methane with 50% excess air.

(3) Rapidaire ovens use air at 85–95°C (say 90°C) with a 95% humidity. Calculate the mass transfer under these conditions using an air velocity of 10 m/s and food surface temperature of 80°C.

## 10.9 REFERENCES

Bengtsson, N. *et al.* (1976), *Journal of Food Science*, **41**, 1047–1053.
Carlheim-Gyllensköld, H. (1970), *Acta Polytechnica Scandinavica*, Chemistry and Metallurgy Series No. 93.
Schaab, M. J. and Dewerth, D. W. (1966), AGA Research Report No. 1415.
Treybal, R. E. (1968). *Mass transfer operations*, (2nd ed.). McGraw-Hill.

## 10.10 FURTHER READING

Matz, S. A. (1965), *Water in foods*. Avi Publishing Corporation.
Sherwood, T. K. *et al.* (1975), *Mass transfer*, Chapter 6. McGraw-Hill.
Van Arsdel, W. B. *et al.* (1973), *Food dehydration*, Vol. 1, (2nd ed.). Avi Publishing Corporation.

CHAPTER 11

# Energy Sources and Control

## 11.1 INTRODUCTION

A chapter is devoted to energy sources and control because the performance of catering equipment is determined to a considerable degree by the performance of the burners and control equipment. Good burner design and its proper use will reduce energy consumption and will improve both the quality of the food and the consistency of its quality. With both gas-fired and electric equipment, safety must also be a prime consideration.

## 11.2 COMPOSITION AND PROPERTIES OF GAS

In many countries, gas-fired catering equipment uses natural gas as a source of energy. Natural gas consists principally of methane but may also contain smaller proportions of ethane ($C_2H_6$), nitrogen and minor proportions of other gases. Hence the behaviour of natural gas when it is burnt reflects the combustion of methane. However, each gas field produces a natural gas of distinctive composition and properties, although these differences are usually small. This means that the burners used in catering equipment must produce a satisfactory flame for the whole range of natural gas within a supply area. Eventually the natural gas fields available will be depleted and even now substitute natural gas (SNG) is needed for peak loads, and will be increasingly required. This gas may be manufactured from oil and/or coal. Again small variations in composition and properties will be apparent.

The other types of gases (liquified petroleum gases, lpg) available to the caterer are usually liquid propane and liquid butane. These are used in remote locations where natural gas is not available or where portable equipment is used, for example outside functions, or as an alternative (cost-wise) to natural gas or electricity. They are available in portable cylinders or, in the case of propane and where large quantities are required for purposes other than cooking, via a road tanker and fixed storage tank. The liquid is vapourised before use. Liquid propane contains about 90% propane ($C_3H_8$), with small proportions of butane ($C_4H_{10}$) and propene ($C_3H_6$).

Liquid butane contains approximately 80 to 90% butane, with smaller proportions of propane and propene. The combustion of these gases is characterised by the combustion of propane and butane respectively.

The gas supplier's aim is to provide a consistent product so as to maintain maximum efficiency of utilisation in appliances and also to give maximum flame stability. A stable flame on the burner is a prime consideration from the point of view of safety as well as convenience. Because of these considerations, a number of measurable and controllable properties have been identified, and these and their significance will be described briefly.

### 11.2.1 Calorific value

Calorific value is a measure of the heat obtained when a gas is burnt. Two definitions of calorific value are used. The **gross calorific value** is that measured in a calorimeter. All the water in the combustion products condense in the calorimeter and therefore the gross calorific value includes the latent heat of condensation of this water. Gas is purchased as a quantity of heat (measured in Joules) but is metered by volume ($m^3$). The gross calorific value is used to convert the meter reading to Joules before the gas is charged to the customer. Therefore the supplier specifies a calorific value for this purpose. **Net calorific value** assumes that the water formed (on combustion of the gas) does not condense and therefore excludes the latent heat of condensation of the water vapour. This is a more realistic calorific value as far as the equipment is concerned because the water in the combustion products does not condense in the equipment and so the latent heat is not available for use. The gross and net calorific values of methane, propane and butane are recorded in Appendix I.

### 11.2.2 Wobbe number and specific gravity

The amount of heat we obtain from a particular burner depends on four factors: the size of the burner, the gas pressure, the calorific value of the gas and the specific gravity of the gas.

Burner size is controlled by burner design and gas pressure is held constant on the appliance. If the calorific value or specific gravity of the gas change however, there will be a variation in heat output from the burner and hence variations in equipment performance. It is therefore important that these two factors are held constant by the gas supplier. They are usually combined together in a factor (often termed the Wobbe No.) which is defined as:

$$\frac{\text{Gross calorific value (MJ/m}^3)}{\sqrt{\text{Specific gravity}}}\,.$$

The specific gravities of methane, propane and butane are recorded in Appendix I. Their Wobbe numbers are 51.0, 76.4 and 85.8 respectively. The gas supplier attempts to keep the Wobbe No. of its gas within certain limits so that the consumer obtains a constant heat output from the burners fitted to his equipment.

### 11.2.3 Flammability limits

A mixture of a combustible gas and air cannot be ignited or burnt if the proportion of gas in the mixture is either too small or too large. The limits within which the mixture will burn are termed the lower and upper flammability limits and are measured by a standard method. The range of flammability may be illustrated as shown in Fig. 11.1.

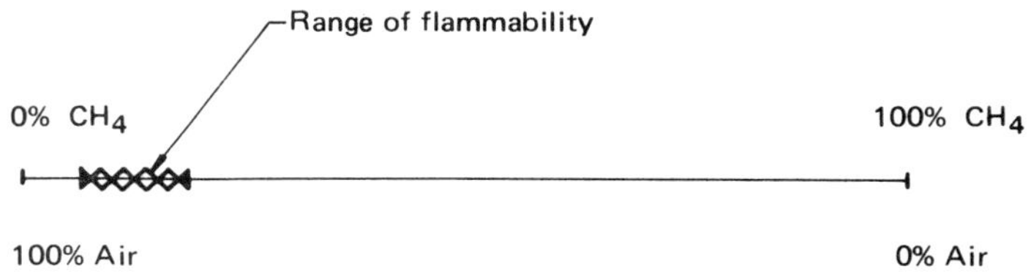


Fig. 11.1 – Flammability limits of methane.

The lower and upper inflammability limits of methane, propane and butane are recorded in Table 11.1.

**Table 11.1 – Inflammability limits.**

| Gas | Lower limit | Upper limit |
|---|---|---|
| Methane | 5.0% | 15.0% |
| Propane | 2.1% | 9.5% |
| Butane | 1.8% | 8.4% |

The limits for these gases are very narrow and burner design must take account of this, otherwise a flame will not be established on the burner.

### 11.2.4 Burning velocity

If a stationary gas/air mixture in a tube is ignited at one end, the flame will travel along the tube at a certain velocity termed the **burning velocity**. The mixture must, of course, have a composition within the flammability limits. The actual speed depends on the proportion of gas in the gas/air mixture, as shown in Fig. 11.2.

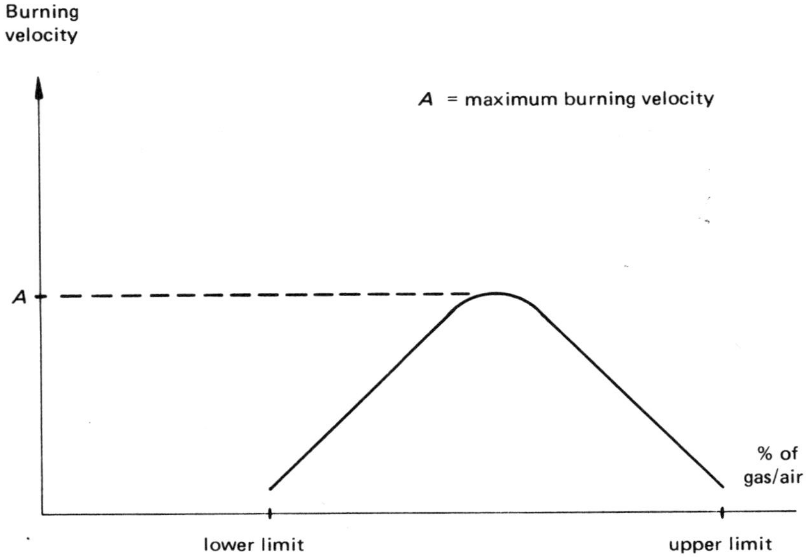


Fig. 11.2 – Variation in burning velocity with mixture composition.

The maximum burning velocities of methane, propane and butane are approximately 0.43, 0.47 and 0.45 m/s respectively. In a flame, the gas/air mixture leaving the burner is moving at a certain velocity, $u$. It is stationary on the burner because the burning velocity ($v$) of the mixture is moving to oppose it as shown in Fig. 11.3. When $v \cos \alpha = u$, there will be a stable flame. If $v \cos \alpha > u$, the flame moves into the burner tube, and if $v \cos \alpha < u$, the flame lifts off the burner.

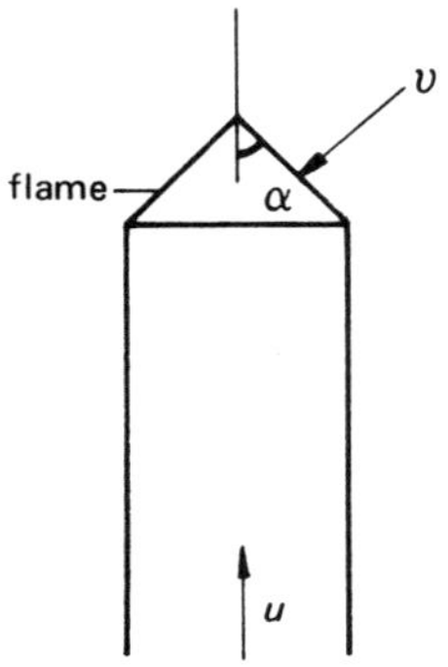


Fig. 11.3 – Stable flame on burner nozzle.

The basic problem with these gases is that they have very low burning velocities and therefore their flames have a tendency to 'lift-off' the burner when the burner is turned up, that is the gas flow is increased. Special care must therefore be taken to ensure that this will not happen because if the flame lifts off the burner, and is thereby extinguished, the appliance would soon be filled with an explosive gas/air mixture.

This explanation of flame stability is very simplified because the range of stability is greater than it would suggest. In fact there is no theory of 'lift-off' which is satisfactory, that is, enables us to specify exactly when it will occur. Various methods are used to overcome this problem of lift-off, as we shall see in the next section.

## 11.3 BURNER DESIGN

Flames, and hence burners, are basically of two types. With a **diffusion flame**, no air is pre-mixed with the fuel gas before combustion. On the other hand, an **aerated flame** has some air pre-mixed with the gas before combustion.

### 11.3.1 Diffusion flames

A diffusion flame is formed when the fuel gas is allowed to issue from the burner without any pre-mixing of the combustion air. The air for combustion is entrained into the gas jet leaving the burner nozzle (Fig. 11.4).

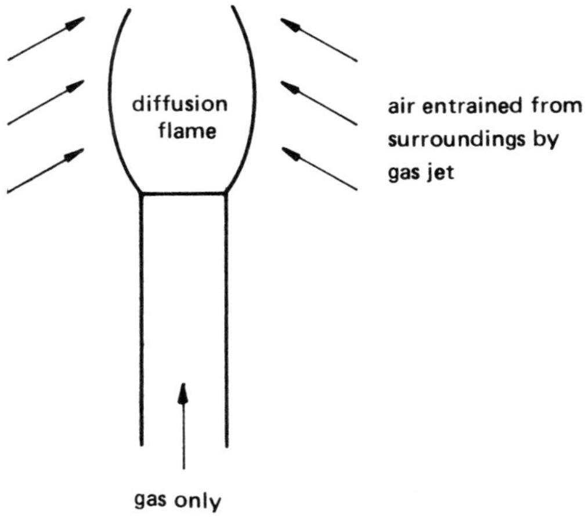


Fig. 11.4 – Diffusion flame.

The process of air entrainment, which involves air mixing with the fuel gas to form a mixture within the inflammability limits, is relatively slow. This means that the gas entering the flame is heated relatively slowly before it mixes with the air and this causes some of the gas to decompose to form soot (carbon). If the flame or a part of the flame containing soot comes into contact with a surface, some soot will be deposited on the surface. This is undesirable if the soot falls either on the food or on the interior surfaces of the oven. The soot particles being solid are relatively slow to burn (oxidise to carbon dioxide) and therefore the flame tends to be voluminous. The soot particles become hot and glow, giving rise to a characteristic yellow flame colouration. This flame colouration makes the flame 'luminous'. Generally speaking a compact flame is required otherwise the combustion space must be relatively large so as to avoid soot deposition.

A further disadvantage of diffusion burners is their marked tendency to give rise to flame lift. Their advantages are that they are silent and they cannot light back. In addition they do not lint, that is they do not have air passing through them which contains dust and so on, which causes blockages in the burner. Therefore if a diffusion flame is used, the aim is to design a burner which will give rapid air extrainment (which implies rapid mixing of fuel gas and combustion air) and therefore a smaller flame.

The **Bray jet** is a common form of diffusion burner. It uses a refractory tip which is designed so that both sides of the gas stream flow inwards and impinge on the middle of the stream. The result is a broad, thin flame (batswing-shaped) with a large surface area for air entrainment. This causes rapid mixing of the air and gas and it also gives a stable flame.

A number of new diffusion burner designs have been developed recently. In one type, the gas issues from opposed orifices and impinges on two stainless steel wings to give two batswing-shaped flames (Fig. 11.5).

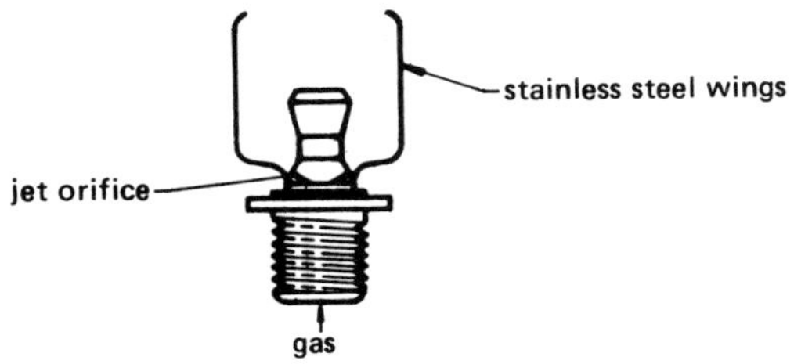


Fig. 11.5 – Diffusion burner.

### 11.3.2 Aerated flames

In an aerated-flame burner, some of the air required for combustion is pre-mixed with the fuel gas so that a gas/air mixture enters the flame (Fig. 11.6). This causes a 'blue cone' to be formed in the flame where the gas is heated relatively quickly. This blue cone is a primary reaction zone where the gas partially oxidises to carbon monoxide ($CO$), carbon dioxide ($CO_2$), hydrogen ($H_2$) and water ($H_2O$) without the formation of soot. On leaving the blue cone these gases entrain more (secondary) air from the surrounding atmosphere and combustion is completed with the CO and $H_2$ burning to $CO_2$ and $H_2O$ respectively.

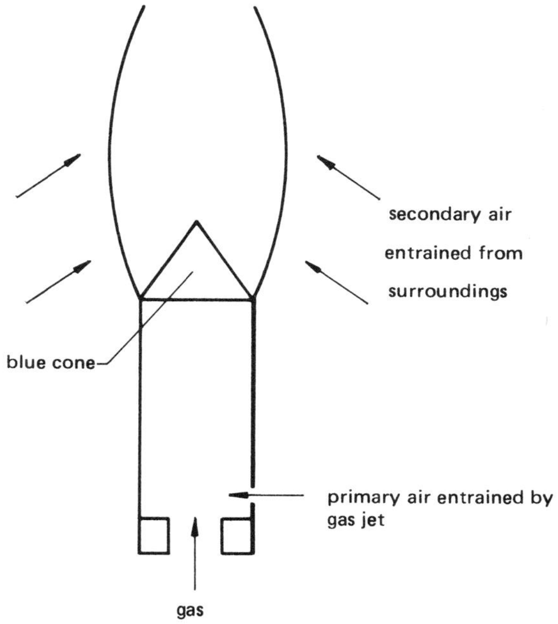


Fig. 11.6 – Aerated flame.

The characteristics of this type of flame depend on the ratio of primary air/gas and the rate of gas consumption (or thermal output of the burner) as shown in Fig. 11.7.

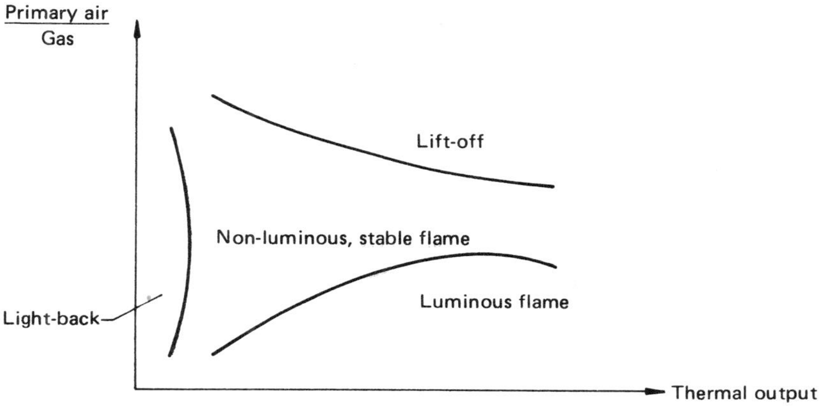


Fig. 11.7 – Combustion diagram.

The limits within which satisfactory combustion are obtained are shown in Fig. 11.7. The limits of a satisfactory flame are light-back, that is the flame moves into the burner tube ($v \cos \alpha > u$), lift-off ($v \cos \alpha < u$) and the formation of a luminous flame (insufficient primary air to prevent soot formation). An aerated flame is smaller than an equivalent diffusion flame.

It should be apparent with both types of flame that if the surrounding air supply is restricted by placing the burner in an enclosure (equipment) then incomplete combustion could easily be the result.

The problems of stabilising aerated flames relate both to lift-off and light-back. Lift-off is overcome by designing burners with some form of stabilisation. Linting may also occur because the primary air may pick-up dust and so on before entering the burner and this may cause partial blockages resulting in a reduction in the volume of air entrained. This could give rise to soot formation in the flame.

In principle, flame stabilisation is achieved by either increasing the size of the burner nozzle, thereby reducing the air/gas mixture velocity and more closely matching it with the burning velocity or using auxiliary flames of low-velocity at the base of the main flame.

### 11.3.3 Bar burners

A single aerated jet is not adequate for catering appliances as it would not achieve the required temperature distribution. It is usual therefore to use bar burners (Fig. 11.8) on some appliances. In these the gas/air mixture for a series of jets is first produced by using a single gas jet, the mixture formed then passing to the series of small jets. In effect a single large burner nozzle is split up into many small jets which aids temperature distribution and flame stability. These small jets may be in the form of many fine (pin-like) jets drilled into the upper surface of a metal box or a smaller number of larger jets (for example Bray jets or something similar) screwed into a piece of tubing or a series of slots in the bar.

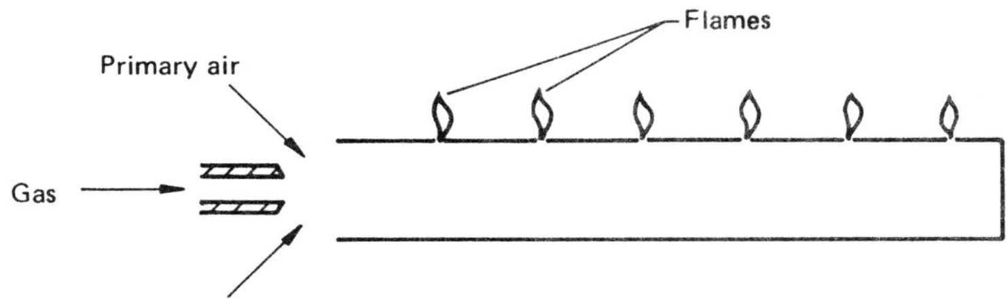


Fig. 11.8 – Bar burner.

### 11.3.4 Gas hob burners

Gas hob burners use either a large number of round-hole jets or slots, sometimes with auxiliary flames, or cause the gas/air mixture to swirl to increase flame stability.

### 11.3.5 Radiant burners

The traditional design of a radiant burner is to use gas jets to heat up a metal fret or ceramic surface which, when hot, radiates heat to the food. This burner has a major disadvantage when used in this way in that, when it is turned down, the release of heat is shifted towards the burner nozzles because the lengths of the flames are reduced. This causes uneven heating over the grill surface and may lead to uneven cooking/browning of the food.

The newer types of radiant burners consist of a porous plate with the gas/air mixture fed from behind. Combustion takes place on the surface of the porous plate in the form of many small jets formed by its pores. As the plate heats up it becomes incandescent and has an even surface temperature. When the gas supply is turned down the whole of the plate temperature is reduced, thus ensuring even heating over its surface but at a lower temperature. The heat is transferred mainly by radiation. This type of burner operates with an air supply much nearer to theoretical than conventional burners. The proportion of heat in the gas generated as radiant heat is of the order of 50%.

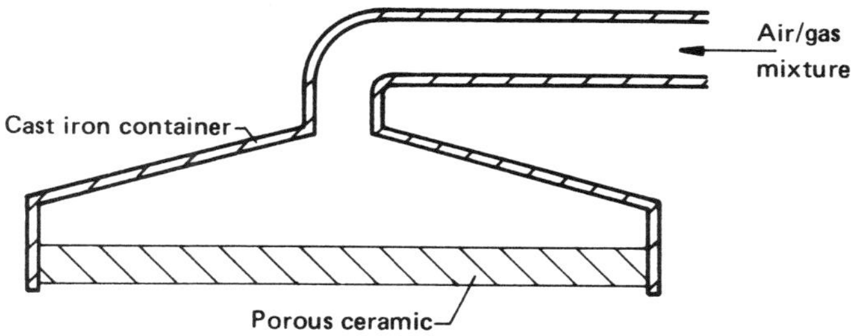


Fig. 11.9 – Porous-plate burner.

Dewerth (1962) has reported on the spectral emission of energy for several types of gas-fired and electric infra-red sources as well as the absorption characteristics of several foods.

## 11.4 ELECTRICAL HEATING

### 11.4.1 Electricity supply

The mains electricity obtained from the supplier is an alternating current (a.c.) supply. Both the current and voltage alternate. The rate of alternation varies somewhat from country-to-country but is typically at 50 Hz (cycles/second), in a sinusoidal wave form. When the power consumption is relatively low, the supply is single-phase (Fig. 11.10). The supply voltage varies for a single-phase supply from country-to-country. It can be as low as 120 V or as high as 240/250 V (with a maximum demand normally of 60 A or about 15 kW.)

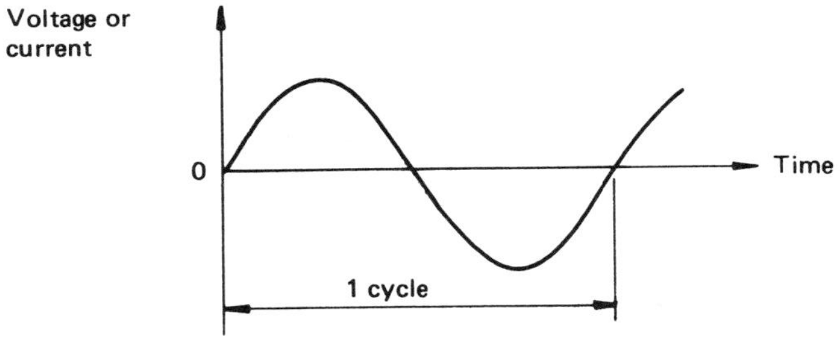


Fig. 11.10 – Single-phase power supply.

With larger equipment however, a single-phase (240 V) supply in inadequate to meet the power demand. In this case a three-phase supply is used, with each phase separated by 120°. A complete cycle represents 360° of rotation of a single conductor in the magnetic field of the generator. In the case of a three-phase supply, there are three conductors, each separated 120° from each other, rotating in the magnetic field. This gives a supply as shown in Fig. 11.11.

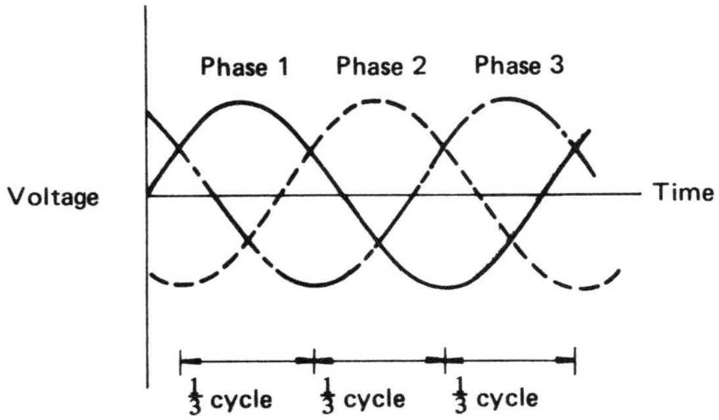


Fig. 11.11 – Three-phase power supply.

In a three-phase motor there are three conductors whilst in a single-phase motor there is only one conductor. For heating purposes, however, the heaters are arranged so they are each supplied from one of the three phases (Fig. 11.12).

In practice, the aim is to distribute the load equally over the three phases. The three-phase supply if usually of the order of 400/450 V and is designed for situations where a single-phase supply cannot provide enough power. If the total loading in a kitchen exceeds the limit of what can be safely obtained from a single-phase supply, then a three-phase supply is used with single-phase appliances being evenly distributed across the phases. In this case however, the equipment on different phases must be physically separated for safety reasons. A qualified electrician should always be used when installing electrical equipment. (Similarly with gas equipment, a qualified gas fitter should always be used for the installation of equipment.)

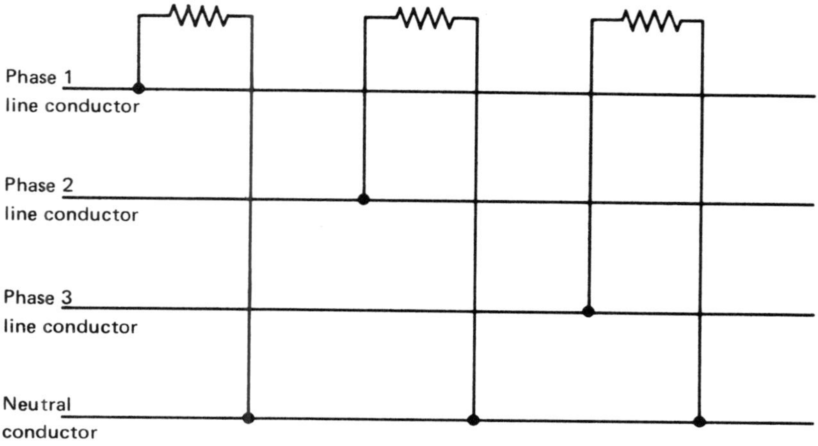


Fig. 11.12 – Heaters arranged on three-phase power supply.

### 11.4.2 Electrical elements

There are three principal types of electrical element. **Immersion elements** consist of a nichrome heater element inside a protective metal sheath (Fig. 11.13). The element and sheath are separated by an electrical insulator, magnesium oxide, so that the sheath is not 'live'. The insulator also needs to be a good conductor of heat in so far as its electrical insulating properties permit this. The length of element will determine the heat output and also whether or not the nichrome is wound in a spiral or is merely straight wire.

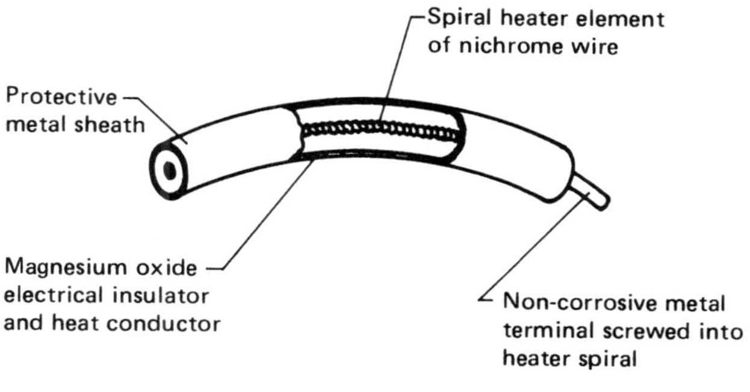


Fig. 11.13 – Sheathed element.

**Infra-red** elements consist of a straight or wound spiral length of wire held down the centre line of a silica tube (quartz glass). The element gets hot and heats the silica tube which then emits the infra-red radiation. They have higher efficiencies than the traditional (immersion) element and take a much shorter time to heat up. The silica tube is fragile and is sometimes protected by a concentric metal tube with holes in it. Alternatively a food loading system may be designed for the catering appliance which ensures that the food packs cannot be pushed into the quartz tubes during loading/unloading.

Radiant heat may also be supplied by an **infra-red lamp**. This is similar to a conventional filament light bulb but with a built-in filter. This filter limits radiation from the bulb to the red end of the visible spectrum and the shorter infra-red rays. There bulbs operate at low power levels and are not used to cook foods but to keep them hot whilst they are standing on the service counter.

### 11.4.3 Magnetrons

A magnetron generates the high-frequency microwaves (2450 MHz) used in microwave ovens. It requires an electrical source of, say, 4 kV d.c. This is supplied from the mains via (a) a transformer which converts the 240 V, 60 Hz a.c. mains to 4 kV a.c. and (b) a rectifier which converts the 4 kV a.c. to 4 kV d.c. The magnetron itself (Fig. 11.14) is a vacuum tube containing a number of components. The essential components are an anode and a cathode. The anode is a

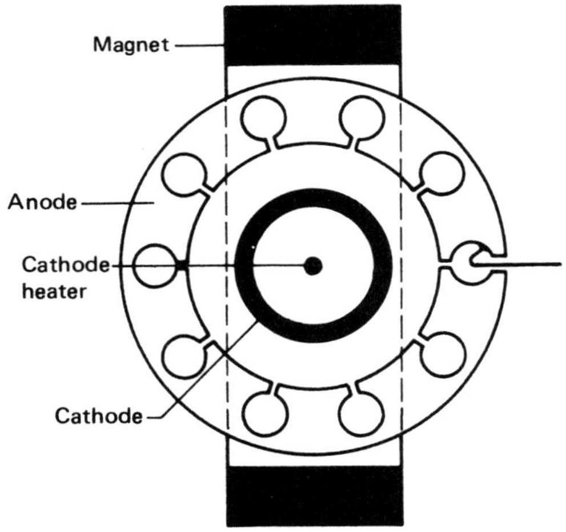


Fig. 11.14 – Magnetron.

ring containing a number of small cavities, while the cathode is a cylindrical tube and is situated centrally in the anode. When heated and when a high voltage (4 kV d.c.) is applied between it and the anode it emits electrons from its surface. In addition, a strong permanent magnetic field is applied perpendicularly to the direction of movement of the electrons, that is along the cathode.

Under the correct conditions the electrons in the anode cavities produce microwaves at the frequency required. These cavities form resonator circuits which generate the high-frequency alternating microwaves. The microwaves are taken out of the magnetron by means of a waveguide (Fig. 11.15), which is effectively a series of resonant cavities connected together.

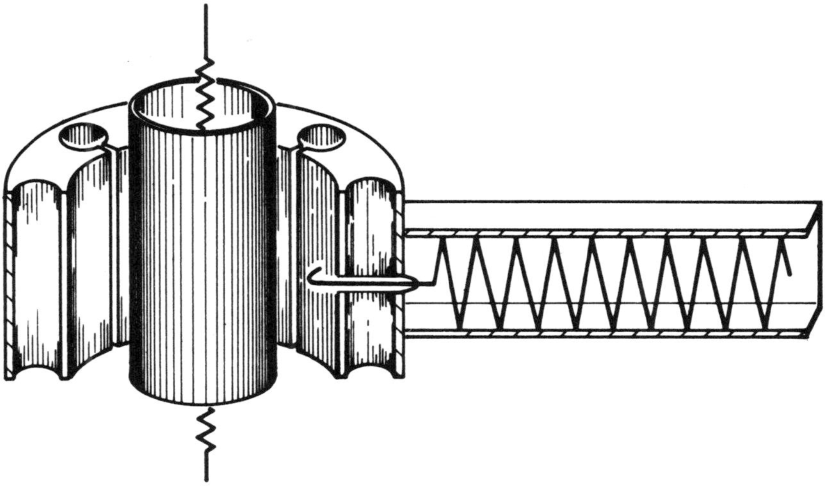


Fig. 11.15 – Magnetron and waveguide.

## 11.5 TEMPERATURE MEASUREMENT AND CONTROL

A wide variety of methods for measuring and controlling temperature are available to the caterer. It is generally recognised that all equipment for heating food or holding it should be controlled, by a thermostat, at a pre-set temperature. The most popular means of doing this is to use either a hydraulic type of thermostat or a rod and tube type, both of which control the power or gas input to the equipment. Sometimes more sophisticated control systems are used and the temperature measuring device is only a part of a more complex control unit. Another important device for measuring temperature is the thermocouple. This finds its major application in research/development studies where temperatures need to be measured at several points in the food and/or when a continuous record of the temperature is required. They are also used in portable devices for checking the temperature of food during production/chilling/freezing by means of a probe and temperature indicator scale.

### 11.5.1 Hydraulic type thermostats

This type of thermostat employs a long, thin phial of volatile liquid as the sensing element. It is connected to the controlling/indicating unit by means of a capilliary tube. The phial is fastened inside the equipment at some appropriate point. The capilliary tube connection allows more flexibility in the positioning of the temperature sensing element (the phial) than is possible with the rod and tube type described in the next section. This is probably the reason why it is the more popular of the two types. Care must be taken in locating the phial and the capilliary tube to ensure that they cannot be damaged when the equipment is in use.

The control device is a flexible metal bellows (Fig. 11.16), the movement of which operates either a gas valve or an electrical contact. When the bulb heats up, the liquid in the phial begins to evaporate or expand. This causes the bellows to increase in length. When the bulb cools the reverse process takes place. This movement of the bellows is used to cause either a gas valve to open or close or an electrical contact to be 'made' or 'broken'. In the case of gas-fired equipment, it is necessary to have a by-pass valve fitted so that a quantity of gas always flows to the burner, otherwise the gas flames would be extinguished when the main gas valve closes.

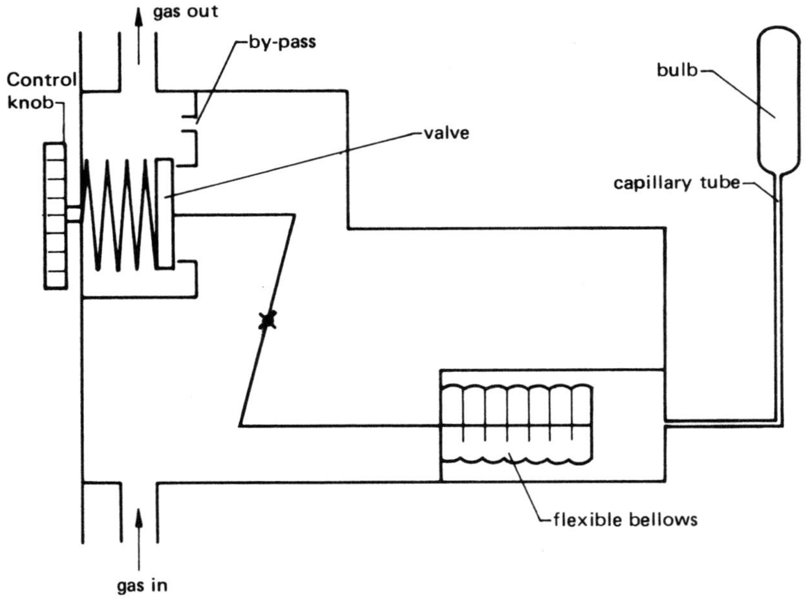


Fig. 11.16 – Hydraulic-type thermostat.

In practice when a piece of equipment is initially started up, the main gas valve will be open so that the equipment heats up rapidly. Once the temperature passes the temperature set on the thermostat, there will be sufficient movement of the bellows to cause the gas valve to close or the electrical switch to be broken. The equipment will begin to cool down and will soon fall below the set temperature. The movement on the bellows unit will then cause the electrical switch to come on or the gas valve to open. A typical oven temperature curve is shown in Fig. 11.17.

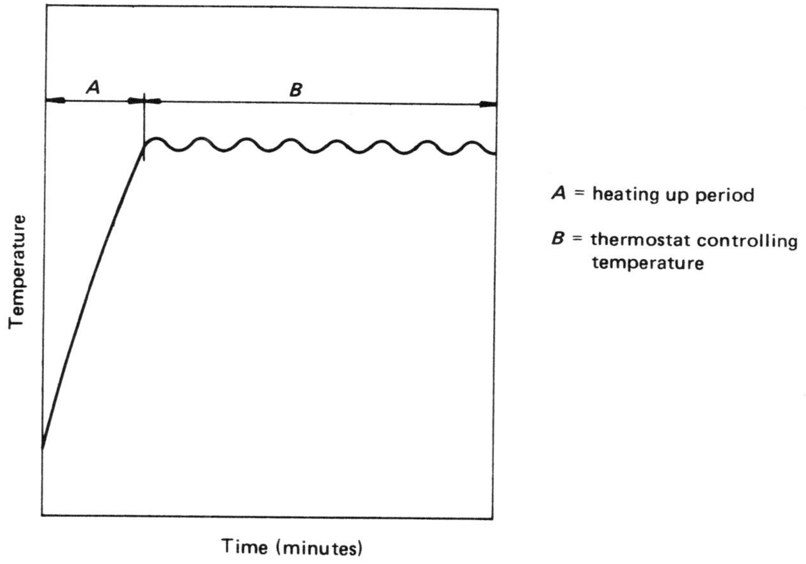


Fig. 11.17 – Oven temperature curve from start-up to control temperature.

Once the oven is up to the control temperature, the temperature will begin to 'cycle' about the set temperature. This cycling occurs because there is a 'lag' in the system between the temperature passing the set temperature and the thermostat actually operating the valve or switch. Careful design is required to minimise the temperature range over which cycling take place, otherwise the food temperature will cycle excessively causing a deterioration in quality. The set temperature may be altered. This is achieved by a control dial which, when moved, affects the point of movement of the bellows at which switching takes place.

Fryers, because of the danger of burning oil, should also be fitted with an 'over-ride' thermostat. This will cut off the gas (electricity) should the temperature of the oil become excessive because of the malfunctioning of the control thermostat. A quick-acting type is required in this case. It is possible to connect a phial to a bourdon tube (a metal tube bent to the shape of a circle) and use it for indicating the temperature on a dial. One end is connected to the phial and is clamped, whilst the other end is sealed but free to move. As more liquid evaporates, the pressure in the bourdon tube increases causing the free end to move. This movement is linked to a pointer moving over a calibrated temperature scale.

### 11.5.2 Rod and tube thermostats

This type of thermostat consists of a steel rod fixed inside and to one end of a brass tube (Fig. 11.18). The steel rod has a very low coefficient of expansion and expands relatively little over the temperature range of the appliance. The brass tube, on the other hand, expands and contracts to a significant extent as

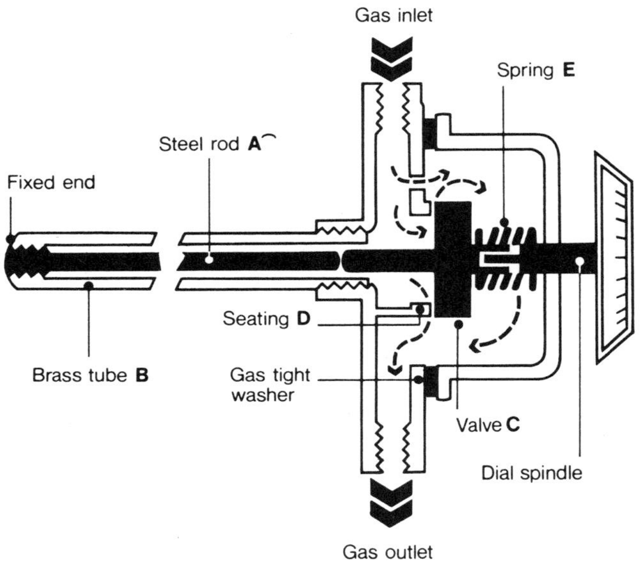


Fig. 11.18 – Expanding rod thermostat.

it heats up and cools down respectively. The brass tube is inserted into the appliance with the control knob just protruding from its front face. As the oven heats up, the brass tube expands to the left in the diagram as its right-hand end is fastened to the appliance casing. As the tube moves it pulls the steel rod with it, thereby closing the gas valve (or breaking an electrical contact). The converse is true as the oven cools down. The sequence of operation is similar to that described above for the hydraulic type. The control knob adjusts the point at which the gas valve closes/opens.

### 11.5.3 Thermocouples

A thermocouple consists of two dissimilar metal wires joined together at each end. The wires may be simply twisted together, but more usually are welded to give a more permanent joint. If one of the end joints is heated whilst the other the wires. An instrument for measuring this current, or more usually the voltage across the junctions may be inserted between the wires at the cold end. The thermocouple will then generate a fixed voltage for a particular temperature difference between the two ends of the thermocouple wires. Often it is advantageous to have the voltage measuring device at some distance away from the junction. In this case, rather than using long termocouple wires, it is possible to use compensating wires (cable) between the two. The compensating cable must be appropriate to the thermocouple wire materials and the thermocouple and compensating wires must be electrically insulated.

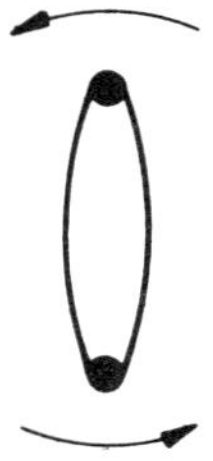


Fig. 11.19 – Thermocouple.

Suitable thermocouple materials should have the following characteristics:

(a) they should develop a relatively large voltage for a given temperature difference;
(b) they should have a stable voltage output at each temperature with time;
(c) they should not change in composition, that is not transfer metal from one to the other at the hot junction;
(d) they should not melt;
(e) they should be reproducible in composition so that the calibration (voltage versus temperature) is always the same for a particular pair of wires.

Thermocouples are produced to standard specifications and those listed are appropriate for use in catering applications. An advantage of the thermocouples listed in Table 11.2 is that they have a linear calibration.

**Table 11.2 – Thermocouple materials.**

| Materials | Tolerance | Temperature range (°C) |
|---|---|---|
| copper versus constantan† | ± 1°C | 0-100 |
| | ± 1% | 100-400 |
| iron versus constantan | ± 3°C | 0-300 |
| | ± 1% | 300-850 |
| nickel/chromium versus constantan | ± 3°C | 0-400 |
| | ± 0.75% | 400-850 |

† Constantan is 40% nickel, 60% copper.

The diameter of thermocouple wires should be small so as to reduce heat conduction along them and to increase the speed of response. The diameter should be large enough to give sufficient mechanical strength. The wires should be long enough to ensure that no part of the compensating cable is above 70°C. The insulation of the thermocouple wires must be chosen so as to withstand the temperatures to which it will be subjected. Maximum temperatures for various materials are recorded in Table 11.3.

**Table 11.3 – Maximum temperatures of thermocouple insulating materials.**

| Material | Maximum permissible temperature |
|---|---|
| PVC | 70-100°C |
| Rubber | 80-100°C |
| Woven glass | 350°C† |
| Woven asbestos | 700°C† |
| Woven silica | 1000°C |

† Must be used in a relatively dry atmosphere.

Thermocouples covered with Teflon are often used to measure the temperature of food in ovens up to 200°C.

The voltage generated by a thermocouple is usually measured using a potentiometer. Since the voltage generated depends on the difference in temperatures between the hot and cold junctions, the cold junction temperature must be compensated for, if this is not 0°C. Because the potentiometer is at room temperature this is always necessary. Usually however, the instrument does this automatically. The potentiometer may give a visual display of temperature on a scale or show it as a digital readout. For many purposes however, it is useful to have a record of the temperature changes and potentiometers with chart recording are available for this purpose. Such potentiometers may have just one channel (for one thermocouple) or they may have several channels so that a record of several thermocouples may be made on the same chart. In this case, the thermocouples may be recorded either in turn, with a time lapse between readings of any particular thermocouple, or simultaneously, giving a continuous record for each thermocouple. Instruments of the second kind are much more expensive than the first type.

If a rapid check of temperature is to be made many times, as in cook-freeze units, then a thermocouple in a metal sheath is used. This makes it easy to use and protects the thermocouple from damage. In this case a single voltmeter may be used, either with a pointer moving over a calibrated temperature scale, or giving a digital readout, both with automatic cold junction compensation. Such an instrument is calibrated only for one pair of thermocouple materials. These are indicated on the scale and also whether or not cold junction compensation is automatic.

## 11.6 PRESSURE CONTROL

### 11.6.1 Governors

The gas supply, as it arrives at a piece of equipment, fluctuates in pressure because the demands of other appliances near to it change. The heat output of the burners in the equipment will therefore fluctuate (Section 11.2.2) unless these fluctuations are smoothed out. There is also a possibility of 'lift-off' taking place if there is a sudden surge in pressure and hence gas velocity to the burners. Therefore it is usually necessary to fit a **governor** to each gas-fired appliance to prevent this happening. An appliance governor is a metal box containing a valve which is attached to a diaphragm. The metal diaphragm is fixed to the sides of the box by means of a flexible seal so that it can move up and down (Fig. 11.20).

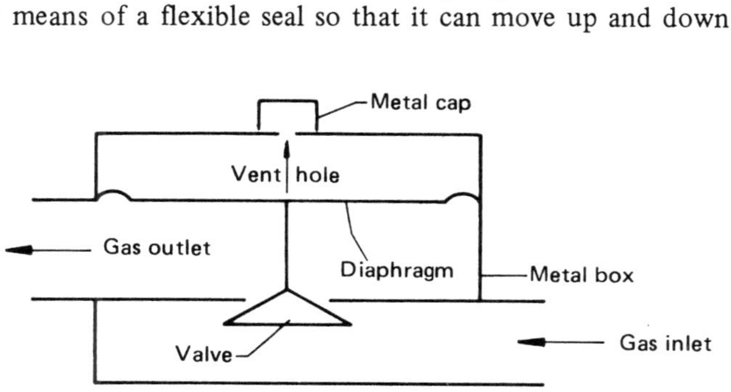


Fig. 11.20 – Appliance governor.

The gas enters the box, passes through the valve and then leaves for the burners. The valve is a restriction to the flow and therefore there is a pressure drop as the gas passes through it. If the inlet gas pressure rises, the diaphragm lifts and the valve closes slightly. The closing of the valve causes the pressure drop to increase, thereby offsetting the rise in inlet gas pressure as far as the outlet gas is concerned. As the inlet pressure falls, the diaphragm moves downwards, opens up the valve and reduces the pressure drop, thereby compensating for the fall in inlet pressure as far as the outlet side is concerned. The upper side of the diaphragm is vented to atmosphere to allow the diaphragm to move freely upwards and downwards.

### 11.6.2 Pressurestats

A pressurestat uses steam pressure to control the input of gas or electricity to an appliance. It is only used therefore on appliances which generate steam. The steam is fed to a bellows unit (Fig. 11.21) which is connected to a spring-loaded valve. As the steam pressure increases, the bellows overcomes the force exterted by the spring and closes the valve. Conversely, as the steam pressure falls, the force of the spring overcomes the force exerted on the valve by the bellows and the valve opens. A small amount of gas may be allowed to flow continuously through a by-pass opening.

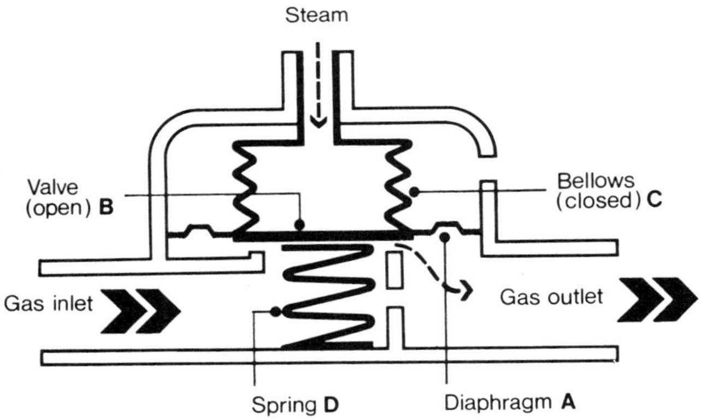


Fig. 11.21 – Pressurestat.

### 11.6.3 Low-level water cut-off valves

These are devices which are used to cut off the energy supply to a water-heating appliance which is in danger of running out of water. If the appliance were allowed to run dry, it could be seriously damaged. The upper section of this unit is a gas valve (Fig. 11.22) which is held open by the pressure of the water operating on it via a flexible diaphragm. If the pressure falls, the weight attached to the diaphragm also falls causing the gas valve to close, the two being attached by means of a rod. The pressure of the water arises from the height of water (potential energy) in the supply tank. If this falls too low then the water pressure falls below the point at which the gas supply is cut off. If more water is added to the water tank, the gas valve will open again. Since this is automatic, it is essential that the burner is fitted with a pilot light (see Section 11.7.1).

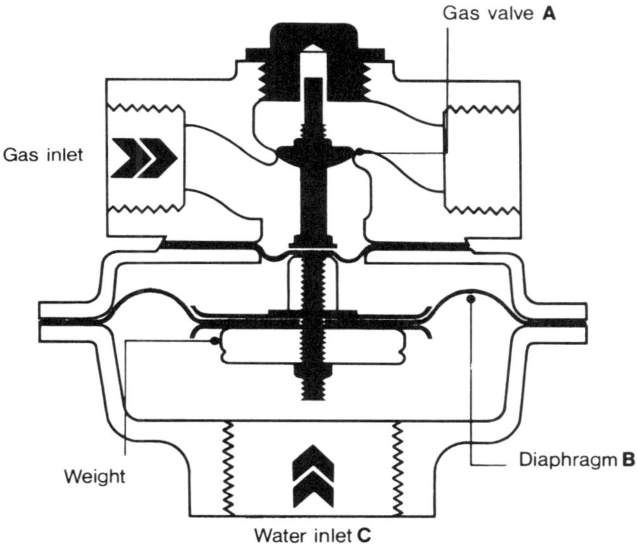


Fig. 11.22 – Low-level cut-off valve.

### 11.6.4 Steam safety valves

All vessels operating under pressure are fitted with a safety valve so as to prevent the steam pressure building up to such a level that an explosion occurs. It is possible to have a valve operating at a fixed pressure or one which is adjustable. They should be designed so that no-one can interfere with them. The pressure is controlled by either a weight or a spring (Fig. 11.23). If the pressure increases above the pre-set value, the valve will lift against either the weight or the spring and release steam to the atmosphere. Once the pressure is below the pre-set level, the valve closes and the appliance continues to operate.

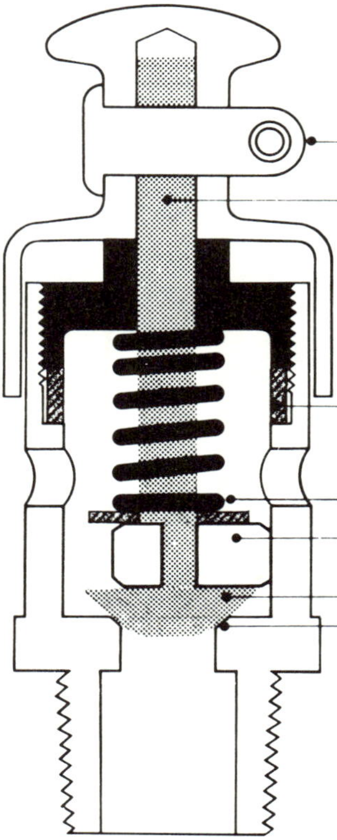


Fig. 11.23 – Safety valve.

## 11.7 GAS IGNITION

Automatic gas ignition is becoming more generally used with gas-fired appliances. Burners are often inconvenient to light manually and automatic ignition is an aid to ensuring that equipment is lit only when it is required, thereby saving energy. A number of methods for automatic ignition are available.

### 11.7.1 Pilot lights

A permanent pilot light is a very small flame burning continuously near to the main flame such that when the main gas tap is turned on the main gas flow will be ignited immediately. Pilot lights must be designed with a number of features.

A very stable flame is required to ensure that it is not extinguished in draughts. The jets must be non-corroding otherwise they will tend to block, thereby preventing the flow of gas to them. In addition they must be prevented from blockage by particles carried along in the gas stream. This is achieved by including a filter in the gas supply pipe. They must be protected from food spillage too, which could cause a blockage in the burner tip. They must also burn satisfactorily in the vicinity of the main burner.

In addition to these design problems, they use gas continuously and this is regarded as a wasteful use of energy.

### 11.7.2 Flashtubes

A flashtube enables a main burner to be lit from a pilot light which is not situated next to it. It is a metal tube connecting the pilot to the main burner. When the main burner is turned on, a gas/air mixture flows back from the main burner towards the pilot light. When it reaches the pilot light it ignites, the flame travels down the tube and ignites the main burner. A single pilot can be used to light the hobs on a gas range as they are required but the tubes are exposed to the heavy duty wear on the top of the range. They may be used to light two grill burners from a single pilot light, where such a design is used.

### 11.7.3 Spark ignition (piezo-electric ignition)

This type of ignition obviates the need to use a permanent pilot light and is push-button operated. It makes use of the fact that when a certain kind of crystal is stressed it produces a high voltage. The spark produced from this high voltage if then used to ignite the gas supply. The crystal is stressed by squeezing it by means of a lever mechanism operated by the push button (Fig. 11.24). The electrode which produces the spark requires protection against damage and spillage. The spark is used to ignite a pilot light which then lights the main gas supply (Fig. 11.25). It is essential to have the correct gas/air mixture arriving at the electrode otherwise ignition will not take place. The pilot is only lit when the main gas supply is in use.

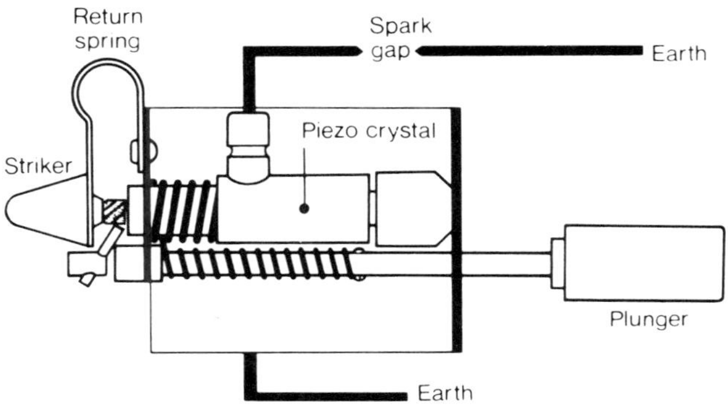


Fig. 11.24 – Piezo-electric igniter.

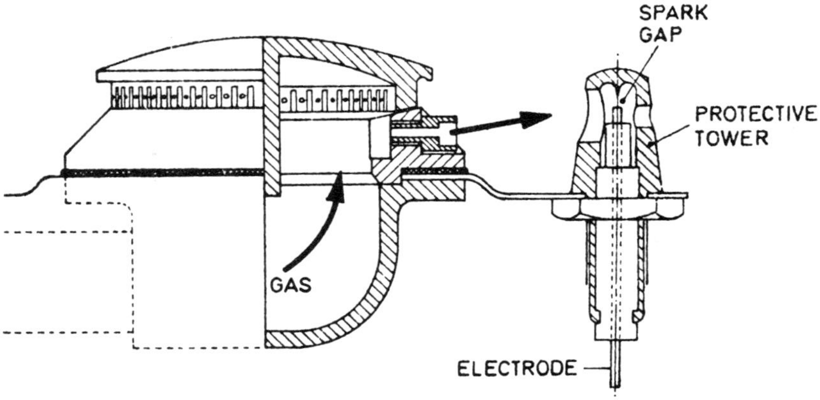


Fig. 11.25 – Spark ingniter attached to a burner.

### 11.7.4 Battery/mains ignition

This is similar in principle to the piezo-electric type except that the energy is supplied from a battery or the main electrical supply. The current is supplied to a platinum wire which is held in the stream of gas issuing from the pilot jet (Fig. 11.26). The current is supplied by batteries or a step-down mains transformer. A push-button valve will supply gas to the pilot and also electrical energy to the igniter (platinum wire). Release of the button cuts off the power supply and the gas supply to the pilot. There is also a form of battery/mains spark ignition which, on pressing a button, produces a continuous chain of sparks.

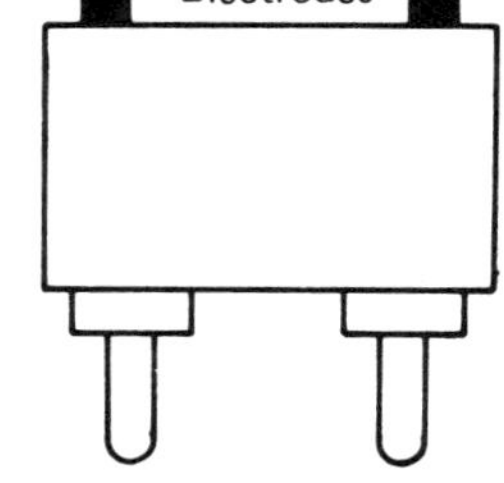


Fig. 11.26 – Battery ignition.

## 11.8 FLAME FAILURE DEVICES

If the flame inside an oven or hot cupboard is extinguished for any reason, a potentially explosive gas/air mixture will build up inside the appliance. If an attempt is then made to re-light the appliance, this mixture could explode. For safety reasons therefore, a flame failure device is often fitted. Should the flame be extinguished, then the gas supply is automatically cut off by this device.

### 11.8.1 Thermo-electric device

In this device (Fig. 11.27), the fact that the burner is lit is sensed by a thermocouple situated in the flame. The thermocouple generates a current which flows through a magnetic coil. The coil generates a magnetic field when the current is flowing and this holds the gas valve open. If the flame is extinguished the current ceases to flow, there is no magnetic field and the gas valve is closed by means of a spring. No gas can flow until the starter button is pressed manually. This button opens the valve against the spring until the flame is lit, a current generated and magnetic field to hold the valve open is set up. The current itself is insufficient to open the valve against the spring; this is carried out manually.

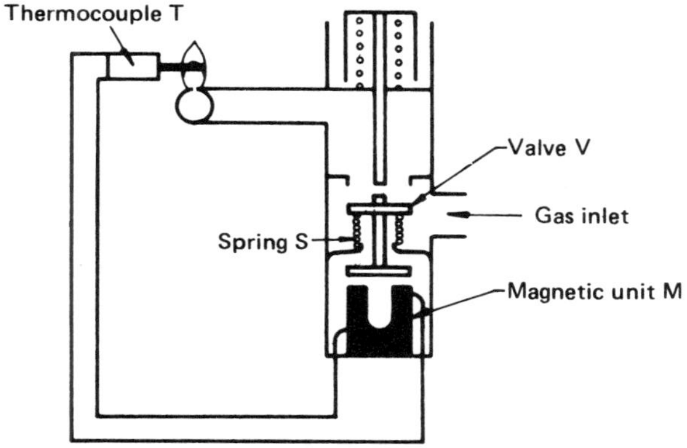


Fig. 11.27 – Thermo-electric flame-failure device.

This device is in part the basis of the pan sensing-device used to cut off the gas supply to a gas ring 30 seconds after the pan has been removed (see Section 5.6.2.1). The other part of the device is a sensor which detects the presence of the pan and allows gas to flow to the burner. When the pan is removed, the gas supply to the pilot light is cut off. The thermocouple (detecting the pilot light is not lit) takes about 30 seconds to cool sufficiently to operate the main gas valve. This lag in the thermocouple control gives the chef 30 seconds to replace the pan before the flame is extinguished. Once the burner shuts down, the pilot light must be re-lit (by automatic ignition) before the main gas supply can be restored.

### 11.8.2 Bi-metallic strip device

A bi-metallic strip device consists of two strips of dissimilar metals fastened together. The two metals expand at different rates when heated/cooled. This differential expansion causes the strip to bend to different extents depending on its temperature. The strip is bent and is situated near the pilot flame as shown in Fig. 11.28.

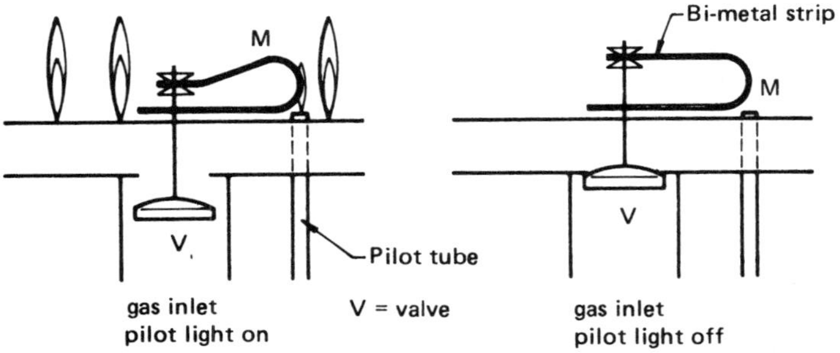


Fig. 11.28 – Bi-metallic strip flame-failure device.

When the pilot is not lit, the main gas valve attached to the end of the strip is held closed. Once the pilot flame is burning however, the strip heats up, expands, bends and causes the main gas valve to open. If the pilot light is extinguished the strip cools, contracts, resorts to its original shape and causes the main gas valve to close.

### 11.8.3 Mercury vapour valve

This operates when mercury vapour, held in a sealed tube, expands as it receives heat from the pilot light (Fig. 11.29). This expansion is transmitted to a beam

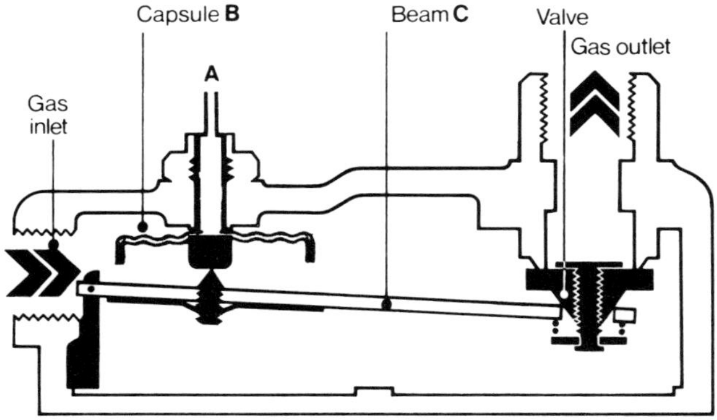


Fig. 11.29 – Mercury-vapour valve.

via a capsule (which is similar to a bellows unit). The beam moves downwards and opens the main gas valve. If the pilot is extinguished, the mercury cools and contracts and this causes the main gas valve to close via the beam mechanism. A method of using a mercury vapour valve is illustrated in Fig. 11.30.

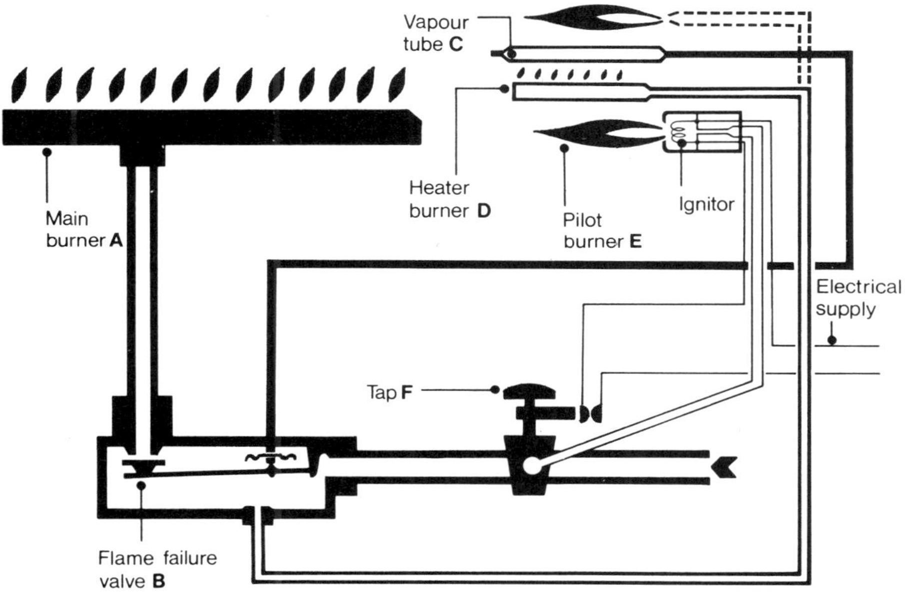


Fig. 11.30 – Vapour tube with electric ignition.

The main burner is connected to the flame failure valve. The vapour tube has its own heater burner. The pilot burner has electrical ignition and ignites this heater burner as well as the main burner. The starting sequence is initiated by pressing the tap. This automatically allows gas to pass through the pilot light and the heater burner and ignites both of them. As the vapour tube heats up it opens the flame failure valve, allowing gas to pass to the main burner.

## 11.9 CONTROL SYSTEMS

There is a trend with certain types of catering equipment towards more automatic control systems. An important component of these systems is often a relay valve, which is used in situations where either the control is remote from the gas supply or the gas supply is too large to be controlled directly. They are used as master controllers and provide a means of controlling an appliance by more than one device. It is possible to control time, temperature, flame failure and the use of electrically-operated mechanisms by these devices. Two types of relay valves are illustrated in Fig. 11.31.

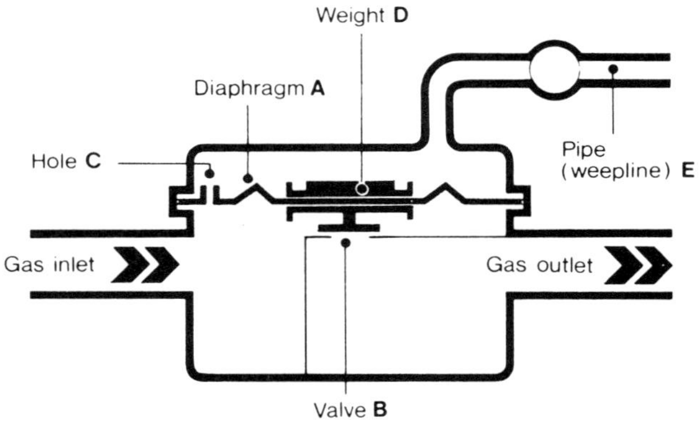


(a) Relay valve on/off.

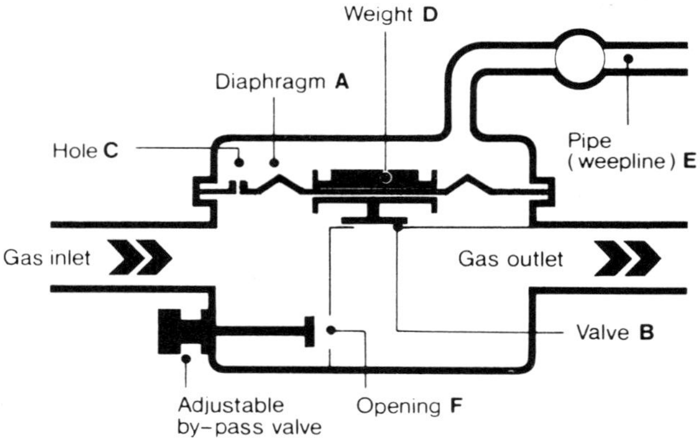


(b) Relay valve high/low.

Fig. 11.31 – Relay valves.

Gas enters the relay valve and fills the space below the diaphragm. A limited amount of gas also passes to the upper side of the diaphragm through a small hole. The diaphragm lifts (by gas pressure), raising the gas valve and allowing gas to flow to the burner. If for any reason the gas cannot leave the upper chamber via the weepline, the pressure on either side of the diaphragm equalises and the diaphragm weight causes the valve to close again. Any ancillary controls are fitted to the weepline, therefore indirectly controlling the main gas outlet valve. All the controls contain a gas valve and if any of these valves close, gas remains in the chamber above the diaphragm and will build up sufficient pressure to cut off the main gas supply. If electric controls are used, a single electrically-operated valve will control the flow of gas from the upper chamber. Gas leaving the weepline is burnt near the main burner.

The high/low relay valve has a by-pass so that some gas always passes to the main burner. The on/off type shuts the gas off completely.

Timer clocks are used with some types of equipment. Fig. 11.32 shows a system used with a forced-convection oven. The timer automatically opens the electrically operated gas valve at the start of the cooking period and automatically closes it at the end. Automatic ignition and flame failure devices are also fitted. The controls could be elaborated further to include a buzzer, a microswitch fitted in the door frame to shut off the burner and fan when the door is opened and indicator lights.

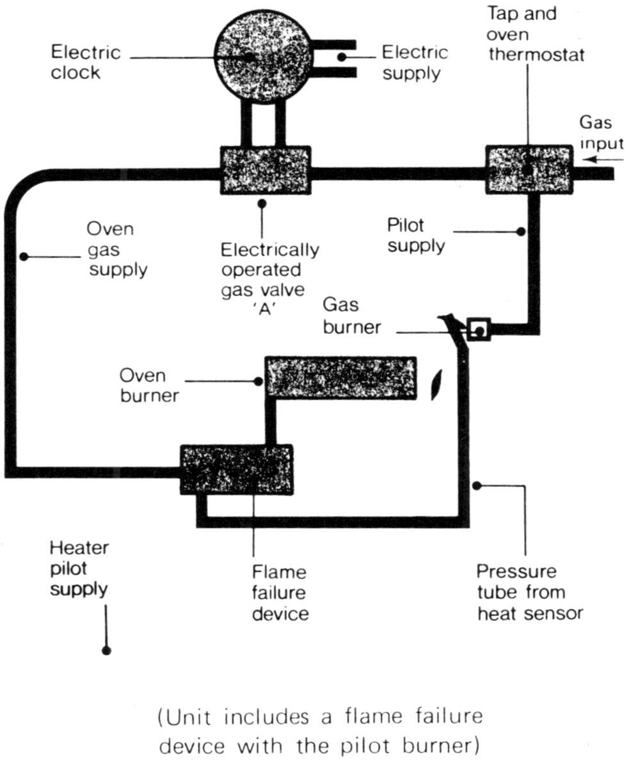


Fig. 11.32 – Time controlled oven (electric clock).

The control circuit for a gas-fired steamer is shown in Fig. 11.33. The gas controls comprise a governor, gas valve (the solenoid type, which is opened/closed electrically), a pilot assembly incorporating an electrical ignition device and a flame failure device operating on the solenoid gas valve. Electrical safety devices are pressure controls, steam safety valve, an exhaust valve, a low water cut-off and a timer.

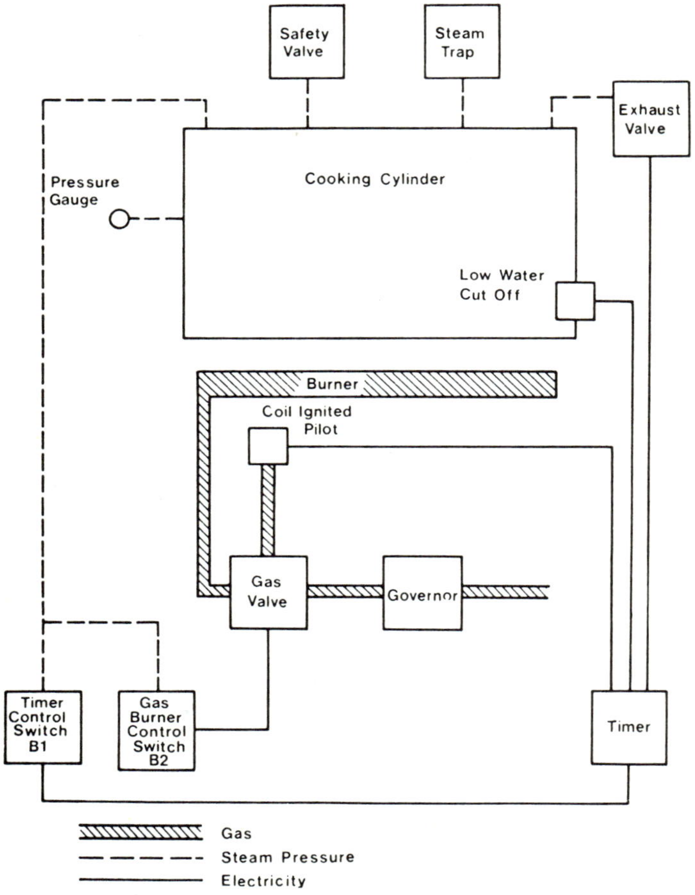


Fig. 11.33 – Steamer controls.

When the timer is turned from the off position to any setting the igniter is energised, the steam exhaust valve closes and the pilot solenoid of the gas valve opens. The pilot is ignited and the flame failure device heated, to make electrical contacts which in turn energise the main gas solenoid valve. The automatic igniter is simultaneously de-energised. Steam is generated within the cooking chamber which displaces the cold air present via the thermostat steam trap. When all the air has been displaced the trap closes automatically. When the pressure in the chamber reaches 0.6–0.7 bar, a pressure switch closes and starts the timer running (the start of the cooking period). When the pressure reaches about 1.0 bar, a further pressure switch opens causing the main gas solenoid valve to close. The pressure switch closes again (and the main gas solenoid opens) when the pressure falls to 0.9 bar. The burner cycles in this way until the end of the cooking period.

At the end of the cooking period the timer causes the main gas supply to be shut off and the exhaust valve to open to release the steam pressure. A buzzer then sounds until it is turned off by the operator. A low water cut-off, which is incorporated into the control system, cuts off the gas supply if the water falls below a predetermined level.

## 11.10 SAFETY

National standards for catering equipment often include requirements for the general constructional features and safety aspects of equipment. A standard for gas-fired equipment may include sections on marking and instructions, gas and water soundness, gas and water connections, construction and design, electrical requirements, heat input, ignition, flame failure devices, pilots, flame stability, surface, floor and wall temperatures, gas cocks and taps, thermostats, governors, flue outlets and draught diverters. A standard for electrically-heated equipment may include sections on finish, voltage rating, marking, protection against electric shock, input, operation under overload conditions, electrical insulation at operating temperature, protection against spillage, insulation resistance and electric strength, mechanical strength, stability, construction, thermal controls, other component parts, supply connections, terminals for external conductors, provision for earthing, screws and connections, creepage distances and clearances and resistance to rusting. Readers should familiarise themselves with the standards, if any, published in their own countries.

Flexible connections with plug-in connectors for gas-heated equipment are available and permissible in some countries. This type of connection enables the user to disconnect the flexible connection without tools. At the same time it automatically shuts off the gas supply by menas of a spring-loaded valve within the socket attached to the gas main. The appliance may then be moved for cleaning. On re-connection, the gas supply is turned on automatically by the plug, which is an integral part of the flexible tube.

Typical guidelines laid down for the installation of flexible connections include the following:

(1) The appliance must be of inherently stable design under all conditions of use and must require deliberate action to be taken before it can be moved.

(2) No strain must be transmitted to the internal pipework during connection and subsequent use.
(3) Tubes and connections should not be subjected to excessive heat by direct exposure to flue products or by contact with hot surfaces.
(4) The plug and socket connection should be easily accessible to the user. The appliance should not be moved whilst connected except for such limited and restrained movement as to reach and operate the plug-in connector.
(5) Tubes and sockets must not be under strain in normal use or whilst being connected or disconnected.

The design, construction, installation and correct use of equipment is essential for the safety of personnel involved in catering. In many countries, legislation has an important bearing on these aspects of equipment design.

## 11.11 REFERENCES

Dewerth, D. W. (1962), AGA Research Bulletin 92.

## 11.12 FURTHER READING

Miles, V. C. and Pinkess, L. H. (1970), *Gas appliance controls.* Walter King.

CHAPTER 12

# Mechanical Preparation Processes

## 12.1 INTRODUCTION

In the previous chapters we have dealt exclusively with mass-transfer, heating and cooling processes. The theory in these areas is well developed and may be applied to catering situations with a fair degree of success. In this and subsequent chapters the remaining processes used by caterers are considered, beginning here with a description of mechanical-preparation processes. These are all the processes which do not involve cooking, heating or cooling of food. Equipment in this area is often very versatile, consisting of an electric motor which is used to drive a range of diverse attachments. Other items are specialist devices which may be used for one purpose only.

## 12.2 SIZE REDUCTION

### 12.2.1 Size-reduction processes

In discussing heat and mass transfer in Chapters 4 and 10, the advantage of having food with large surface areas was mentioned. This is utilised in a number of processes in catering, for example grinding coffee beans to speed up the extraction of the coffee and dicing or slicing vegetables prior to boiling to give short heating times.

The advantage of size reduction in terms of surface area may be shown by the effect on surface area of progressively cutting up a cube into smaller cubes. This can be illustrated by starting with a single cube 10 cm × 10 cm × 10 cm and progressively reducing it in size as shown in Table 12.1.

In some operations, such as dicing and shredding, the final shape of the product can be controlled. In grinding however, it is not possible to do this and the product covers a range of sizes. Much research has been carried out on grinding processes and therefore a reasonable theory has been developed. In the case of the other processes no such theory exists and in many catering operations these other processes are often carried out manually although mechanical aids are often available nowadays. Lowrison (1974) has classified the methods of size

**Table 12.1 – Size reduction and surface area.**

| Size of cube | No. of cubes cut from initial cube | Surface area of each cube† | Total surface area |
|---|---|---|---|
| 10 cm X 10 cm X 10 cm | 1 | 600 $cm^2$ | 600 $cm^2$ |
| 5 cm X 5 cm X 5 cm | 8 | 150 $cm^2$ | 1200 $cm^2$ |
| 2 cm X 2 cm X 2 cm | 125 | 24 $cm^2$ | 3000 $cm^2$ |
| 1 cm X 1 cm X 1 cm | 1000 | 6 $cm^2$ | 6000 $cm^2$ |
| 0.5 cm X 0.5 cm X 0.5 cm | 8000 | 1.5 $cm^2$ | 12000 $cm^2$ |

† A cube has six sides.

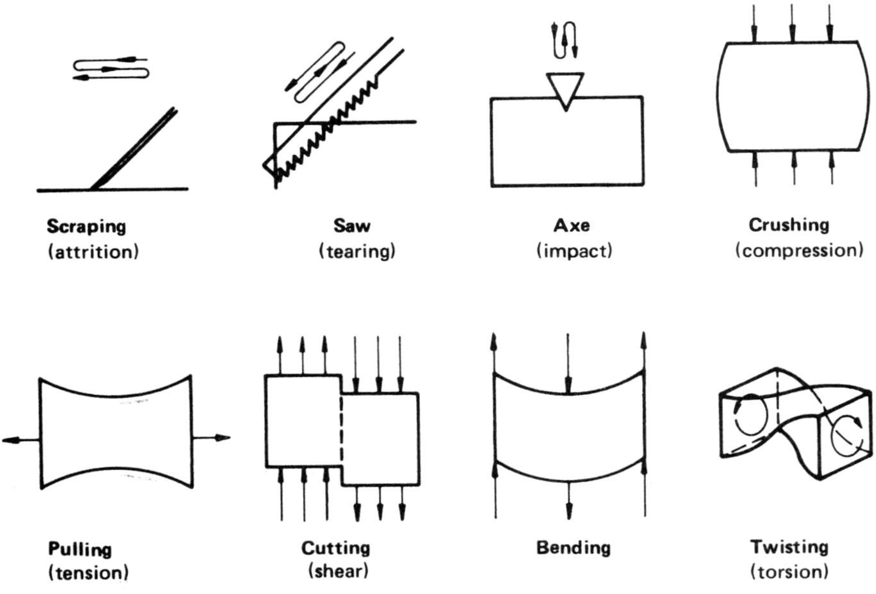


Fig. 12.1 – Forces required to break materials.

**Scraping** (attrition) – the solid is worn away with another solid.
**Saw** (tearing) – the solid if torn with another solid.
**Axe** (impact) – the solid is impacted by another solid.
**Crushing** (compression) – pushes it in.
**Pulling** (tension) – pulls it apart.
**Cutting** (shear) – subjects solid to shear by another solid.
**Bending** – subjects it to flexure.
**Twisting** (torsion) – twists the solid.

reduction by the kind of forces required to break the material. To break a solid, it is necessary to change its shape beyond certain limits. The processes used to do this are illustrated in Fig. 12.1.

In the processes of crushing, pulling, cutting and twisting a stress is applied to the solid. This is measured as the force applied divided by the area over which it is applied. The solid deforms by an amount called the **strain**. The strain is measured as the amount of deformation divided by the original size of the material. The definition of strain in several breakage situations is illustrated in Fig. 12.2.

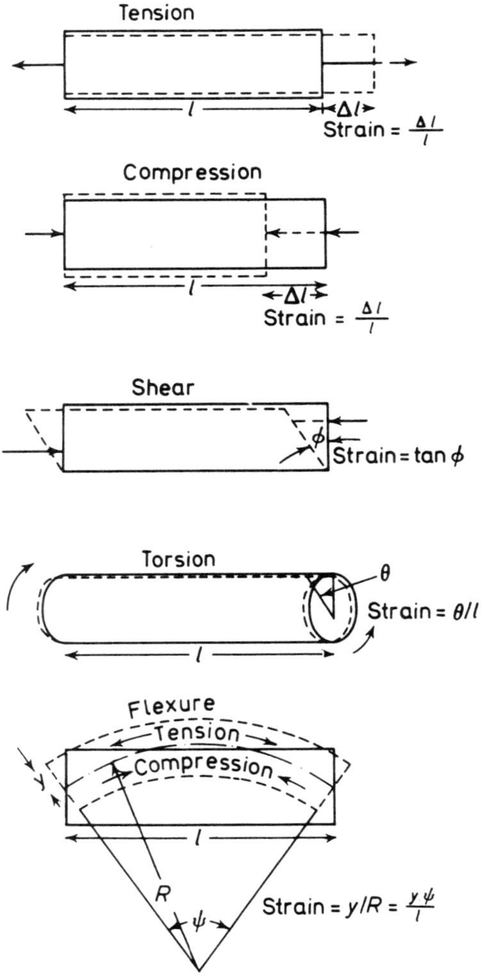


Fig. 12.2 – Stresses and strains in breakage processes.

If stress is plotted against strain a curve of the type shown in Fig. 12.3 is obtained.

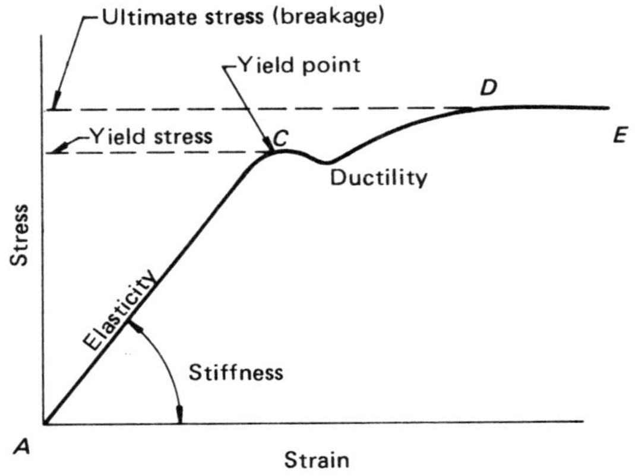


Fig. 12.3 – Stress/strain diagram for a solid.

Catering operations involving size reduction may be defined in terms of these size-reduction processes:

(a) **grinding** – uses crushing and attrition;
(b) **dicing, slicing, shredding** – uses cutting;
(c) **mashing** – uses crushing and attrition;
(d) **sawing** – uses sawing;
(e) **mincing** – uses cutting;
(f) **vegetable peeler** – uses attrition.

These operations are considered briefly below.

### 12.2.2 Crushing and attrition equipment

In grinding and mashing, crushing (compression) is carried out by means of two solid surfaces pressing together on the food while the grinding takes place between pieces of food in contact with each other and between the pieces of food and the parts of the machine. Both processes take place together, and it is difficult to separate the effects of them.

Most research on crushing/grinding is of a purely empirical nature. The best relationship available which relates power input to feed and product sizes is that due to Bond:

$$W = 10\,W_1\left(\frac{1}{\sqrt{P}} - \frac{1}{\sqrt{F}}\right)$$

where $W$ = energy input (kWh/ton), $W_1$ = work index (kWh/ton), $P$ = product size ($\mu$m) and $F$ = feed size ($\mu$m).

The work index is a property of the material being crushed. The feed and product sizes ($F$ and $P$) are those for which 80% of the feed and product pass respectively. The effect on power requirement of product and feed sizes may be demonstrated by simple calculation. A comparison will be made of the power requirements to grind a material with a work index of 2 kWh/ton from a size of 10 mm to a size of 1 mm, 1 mm to 100 $\mu$m and 100 $\mu$m to 10 $\mu$m.

For size reduction from 10 mm to 1 mm:

$$\begin{aligned} W &= 10 \times 2\,(1/\sqrt{1000} - 1/\sqrt{10\,000}) \\ &= 0.43 \text{ kWh/ton.} \end{aligned}$$

For size reduction from 1 mm to 100 $\mu$m:

$$\begin{aligned} W &= 10 \times 2\,(1/\sqrt{100} - 1/\sqrt{1000}) \\ &= 1.37 \text{ kWh/ton.} \end{aligned}$$

For size reduction from 100 $\mu$m to 10 $\mu$m:

$$\begin{aligned} W &= 10 \times 2\,(1/\sqrt{10} - 1/\sqrt{100}) \\ &= 4.32 \text{ kWh/ton.} \end{aligned}$$

Consequently the finer the product size the greater the energy requirement and this increase is more rapid the finer the product size. In practice, the input to the grinder will be constant and therefore the extra energy input is reflected in longer grinding times. Coffee is normally ground in between two cylindrical surfaces, the inner one rotating, the outer one stationary. The surface may be covered by ridges. Potatoes are mashed in a vertical mixer (see Section 12.3). Milk should only be added at the end of mashing since if added early, it lubricates the surfaces of the potatoes, thereby reducing attrition between the potato pieces and between the potato pieces and the machine.

**Mincing machines** work on a rather different principle to crushers, although they reduce the size of meat and so on. The meat is first extruded into a series of small holes, this process causing some breakup of the meat structure. Immediately before leaving the holes the meat is cut by a rotating knife, thereby reducing its size further. The feed material is carried towards and is forced through the holes by means of a screw conveyor, rotating inside a cylindrical casing. The screw conveyor is turned by an electric motor. The rotating knife is also attached to the rotating shaft, whilst the plate containing the holes is held fixed by the casing of the machine. For safety reasons the rotating knife is mounted inside the machine.

### 12.2.3 Cutting machines

In the mechanism of cutting the finest edge of the knife, say, penetrates as a result of plastic deformation of the material being cut. The material is first displaced without rupture. If the cutter is moved, abrasion also takes place. This is because, at high magnification, the knife edge consists of peaks and troughs which interact with similar ones on the food surface so that the 'hills' break away. The relative influences of plastic deformation and abrasion depend on the nature of the knife and the material. A shape knife passing through butter is almost wholly caused by plastic deformation, whilst the action of a saw through bone is almost wholly caused by abrasion. Knives and shears are usually used with plastically-deformable material, whereas saws are applied to more brittle materials. There is no theory available to calculate the energy required during cutting.

Several types of cutting machine are available and include knives (with one cutting edge), shears (which have two cutting edges) and saws (with many cutting edges). They may be operated manually or by incorporating them into a machine. Various designs of cutters/slicers are available for different purposes. The blade(s) are attached to a disc: the distance between the blade and the disc determines the thickness of slice and usually a range of sizes (thicknesses) of each type is available. This is particularly important in catering applications because the shape of the product required is closely allied to customer satisfaction. For meat a cutter-mixer is also available: the cutting blades are mounted on a rotating shaft within a mixing bowl. A bandsaw consists of a flexible metal blade with saw teeth cut into one edge. The blade moves downwards and cuts the food as it is held on a metal table. Tenderisers have knives fitted to a shaft which is rotated by a motor. The knives cut the surface of the meat. In general, cutting is a slow process and (if it is a limitation) the minimum size of the product is large compared with other size-reduction processes. Some cutting machines are often difficult to guard for safety.

Machines are available for slicing, chipping and dicing vegetables. Slicing involves cutting in one direction, chipping in two directions and dicing in three directions. Machines with interchangeable blades are available for performing all of these processes for a range of vegetables. Alternatively, attachments for mixing machines which can be used in a similar way can be obtained. Where large throughputs of diced meat are required machines are available to do this.

**Slicing machines** are used for meat and bread. Both operations are time consuming if carried out manually. The machine also gives slices of regular thickness. Usually a feed tray holds the food firmly against a circular cutting knife. The tray usually slopes towards the knife so that the food is automatically fed by gravity for slicing. The knife may be rotated and advanced manually or mechanically or by a combination of both. A calibrated scale is fitted to control the thickness of the slices. It is also possible to have an automatic and combined bread cutting and buttering machine.

**Vegetable shredding machines** may be specialised, such as potato chippers, or versatile, where a number of interchangeable cutting discs are used. The operations performed by such machines include shredding, slicing, grating, cutting straws and chips and making breadcrumbs. Potato chippers first cut the potato into slices and then the slices are cut again to form chips. Alternatively they may be cut to chips in one single cutting action. Shredders and so on are rotating discs with cutting edges pressed out of the metal disc just above its surface. These edges may be the length of the disc's radius, thereby forming a slicer, or they may be in the form of many small cutting edges, therby forming a shredder/grater. A grid of knives combined with a slicer is used for dicing materials.

**Liquidisers** contain several knives rotating at high speed inside a container and are capable of producing food material of fine size.

**Waste disposal units** pulverise kitchen waste so that it can be passed into the drains. Such a machine consists of a series of steel knives rotating at high speed inside a metal tube. A feed hopper is attached. Most units are fitted with a device to prevent cutlery falling into the grinding chamber.

### 12.2.4 Vegetable paring machines

These machines are used for cleaning potatoes and other root vegetables. They operate by bringing the vegetable into contact with a rough surface which rubs off the outside layers of the vegetable by attrition. The container is lined with carborundum and has inside its base a rotating disc also covered with carborundum and driven by a motor. The disc rotates at moderate speed and in doing so throws the potatoes against the rough surfaces and removes the peel. The disc usually has an undulating surface to improve its efficiency, that is its ability to bring the surface of the potatoes into contact with the carborundum surfaces. Dirt and scrapings are removed with the aid of running cold water. The machine is designed so that the water is directed onto the surfaces of the vegetables and so that the cleaned vegetables can be removed from the machine separate from the water and peelings.

## 12.3 MIXING

In catering a wide variety of mixing operations are carried out. It is usual to classify these operations in terms of the types of materials being mixed, that is liquids of low or moderate viscosity, high-viscosity pastes and plastic solids (including non-Newtonian liquids) and dry solids.

Mixing machines appropriate to each class of mixture can be designed, thereby minimising costs for the easier mixing operations and only using the most expensive machines for the most difficult situations. In catering applications however, the utilisation of mixing equipment is so low that it is not possible to have several machines of varying costs, instead one versatile machine which will cover all eventualities is needed. Of the materials listed, it is the high-viscosity pastes and plastic solids which are the most difficult to mix and therefore the mixing machines used in catering are designed for this situation.

Some operations involving mixing have a secondary purpose, for example incorporating air into a sponge mixture (mixing plus aeration), potato creaming (mixing plus attrition) and making mayonnaise (mixing plus emulsification). In other operations, such as the mixing of bread dough, considerable amounts of mechanical energy are required to produce the changes from flour to dough. Specialised high-speed mixers have been evolved for the mechanical development of dough by the Chorleywood process: in this the measurment of energy input is an important control parameter.

### 12.3.1 The measurement of mixedness

Ideally, a mixing process begins with the components grouped together in some container but still separate as pure components. If small samples are taken throughout the container they will almost all consist solely of one pure component, the frequency of occurrence of these components being proportional to their fractions in the whole container. As mixing proceeds, samples will increasingly contain more of the components, again in proportions approximating to their overall proportions in the whole container. Complete mixing could then be defined as that state of affairs in which all samples are found to contain the components in the same proportions as in the whole mixture. This is an improbable result for any practical mixing process. Another approach can then be made, defining the perfect mixture as one in which the components in samples occur in proportions whose statistical chance of occurrence is the same as that of a statistically random dispersion of the original components. Such a dispersion represents the best that random mixing processes can achieve. The standard deviation of sample compositions from the mean composition of the overall mixture represents a measure of the mixing process. Hence standard deviation can be used to represent the degree of mixedness with multicomponent mixtures.

A standard deviation must be calculated for each component in turn, with all the other components grouped together. Therefore, with multicomponent mixtures, component A can be well mixed with (B, C, D, E and so on) taken together, but B may be poorly mixed with (A, C, D, E and so on) taken together. For a perfect mixture, the standard deviation for each component would be zero. As mixing proceeds, the standard deviations will all decrease. Unfortunately, the standard deviation measured will also be dependent on the size of the sample taken, larger samples tending to even out variations, giving lower standard deviations than smaller samples removed from the same mixture. In the extreme, samples that consist of just one particle would lead to the conclusion that no mixing had occurred.

In practice the sample size used should correspond to a **characteristic**

**volume** with which all the components must be present in approximately the correct proportions, but not necessarily well mixed. This volume is determined by the use for which the mixture is intended, and in catering will be related to the organoleptic (those qualities appealing to the customer's sight, taste and smell) properties of the end product. Thus in Cornish pasties using a potato/minced-meat filling, the customer will need to taste *both* potato and meat when he takes a bite of the pastie and not one or the other alone. **A mixing index** ($A$) may be defined by:

$$A = \frac{\log S_0 - \log S}{\log S_0 - \log S_R}$$

where $S$ = the standard deviation at any time during mixing, $S_0$ = the standard deviation in the unmixed state, and $S_R$ = the standard deviation for the ideal mixture. Consequently the denominator represents the path from unmixed to ideally mixed whilst the numerator represents how far mixing has proceeded along this path. No values of mixing index are available in the literature for catering situations.

### 12.3.2 Mixer design

The most widely used mixer consists of a pan containing the materials to be mixed and a mixing element which is used to cause the displacement of materials relative to each other. The performance of mixers for high viscosity pastes depends on direct contact between the mixing element and the materials of the mix. Therefore the mixing element must travel to all parts of the mixing vessel. If it does not do so then 'dead spaces' of unmixed material will occur in the pan. In practice this means that there must be a very small clearance between the mixing element and all parts of the pan surface. Good contact is achieved with the mixing element itself rotating and also moving in a planetary path within the pan as illustrated in Fig. 12.4.

The local mechanisms responsible for mixing include kneading (where material is pressed against adjacent material), folding (where fresh material is enveloped by already mixed material), shearing and stretching and tearing. Several designs of mixing element are used to make the mixer more versatile (Fig. 12.5). The beater is used for creaming butter and sugar and for mashing and beating potatoes. A wire whisk is used to thicken mixtures and dough hooks are used for rubbing fat into flour and for kneading doughs.

The various operations are carried out at different speeds and therefore the machine has three or four gears or, for smaller machines, a continuously variable speed control. On some machines the same motor can be used for some size-reduction operations thereby making the machine even more versatile.

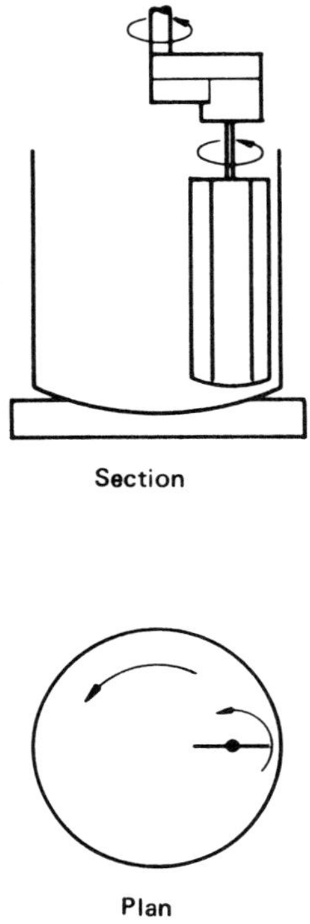


Fig. 12.4 – Food mixer.

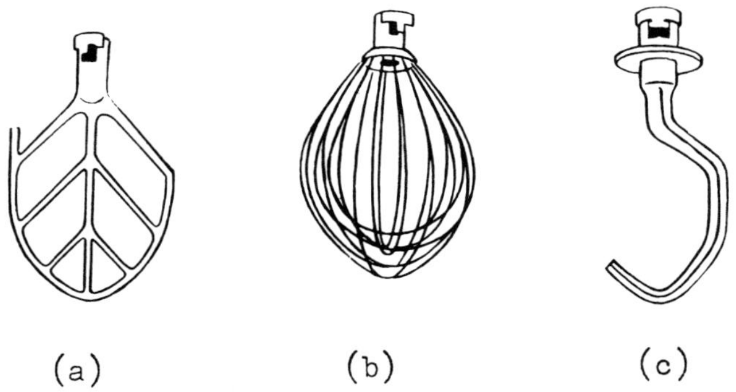


Fig. 12.5 – Mixing elements.

It should be noted that mixtures can be over mixed. With dry solids, this can cause segregation to take place. In the case of certain baking mixtures, over mixing will cause a deterioration in organoleptic properties.

A mobile turbo-mixer is also available for mixing large quantities of liquid/ semi-liquid materials. The mixing head is fixed on the end of a flexible shaft and acts in a similar manner to a pump. Its pumping action causes the liquid to be stirred rapidly, thereby causing it to be mixed.

### 12.3.3 Power requirements for mixing

The variables affecting the power requirement of a simple mixer for mixing Newtonian fluids are:

| | |
|---|---|
| power required for stirring (W) | $Pr$, |
| absolute/dynamic viscosity of mixture (kg/m.s) | $\mu$, |
| density of mixture (kg/m$^3$) | $\rho$, |
| stirrer speed (1/s) | $N$, |
| stirrer diameter (m) | $D$, |
| acceleration due to gravity (m/s$^2$) | $g$. |

Dimensional analysis give the following functional relationship:

$$\underset{\text{Power number}}{\frac{Pr}{\rho N^3 D^5}} = f\left(\underset{\text{Reynolds number}}{\frac{\rho N D^2}{\mu}}, \underset{\text{Froude number}}{\frac{N^2 D}{g}}\right).$$

This equation assumes that the geometry of the mixer remains constant. In the Reynolds number, the velocity is $ND$, which is the velocity of the tip of the stirrer, (strictly it should be $\pi ND$). The Froude number may be relatively unimportant in most catering applications because gravitational effects are not likely to be significant.

If two immiscible liquids are being mixed, the interfacial tension between them will also be important and this would lead to a fourth dimensionless group:

$$\text{Weber number} = \frac{N^2 D^2 \rho}{\sigma}$$

where $\sigma$ = surface tension (force/unit area) between the two liquids (kg/m.s$^2$).

For non-Newtonian fluids the non-Newtonian constants must be introduced into the analysis, which would lead to further dimensionless groups. For pan mixers the mixing element's speed of rotation in its planetary path would also be required.

Data available for simple mixers suggest that the Power number decreases as Reynolds number increases up to $Re = 10$, that is other things being equal, power consumption increases as viscosity increases or as stirrer speed and density decrease. At $Re = 1000$ to $10\,000$, Power number is little affected by further increases in Reynolds number. A transition region exists between $Re = 10$ to 1000/10 000 where there is a lower dependence of Power number on Reynolds number than for $Re < 10$ though it is not independent of it. Unfortunately no studies have been carried out on the types of mixers used in catering, but these general comments may be applicable to such equipment.

### 12.3.4 Rates of mixing

If it is assumed that the rate of mixing at any time, under constant working conditions, is proportional to the extent of mixing at that time, then

$$\frac{\mathrm{d}A}{\mathrm{d}t} = -KA$$

where $A$ = the mixing index (Section 12.3.1), $t$ = time (s) and $K$ = a constant (1/s) dependent on the mixing machine and the mixing conditions.

Integrating this equation gives:

$$\frac{A}{A_0} = e^{-Kt}$$

where $A_0$ = the mixing index at the start of mixing.

The mixing index changes from $A_0$ to $A$ as time increases. This equation shows that the rate of mixing has an expotential relationship to time, which has been found to apply in many cases for the mixing of dry solids. It means that increasing the degree of mixing takes longer and longer the more the food is mixed. In the case of Newtonian liquids, mixing time, other things being equal, decreases as Reynolds number increases for $Re < 1000$ and is more or less independent of it for $Re > 1000$.

## 12.4 FORMING PROCESSES

Forming implies the shaping of a product. This process is required by two principal types of products: pastry and meat products. Bulk pastry is rolled into thin sheets and these thin sheets of pastry are formed into pie bases and so on. Minced meat is formed into patties or burgers and is forced (formed) into a skin to make sausages.

Pastry rolls may be hand or machine operated. The pastry is carried by a conveyor belt between two rollers. The distance between the rollers is adjusted to obtain the correct thickness of rolled pastry.

Pie bases and so on are moulded between two metal dies. The machine also cuts off surplus pastry so as to produce a correctly-sized base of the required thickness. A range of dies are supplied with each machine. Small machines are manually operated whilst the larger ones have automatic pastry feed from a hopper. Some machines may also be used to fill and lid the pies. Specialised devices for forming cornish pasties for example, are also available.

Patties and burgers may be moulded to shape automatically from a continuous meat feed. They may be delivered from the machine onto a paper base and are cut automatically to the desired weight. Sausages are filled through a nozzle, with the skin being fed concentrically over it. The sausage meat is pumped through the nozzle by means of a piston.

Another piece of equipment which uses forming is an automatic doughnut maker. The mixture is fed from a hopper through a forming device before it drops into the frying oil. It may be batch-fried or carried through the oil on a wire-mesh conveyor belt.

## 12.5 NOMENCLATURE

| | |
|---|---|
| $g$ | acceleration due to gravity ($m/s^2$) |
| $t$ | time (s) |
| $A$ | mixing index (–) |
| $D$ | stirrer diameter (m) |
| $F$ | feed size ($\mu m$) |
| $K$ | mixing constant (1/m) |
| $N$ | stirrer speed (1/s) |
| $P$ | product size ($\mu m$) |
| $Pr$ | power required for stirring (W) |
| $Re$ | Reynolds number (–) |
| $S$ | standard deviation (–) |
| $W$ | energy input (kWh/ton) |
| $W_1$ | work index (kWh/ton) |
| $\mu$ | absolute/dynamic viscosity (kg/m.s) |
| $\rho$ | density ($kg/m^3$) |
| $\sigma$ | surface tension ($kg/m.s^2$) |

## 12.6 REFERENCE

Lowrison, G. C. (1974), *Crushing and Grinding*. Butterworths.

## 12.7 FURTHER READING

Skelland, A. H. P. (1967), *Non-Newtonian Flow and Heat-Transfer*, Chapter 9, Wiley-Interscience.

CHAPTER 13

# Food-Holding Equipment

## 13.1 INTRODUCTION

In this chapter equipment used to store food at various temperatures and therefore for varying lengths of time is considered. The temperatures involved may be classified as follows:

(a) **Ambient temperatures** (10–20°C) where no means of controlling the temperature are used. This type of storage is designed by an architect and does not constitute equipment for the purposes of this book. These temperatures are suitable for the storage of 'dry goods'.

(b) **Chilled storage** (0–10°C): foodstuffs stored in this temperature range include the following:

   (i) Unprepared or partially-prepared raw materials such as raw and cooked meat, fish, poultry, vegetables, fruits, salad materials and dairy products. These are stored in either cold rooms or refrigerators. Regulations often stipulate a maximum storage temperature of 10°C for cold foods.

   (ii) Meals chilled as part of a cook/chill system: a cold room (below 5°C) is used.

   (iii) Prepared salads and cold desserts on display to the customer: these are held in refrigerated display units.

   (iv) Beers should be held in refrigerated cellars or store rooms at about 12–13°C; in cellars this is achieved by the installation of a refrigeration unit, the walls of the cellar acting in the same way as the walls of a cold room.

   Vending equipment, which also holds chilled food, will be considered in the next chapter.

(c) **Frozen storage** (−20°C) is used for bought-in frozen foods or for holding foods as part of a cook/freeze system. The equipment used for this purpose is either a frozen-food cabinet (movable) or a walk-in frozen-food store. Ice cream conservators operate at similar temperatures.

(d) **Warm holding** (60–80°C) is used to keep food and crockery warm for short periods before and during service. Regulations often stipulate a minimum storage temperature of about 60°C for warm food. Equipment used for this purpose is either a *bain-marie* or a hot cupboard, and is for the static, bulk storage of food. In the next chapter, portable equipment used to keep food warm as part of a food-handling system will be considered.

## 13.2 CHILLED AND FROZEN STORAGE EQUIPMENT

All equipment for chilled and frozen storage works on the principle of the refrigeration cycle discussed in Section 8.4. The temperature inside these units is below ambient and therefore heat is transferred from the surroundings into the cooler storage space. This heat must be continuously removed to maintain the correct temperature in the storage unit. Hence energy needs to be expended to remove this heat. The equation governing this rate of heat-transfer was considered in Chapter 5 (see Problem (3)).

It is clear from this equation that the temperature difference will be much greater for frozen storage (about 40°C) than for chilled storage (in the range 10–20°C), assuming an ambient temperature of 20°C. Therefore there is a greater need for insulation in the case of frozen storage, if excessive heat removal is to be avoided. Foamed urethane is the most effective insulating material available for this purpose. Similar cost considerations apply to those discussed for other insulation in Section 4.4. This material also enables rigid wall and so on sections to be built and so modular construction methods may be used to give a wide capacity-range for walk-in units. Energy expenditure may also be conserved by observing several operational factors. First the thermostat setting should be as high as food quality considerations will allow. Then the door should be opened as little as possible, as this allows warm air to enter the cold space, thereby increasing the quantity of heat to be removed and causing excessive ice formation on the cooling surface. The cooling surface should be defrosted regularly since excessive ice deposition on it acts as an insulator between the cooling surface and the air being cooled by it, thereby reducing the rate of heat abstraction from the air. Finally the food should be packed so as to allow the free circulation of air around it, thereby ensuring it is always exposed to cold air. It should also be noted that fluctuations in temperature in the cold space cause loss of product quality. These fluctuations may be caused by opening the door, thermostat cycling, defrost cycles and placing warm food in the unit.

The size range of equipment for storing chilled and frozen foods varies considerably. Among the smallest are small refrigerated bar units, refrigerated display units and domestic-size refrigerators. The larger units include large chest-type frozen-food cabinets and walk-in chilled- and frozen-food stores.

Chilled units with a storage space of less than about $0.25m^3$ may use an absorption-type refrigeration cycle based on gas or electricity. The principal advantage of this cycle is that it is silent and is suitable for mini-bars situated in hotel bedrooms and similar applications (for a description of the principle of this cycle see Electrical Association for Women, 1971). For larger chilled-storage units and all frozen-food storage however, it is necessary to use a compression cycle which requires an electric motor. The equipment used is illustrated diagrammatically in Fig. 13.1.

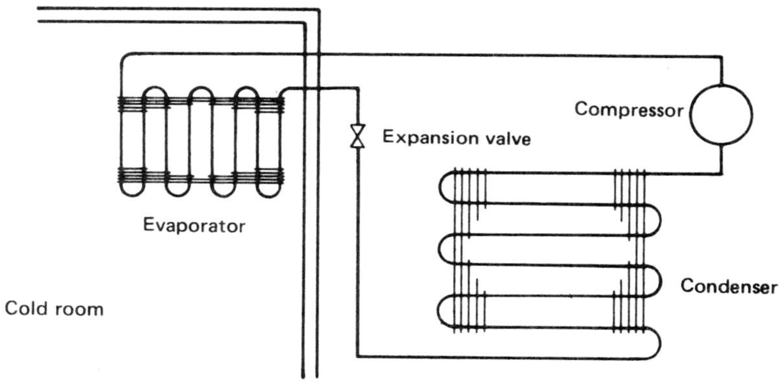


Fig. 13.1 – Compression-cycle refrigerator.

The refrigeration unit is sealed and contains a refrigerant, for example methyl chloride of dichlorodifluoromethane (Freon 12). The cycle of operations is as follows. The refrigerant leaves the evaporator as a vapour. The (electric) motor-driven compressor is used to increase its pressure and this compression also raises its temperature above ambient. On passing to the condenser, which is a series of pipes, the vapour is cooled by the ambient air and condenses to a liquid. This convective heat-transfer process is aided by fitting fins (Section 4.5) to the condenser pipes, and, in addition, air may be blown over the surface (natural convection changed to forced convection). It is advantageous, from a heat-transfer point of view, to have the cooling air at as low a temperature as possible. The condenser should not therefore be situated near to cooking equipment or warm pipework. Condensers are often sited on top of refrigerator cabinets although this is not the best place because warm air will accumulate at this height in the kitchen, so lowering the coefficient of performance of the refrigeration cycle (Section 8.4).

The liquid then passes through an expansion valve into the evaporator. In the evaporator, the liquid extracts heat from the cold space as it evaporates to vapour. This process is aided by fitting fins to the evaporator tubes and in large units this is supplemented by blowing air over the tubes. This also ensures good circulation of the air in the cold space. The air in the cold chamber is at a higher temperature than the surface of the evaporator tubes (temperature drop through the boundary layer). Because the water in contact with the tube surface cannot hold all its moisture content on cooling, water is deposited on the evaporator surface. The temperature of the surface is below 0°C (even in a chill-food store) and the deposited water freezes. Thus, periodic de-frosting is required and this is achieved automatically on large units by using an electrical heater situated on the cooling surface. Adequate drainage for the water formed must be provided.

The demand for energy conservation has led to the use of the heat rejected from the condenser unit for space or water heating. The general considerations relating to such usage of the waste heat from refrigeration cycles has been discussed by Catterall (1978).

## 13.3 WARM-HOLDING EQUIPMENT

### 13.3.1 *Bain-maries*

These are used to keep liquid foods warm. The liquids foods, for example soups and sauces, are held in metal containers which are kept hot by immersing them in hot water, warm air or steam. There are two principal types. The **open type** is simply a tank of water, heated by gas or electricity, and containing a perforated plate on which to stand the metal containers (Fig. 13.2). In the **closed type** the food containers are held within a metal top plate so that the method of heating is not visible (Fig. 13.3).

The method of heating may take one of three forms. The containers may be immersed in hot water heated by gas or electricity; immersed in warm air heated by an electrical element or by combustion products generated from a gas burner or thirdly the containers may be immersed in steam generated from a shallow layer of water situated below the metal containers and heated by gas or electricity. The empirical design of gas-fired *bain-maries* was considered by the American Gas Association (1948 and 1949).

Equipment standards may contain specifications for *bain-maries* along the lines already discussed in Sections 1.4, 2.7 and 11.10. Readers are advised to study the standards, if any, pertaining to their own countries.

### 13.3.2 Hot cupboards

These are used to heat plates and to keep cooked food warm. They may be **directly-heated by gas** with a burner sited below a baffle plate in the heating space. This design is recommended for plate warming at 60°C. Alternatively, they may be **indirectly heated by gas** (Fig. 13.4) with either combustion products circulating around the heating space in channels but not entering it or with steam generated in a well in the base. This type is recommended for keeping food hot at about 80°C.

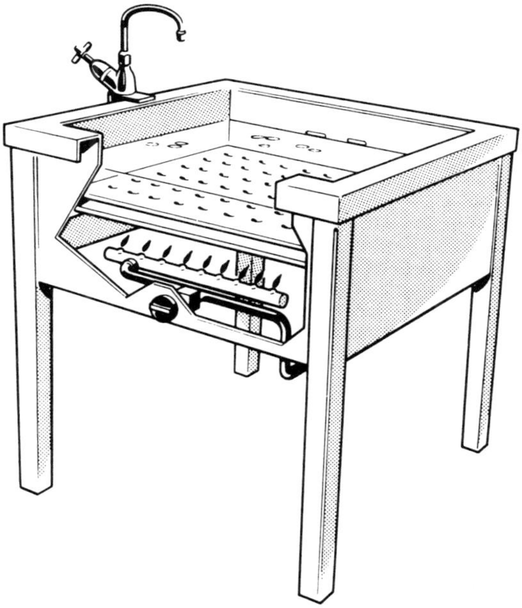

Fig. 13.2 – *Bain-marie* (open).

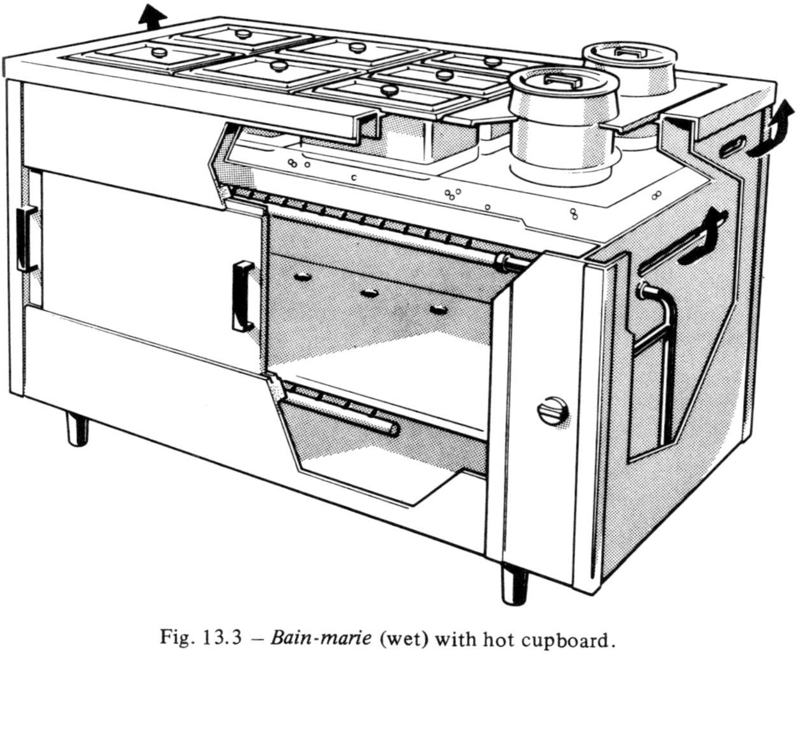

Fig. 13.3 – *Bain-marie* (wet) with hot cupboard.

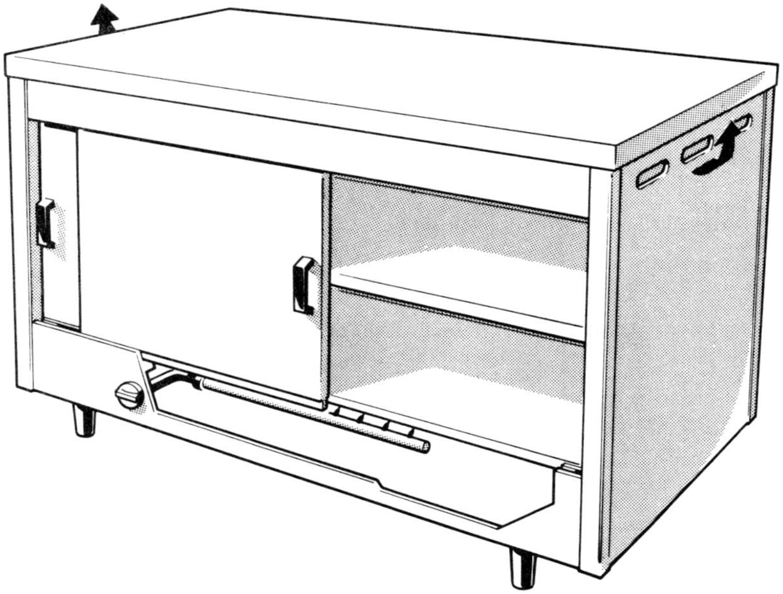

Fig. 13.4 – Hotcupboard.

**Electrically-heated hot-cupboards** have heating elements in the heating space at the appropriate temperature. They may also be **heated with steam**, the steam passing through long lengths of pipe in the base of the unit.

It should be noted that hot cupboards may be obtained with either open or closed top *bain-maries*. Equipment standards are available for hot cupboards in several countries.

## 13.4 REFERENCES

American Gas Association (1948), AGA Research Report No. 1112.
American Gas Association (1949), AGA Research Report No. 1130.
Catterall, F. L., (1978), *Catering & Hotel Management*, January, **17**, 36.
Electrical Association for Women (1971), *The EAW Electrical Handbook*, 9th ed., Chapter 14. English Universities Press.

CHAPTER 14

# Food-Handling Equipment

## 14.1 INTRODUCTION

Food-handling equipment is considered under a number of headings in this chapter. Kitchen equipment is discussed first: in small installations most of the food is moved around the kitchen manually but in larger units some efforts have been made to mechanise food handling. Next we look at containers and packaging: standardisation of food containers and the consequent standardisation of cooking and other equipment sizes to take them has resulted in a reduction in the number of times food is transferred from container to container during production and service. Modern systems of operation, for example cook/freeze and vending, require the food to be packaged.

A discussion of serveries follows: a service counter is normally required whether or not waiter/waitress service is involved. These counters are usually made up of food-holding equipment (Chapter 13) plus other special items. Mobile servery units are also used to distribute food over a distance before service.

Vending machines are special machines which allow the customer to purchase food/beverages by placing a coin in a slot. Finally tray distribution systems are considered: these are normally used in hospitals and similar situations, where the food is plated-up in the kitchen and transported to the wards for the patients or to other types of customer, for example passengers in an aircraft.

Modern catering operations making good use of food-handling equipment have been described by Brunnenkant (1977), Doyen (1977), Gallo (1977), Passmore (1977) and Van Pottelsberghe (1977).

## 14.2 KITCHEN EQUIPMENT

The simplest form of handling equipment consists of racks which hold standard-size food containers (Section 14.3) fitted with a base frame running on four wheels. If standard-size equipment is used, then it is possible to load the food into and out of the cooking equipment, freezing/chilling equipment, storage and re-constitution equipment without the need to transfer it from the original containers. Bins mounted on wheels are also available for moving larger quantities of food around the kitchen. Examples of the use of these are shown in the diagrams of continuous cooking equipment in earlier chapters. As an alternative, an overhead crane may be used for handling the bins (Anon, 1975).

Conveyors of several kinds find applications in catering. **Roller conveyors** consist of a series of steel rollers held in a metal framework. These are suitable for transporting boxes of food supplies from the delivery point to storage. The boxes move by gravity if the conveyor is inclined or they may be pushed along if it is horizontal. This type is also used to transport trays of crockery and so on in wash-up units. **Belt conveyors** consist of a belt driven by an electric motor and travelling over a metal support plate or on rollers. Belts supported on rollers need to be much stronger than those supported on a plate. These operate horizontally or up/down an incline. They are used to convey food materials and crockery, to transport trays and crockery away from the dining area to the wash-up as well as for tray assembly in hospitals and so on. In some applications, including tray assembly, the belt speed is variable. **Screw conveyors** consist of a continuous screw rotating in a casing and driven by an electric motor. A typical use for this equipment would be a vertical unit lifting loose potatoes from a loading bay to a kitchen on a higher floor or to a bin situated above potato paring machines; these can then be fed by gravity from the bin.

Doyon (1977) has described an automated handling system for the recovery of frozen food packs from store, passing through tempering refrigerators and then transferring them to the tray-assembly area. Foods to be heated are automatically fed through tunnel microwave ovens prior to assembly.

In situations where the kitchen and dining area are on different floors, it is usual to convey food between them by means of lifts.

## 14.3 CONTAINERS AND PACKAGING

There is an international standard for the size of catering containers based on the gastronorm size of tray. It specifies the sizes of containers for cooking, serving, preparing, storing and transporting food within and from the kitchen. The sizes specified are based on a module of 530 mm × 325 mm, with ten fractions of the module and one multiple of it. Dimensions for flanged and non-flanged containers are available.

Dorney *et al.* (1974) discussed suitable packaging materials for cook/freeze systems, advocating re-usable heavy-gauge aluminium and stainless steel and disposable aluminium foils. For single portions, foods can be packed in 'boil-in-the-bag' pouches of high density polythene, nylon or nylon-polythene laminate. The development of large central production units for the production of packaged meals has led to the use of automatic portioning. The meal containers are transported along conveyor belts situated underneath the filling devices. Van

Pottelsberghe (1977) discussed the problems associated with the automatic portioning (filling) of liquids, granular products (2 mm cubes to 55 mm spheres), semi-solids and precooked frozen foods.

Liquids are portioned from a hopper through a pump and valve. The valve is opened and closed automatically, the open time determining the quantity of liquid dispensed. Granular products, which are too soft to be handled mechanically and automatically, are portioned using a horizontal rotary table containing measuring pockets. The food is fed from the hopper into these pockets before the food is dispensed into the food container. Semi-solids are usually portioned with a piston filler which measures out the volume to be dispensed.

After portioning (filling) the container needs to be closed. This can be achieved automatically by hooding (the lid is crimped onto the container), lidding (the lid is folded with the container sides) or heat-sealing (the lid is sealed onto the container by the application of heat). The method used depends on the materials. Descriptions of high-speed filling lines as used in catering have been given by Van Pottelsberghe (1977) and Gallo (1977).

Boil-in-the-bag meals are packed in plastic pouches under vacuum. The plastic pouch containing the food is placed in a chamber, its top positioned so that when the lid is closed it is clamped against a heating element. The vacuum pump operates to evacuate the air from the chamber and the pouch and then the heater forms a seal at the top of the bag. The vacuum is then released from the chamber and the lid is opened to enable the package to be removed.

Sandwiches and similar items are packed in plastic film. The film is usually fed from a roll and a hot plate is used to heat-seal the ends of the film together to give a partial barrier against air and moisture migration. For large throughputs, automatic machines which are also capable of weighing and labelling the pack can be bought. In this case shrink-wrap film may be used. The film is wrapped around the product and on heating in warm air the film shrinks to the size and shape of the product and is sealed.

The gas and moisture permeability of a plastic film is often of importance in maintaining the quality of the food. Some foods require permeable film to allow them to 'breath' whereas others need a moisture-proof film to prevent drying-out or moisture-absorbtion.

## 14.4 SERVERIES

Serveries are used to serve the customer directly by service staff, allow the customer to serve himself, or act as a point for service staff to pick up the food and crockery they require to serve the customer at table. Service counters are usually designed individually for each operation. They are however, made up of a number of either standard units or purpose-built units. Some of these units have been discussed already in previous chapters such as *bain-maries*, hot cupboards, chilled/refrigerated storage units and units holding café sets with a drain to take away liquid waste. Other units which have not been discussed include unheated cabinets for the storage of plates and so on, cashier units, tray rails for customers to slide their trays over, shelves and enclosed display units, tray and cutlery stands and lowerators. Lowerators hold plates, trays and so on at service level. The plates, say, are stacked on a metal base plate with a spring(s) underneath it. When a plate is removed, the loading on the spring diminishes and this allows it to extend sufficiently to bring the next plate in the stack into the position occupied by the previous plate. Heating elements may be incorporated into the unit to keep the plates warm.

All of the above units and their layout are designed mainly from an ergonomic viewpoint rather than from a process engineering viewpoint. Display sections for hot food usually incorporate overhead electric infra-red lights as well as heating from the underside. The development of a gas-fired overhead radiant food warmer was described by Dewerth (1965).

Serveries are usually made of stainless steel, sometimes incorporating coloured panels. Back bars are also used for small operations based on convenience foods. These incorporate small items of cooking equipment in addition to servery equipment. Usually they consist of shelf-mounted equipment such as fryers, griddles and *bain-maries.* They can be built into a unit with its own extraction system.

A newer type of servery is in the form of the carousel. It is used for self-service operations and consists of a set of food display units which revolve around a central axis, driven by an electric motor. The display units may be heated or cooled as with a conventional servery. The units are re-filled as they rotate and this can be achieved automatically in some instances. In this case photo-electric cells can be used to monitor for empty spaces on the displays and the electrical signal from them can be used to actuate pusher devices which may load, for example, a new tray onto the display unit. In the normal servery unit, the food is static whilst the customer moves. With a carousel the converse is true. Coetze (1977) has described such a unit for student catering.

Mobile equipment is used to transport and/or serve food where the point of service is remote from the production centre. Units for carrying trayed meals will be dealt with in Section 14.6, whilst here we concern ourselves with bulk-food delivery. Sometimes mobile equipment is used in conjunction with serveries which may be adjacent to the kitchen whilst in other cases the journey may be several miles by road vehicle.

Some units are combinations of hot cupboards, *bain-maries* and refrigerated units but they may be as simple as portable insulated boxes. This latter type of food conveyor may be specified in national standards in terms of temperature distribution and temperature-holding performance.

Insulated boxes are also available for transporting frozen and chilled foods without resorting to the use of refrigerated vehicles. Reasonably long storage times can be achieved with modern insulating materials which also have the advantage of being light in weight.

## 14.5 VENDING MACHINES

Vending machines enable a customer to be served with food/beverages by placing coins in a slot. They are able to provide the following food and beverage items:

(a) hot drinks, including coffee, tea, chocolate and soup;
(b) cold drinks, including carbonated and non-carbonated in several flavours;
(c) snacks, such as biscuits, chocolate, sweets, sandwiches, desserts and hot foods;
(d) pre-cooked meals (main courses and sweets) which are re-heated by the customer in a microwave oven; some make use of a time-token to control the re-heating time whilst others have push-buttons giving a pre-set heating time.

Hot drinks machines (Fig. 14.1) have separate containers for each of the dry ingredients. Hot water is supplied from a tank by means of a pump. The machine is designed to ensure that the dry ingredients are not only flushed into the receiving cup but that the feed line is thoroughly cleaned each time the machine is operated. The shape of the mixing bowl, causing the water to swirl, ensures thorough mixing. Alternatively, the ingredients and water may be whipped together. Machines using ground coffee or tea leaves are available. The coffee or tea is held on filter paper (fed automatically from a roll) whilst boiling water is pumped over them.

Cold drinks machines (Fig. 14.2) have separate containers for each of the flavour syrups or powders. A chilled water supply is used for dilution. For carbonated drinks, carbon dioxide is mixed with the water in the cooler. The syrup may be fed by either carbon dioxide under pressure or by a pump.

Pre-cooked meals are supplied from one of two types of machine. **Rotating drum machines** consist of a series of shelves one above the other. Each shelf is divided into compartments. The meal is removed directly from the compartment by the customer, the shelf revolving each time a pack is taken from it. Each shelf has its own door through which the food is vended. **Lift machines** consist of a series of horizontal rectangular shelves, at each end of which is a lift. The lifts asend to the correct shelf. One lifts pushes an empty tray on to the shelf which moves all the other trays sideways, causing the end (full) tray to be pushed on to the other lift. The lifts descend so that the customer can remove his tray. Only one vending door is required on this type of machine. Hot can machines are also available. These eliminate the need for a microwave oven as the cans are kept hot in the vending machine itself.

The coin mechanisms on vending machines must ensure that 'false' coins are not accepted. Usually several parameters, for example size, weight, magnetic properties and thickness, are checked before a coin is accepted. Further details about vending machines have been published by the Automatic Vending Association of Britain (1972) and the Tea Council (undated). In some countries, standards are available covering general requirements and tests for coin-feed food and drink vending machines.

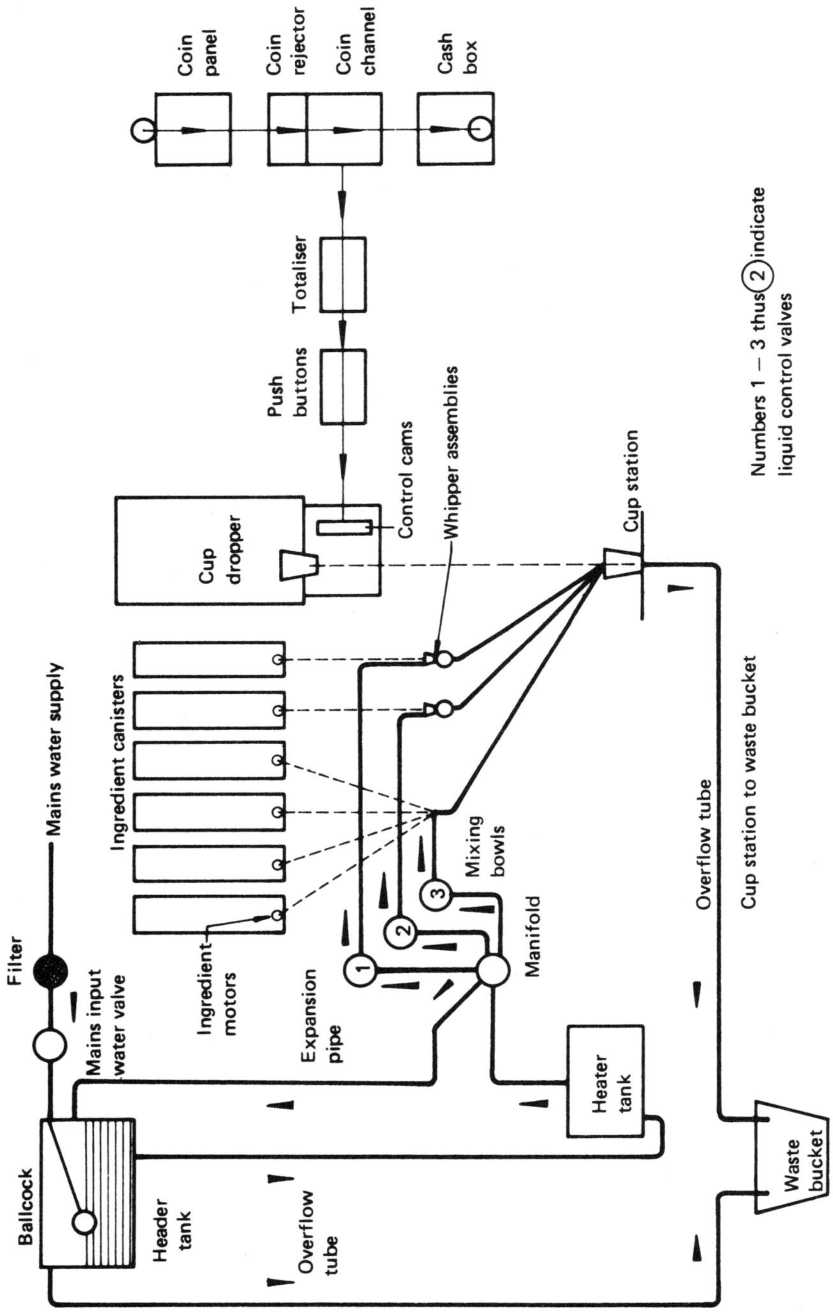


Fig. 14.1 – Vending sequence – hot drinks machine.

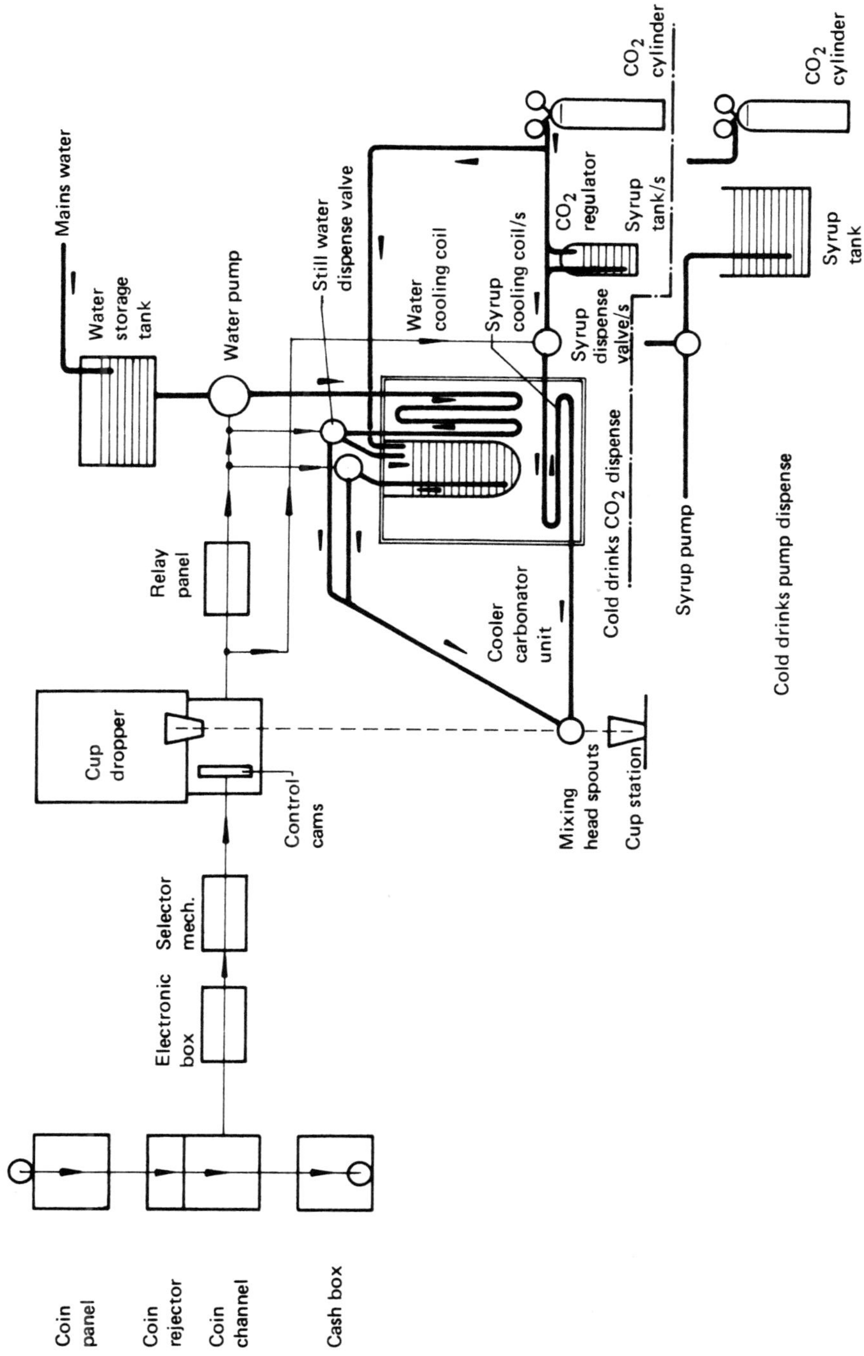


Fig. 14.2 – Vending sequence – cold drinks machine.

For a refrigerated food machine it may be typically specified that the temperature shall not exceed 10°C; if it does then it must go 'off sale' until all vulnerable products are replaced. For a heated food machine it is typically specified that the temperature of the food compartment shall not fall below 66°C; if it does than it must go 'off sale' until the vulnerable products are replaced. However, a period of 30 minutes is usually allowed after loading of food machines to comply with the temperature requirements. Details of these tests must be obtained by reading the appropriate national standards.

## 14.6 TRAY DISTRIBUTION SYSTEMS

The traying-up of meals is used extensively in hospitals and in some other catering areas. The method involves using a belt conveyor (see Section 14.2) down each side of which is situated mobile service equipment, for example lowerators and *bain-maries* (see Sections 13.3.1 and 14.4). At the beginning of the belt, operatives place trays, each having a menu card with choice indicated, and crockery on them. At subsequent work stations the food is portioned on to the crockery from the mobile *bain-maries*. After checking, the food is transported to the customer. These should ensure that the food arriving at the customer is at the correct temperature.

The trays may be made up in one of three forms. The first type utilises pre-heated plates on the trays. The trays are transported in heated trolleys or on hot shelves in a trolley.

In the second type pre-heated plates, seated on a hot pellet with a large heat capacity and covered with a lid, are transported in unheated trolleys. The third method is to have food contained in thin, disposable plastic containers, the containers themselves being held in a plastic insulated box, with transportation in unheated trolleys (Harvey, 1977).

Shea (1974) carried out temperature loss tests on sixteen types of tray system using meat loaf, mashed potatoes and diced carrots. She measured the time which elapsed before the temperatures of the meat loaf dropped to 66°C and the potatoes and carrots dropped to 71°C. The meat loaf and potatoes were also tested to 60°C. The results supported the use of a pre-heated plate, a plate cover (especially in combination with a heated plate) and of plate covers made of stainless steel. The hole in a plate cover appeared to have no effect on heat retention. Of the three insulated systems tested, insulated glass fibre was the best. The heated disc system showed relatively good heat retention for the first eighteen minutes but insulated glass fibre was more effective over a longer time period.

Jonsson *et al.* (1977) compared three distribution systems: two had heated stainless steel underplates, one with a plastic lid with a hole and the other with a stainless steel lid. The third system was of the plastic insulated box type. The differences in the time-temperature curves for the first two systems were generally not very great with both hot and cold foods. The third system was superior to the other two in its ability to retain the temperature of the food after about 15 minutes from the start of an experiment, for both hot and cold foods. This difference was also observed in a transportation test over 10 km. The authors pointed out that factors other than temperature also need to be taken into account in selecting the system, for example nutritive losses, cost, durability, handling time and problems relating to washing up.

The trays of food are distributed to the customers in mobile trolleys which are either moved manually or travel in a small vehicle. More sophisticated transporation systems are available in a few hospitals. These make use of miniature railway systems or monorails. The cost of these systems, which may be automated, can only be justified in the context of a general hospital transportation system.

## 14.7 REFERENCES


Anon. (1975), Staff and Welfare Caterer, October, 47.

Automatic Vending Association of Britain (1972), *Vending*.

Brunnenkant, W. (1977), paper in Glew, G. (ed.), *Catering equipment and systems design*. Applied Science Publishers.

Coetze, R. M. (1977), paper in Glew, loc. cit.

Dewerth, D. W. (1965), AGA Research Report No. 1390.

Dorney, D. C. *et al.* (1974), *Hospitals, JAHA*, **48**(13), 81-83.

Doyon, P. R. (1977), paper in Glew, loc. cit.

Gallo, C. (1977), paper in Glew, loc. cit.

Harvey, J. (1977), paper in Glew, loc. cit.

Jonsson, L. *et al.* (1977), paper in Glew, loc. cit.

Passmore, F. (1977), paper in Glew, loc. cit.

Shea, L. A. (1974), *Jnl. American Dietetic Association*, **65**, 430-436.

Tea Council (undated), *Manual on tea vending*.

Van Pottelsberghe, P. (1977), paper in Glew, loc. cit.


## 14.8 FURTHER READING (on vending)


Beckett, T. W. (1977), paper in Glew, loc. cit.

Hannay, N. J. (1977), *The Cornell H. R. A. Qtly.*, November, 39-45.

Hartley, D. E. (1976), *J. Milk Food Technol.*, **39**(1), 59-62.

Yule, P. *et al.* (1974), *Catering and Hotel Management*, April, 14, 17, 18, 21 and 22.

CHAPTER 15

# Dish Washers

## 15.1 INTRODUCTION

Although the primary purpose of the dishwashing area is to wash crockery, cutlery, cooking utensils and trays, it is also an area where handling is important. All dirty items have to be removed from the dining area, sorted, loaded, washed and sterilised, and stacked ready for re-use. They can be conveyed to the wash-up area manually or by means of a conveyor belt (Section 14.2). Sorting these items and loading them into trays prior to washing is usually a manual operation and an unpleasant one. Attempts have been made to design automatic sorting machines but these are not widely used because not only are they very costly, but they have a tendency to cause breakages and misplaced items in the machine can cause breakdowns. Cutlery, if it has magnetic properties, can readily be separated from crockery and trays by using a magnet situated behind a moving belt. As the trays pass on the main belt underneath the magnet, the cutlery is pulled clear and held against the cutlery belt. This latter belt is moving, and when the cutlery passes out of the influence of the magnet it is released, falling into a collection bin. The sorting of cups, plates of several sizes, saucers and trays from each other is however a much more difficult mechanical process. For smaller throughputs the cost of the machine would not be justified in terms of savings in labour costs, but in large operations there is a possibility that such a machine would be economic. Another reason for its installation may be the removal of bad working conditions, even if the overall cost is higher than with manual sorting.

Waste must be disposed of in the sorting area. This may be cleaned off the trays and plates into a sack but common practice is to use a waste disposer (Section 12.2.3). Larger waste (tins, packages, bottles and so on) may be disposed of by means of bins but increasing use is made of waste compaction units. These use a ram driven by an electric motor to compact the material into a metal bin which reduces the volume of waste considerably. The bin is fitted with a strong lining bag and this serves as a container for the compacted waste.

The washing equipment itself may be classified under three groupings. In the first, a series of **sinks** is used for the washing of small quantities of crockery and cutlery or for pan washing. A sterilisation sink is required (in addition to the washing sink) to sterilise cleaned items. This needs to be heated so as to maintain the water at 82–88°C which sterilises the items. It also allows them to get sufficiently hot and retain enough heat to dry off the surface moisture when the items are left standing for a short period of time.

In **batch dishwashers** the crockery/cutlery is loaded into the machine on tray(s) and remains stationary whilst it undergoes the several stages of the wash cycle, for example pre-wash, cold wash, hot wash, rinse and drying. Most machines pass through the cycle automatically.

Finally, in **continuous ('flight') dishwashers** the trays of crockery are carried through the machine continuously on some form of conveyor. The machine is zoned, each zone representing a stage in the wash cycle. As it passes through the machine, each tray undergoes the complete wash cycle as it passes from zone to zone.

The stages in the wash cycle are basically the same although the larger continuous machines may be more sophisticated in their time/temperature treatment of the soiled ware. It is usual to clear the soiled items manually of gross waste before they are loaded into the machine so as to reduce the quantity of soil to be removed from the machine itself. Once in the machine, the cold-water pre-wash is used so as to remove the other easily-removeable soil. With sterilising sinks, this is carried out using a hand spray attached to the sink. In the case of pans, rotating wire brushes on flexible leads can be used to loosen burnt-on food.

The next stage is to wash the load with hot water containing a detergent. The wash water should be at a temperature of at least 60°C and this is achieved by using thermostatic control in conjunction with water heaters. On most machines the detergent additions are made automatically because correct dosage is essential for consistent washer performance. This wash stage should remove all the remaining soil from the ware surfaces. Sometimes two hot wash stages are used.

After washing, the load is rinsed to remove the wash water and is sterilised. Sterilisation in a machine must take place with the water at a temperature of at least 80°C. It is usual to add a rinse aid (automatically) to the water at this stage. This minimises the wetting effect of the water, that is it encourages it to 'run off' the surface. This effect coupled with the heat held in the ware causes rapid drying on standing in air for a short period. Usually the ware is held in the machine to allow the removal of the steam evolved during this drying period. The water used at each stage is kept separated from that used in the other stages by using several sinks, by pumping the washer empty after each stage or by zoning the machine. The flow of water may in some cases be counter-flow so that sterilising water is re-used for washing.

In hard water areas, it may be necessary to soften the incoming water supply to prevent the build-up of scale on heating surfaces, which impairs the efficiency of the heaters.

## 15.2 SINKS

Sinks used for washing and sterilising crockery are made in a range of sizes appropriate to the size and quality of items to be washed. A hand spray is often fitted for pre-washing and detergent additions are made manually. The individual items are handled seperately.

## 15.3 BATCH DISHWASHERS

Three washing actions are available for removing soil in batch dishwashers. The first involves the immersion of the soiled ware in water. An agitator, driven by an electric motor, is used to create water movement over the soiled surfaces. This movement of the water provides a scrubbing action, or frictional force, on the soil adhering to the surface. If this frictional force overcomes the adhesive force between the soil and the surface, then the soil will be removed. The size of such a unit is limited because of the problems of designing a large agitator which is mechanically strong.

In the second type of batch dishwasher the surface of the crockery is scrubbed with sprays of water (Fig. 15.1). The water sprays impinging on the surface of the ware act in a similar manner to the water movement achieved by the agitator method described above. The sprays of water are produced by pumping the wash water through a series of nozzles. The nozzles are located in pipes (arms) which are free to rotate or, sometimes, in fixed pipes. The action of pumping the water through the nozzles causes the arms to rotate thereby giving better coverage of the surfaces to be cleaned. The design of the nozzles varies from manufacturer to manufacturer. In some machines the tray of crockery also rotates. This design can be used for large units and is employed therefore in pot washers which are used for cleaning large cooking utensils. The utensils used in a kitchen are of widely varying shapes and sizes and this means that to design a machine to wash them all efficiently is difficult. The major problem is to ensure that the water reaches all the surfaces and that they are all adequately washed. The adhesion between the soil and the surface is also higher for cooking utensils than for plates and so on because they have been used for cooking.

The third and most positive design of dishwasher is the use of revolving brushes to mechanically remove the soil from the surface. The brushes are situated partly below the water surface in a sink or they may be supplied continuously with water. This design is limited to situations where the soiled crockery is of fairly consistent shape and it requires manual feeding. Glass washers are specialised versions of this design. Separate sterilisation facilities are required.

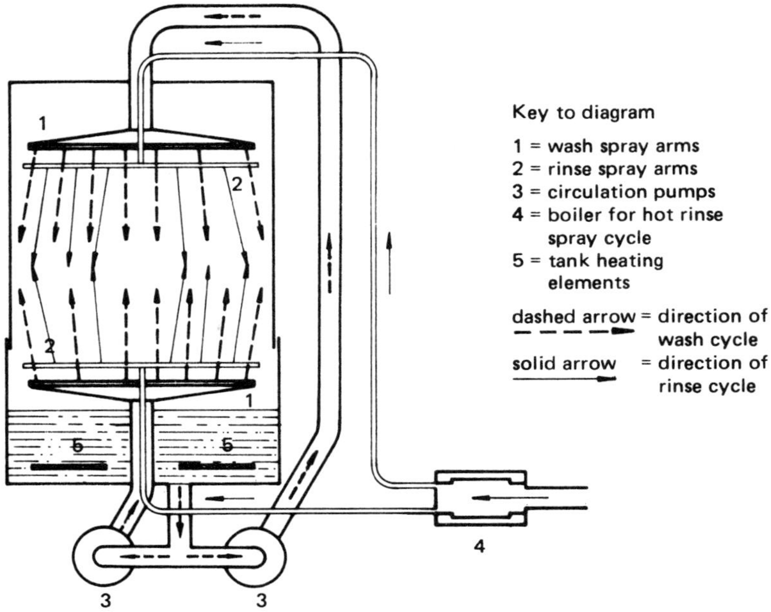


Fig. 15.1 – Batch dishwasher with spray action.

## 15.4 CONTINUOUS DISHWASHERS

In a continuous dishwasher, as explained in Section 15.1, the trays of dirty items are loaded at one end of the machine and emerge clean and dry at the other end (Fig. 15.2). Each zone of the machine is independent, with its own water tank, heaters and pump. Continuous machines use water sprays in each zone for cleaning off the soil and for rinsing purposes. It should be noted however that each manufacturer has its own variations of the basic design, for example his own type of spray nozzle, location of spray arms and whether they are fixed, rotary or both, number of zones and type of conveyor. In general the higher the throughput of the machine the more sophisticated it will be in terms of the number of zones, wash temperatures and times.

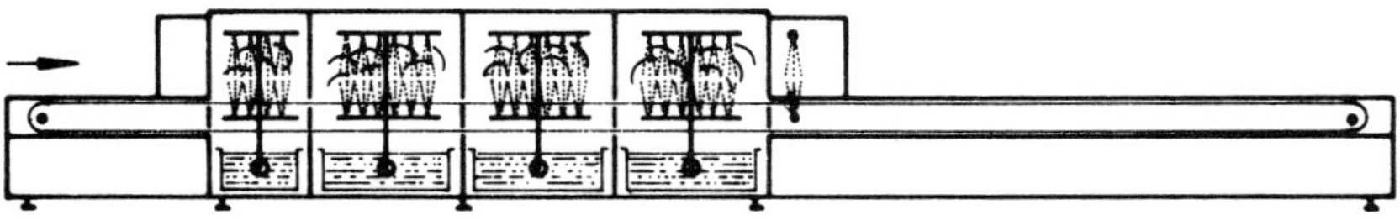

Fig. 15.2 – Continuous ('flight') dishwasher.

Arrangements for loading and unloading continuous machines vary considerably. In the larger installations, subsidiary conveyors and tabling are used to facilitate sorting, loading, unloading and handling of the ware. Where there is less space, a machine with a conveyor in the form of a continuous loop may be used. The conveyor passes through the machine and returns along one side of it, giving a length of conveyor (about the same length of the machine) for loading/unloading. The simplest case is where there is a small area at each end for loading and unloading. This would however pre-suppose other facilities for sorting and so on.

Newer designs of continuous dishwashers emphasise the need to conserve water and energy as well as detergent and rinse aid. Make-up water is usually added only at the rinse stage and there is a flow from there through the various zones to the first zone in the machine as the quantity of water in each zone is depleted. Potentially a large amount of energy is leaving the machine in the hot dry ware and in the steam evaporating from the ware after it has passed through the rinse zone. Steam is also evolved in the wash zone. These two factors also give rise to unpleasant working conditions (heat and moisture). Several manufacturers now fit equipment for recovering this heat and re-cycling it back to the machine. In one design, the steam leaving the machine is condensed and is returned to the rinse water tank. Another design recovers heat from the surrounding atmosphere (heat lost through walls of machine by conduction) and returns it to the washer.

## 15.5 CONSUMABLES

The consumption of detergents and chemicals in dishwashers has been studied by McCaughey and Montag (1974) whilst Molzahn and Montag (1974) studied the consumption of electricity, steam and water. The former authors found that machine setting had the greatest influence on cost; other factors affecting cost were water hardness, condition of the machine, thoroughness of dish scraping, temperatures of wash and rinse water and machine running time. Their results also showed that a high amount of detergent did not necessarily result in clean dishes. Factors affecting cleanliness included training of operatives and machine maintenance.

Day (1975) has studied the strength of bonding between food soils and dishes. His results show the effects of detergent concentration, wash temperature, treatment time, soaking, water hardness, detergent type, energy treatment and ware material on soil removal. The results suggest that there is a maximum level of detergent concentration beyond which no further soil is removed, the value of the maximum depending on the hardness of the water and its degree of ionisation. The soil removal rate is doubled for each 10.5°C rise in wash water temperature, with 100% removal at 90°C under the conditions of the test. Ease of soil removal increased with both soaking temperature and soaking time.

These results support the general features of dishwasher design and operation. How far his tests correlate with dishwasher performance is not known.

## 15.6 REFERENCES

Day, J. M. (1975), *Jnl. American Oil Chemists' Society*, **52**(11), 461–464.

McCaughey, P. E., and Montag, G. M. (1974), *Jnl. American Dietetic Association*, **65**, 548–553.

Molzahn, G. M. and Montag, G. M. (1974), *The Cornell H. R. A. Qtly.*, May, 98–106.

# Fluid-Flow Charts and Data

## A.1 FRICTION FACTOR CHART

Fig. A.1 is used to calculate the friction factor ($C_f$) in fully-developed, turbulent flow using the Reynolds number ($Re$) and relative roughness ($\epsilon$) of the pipe

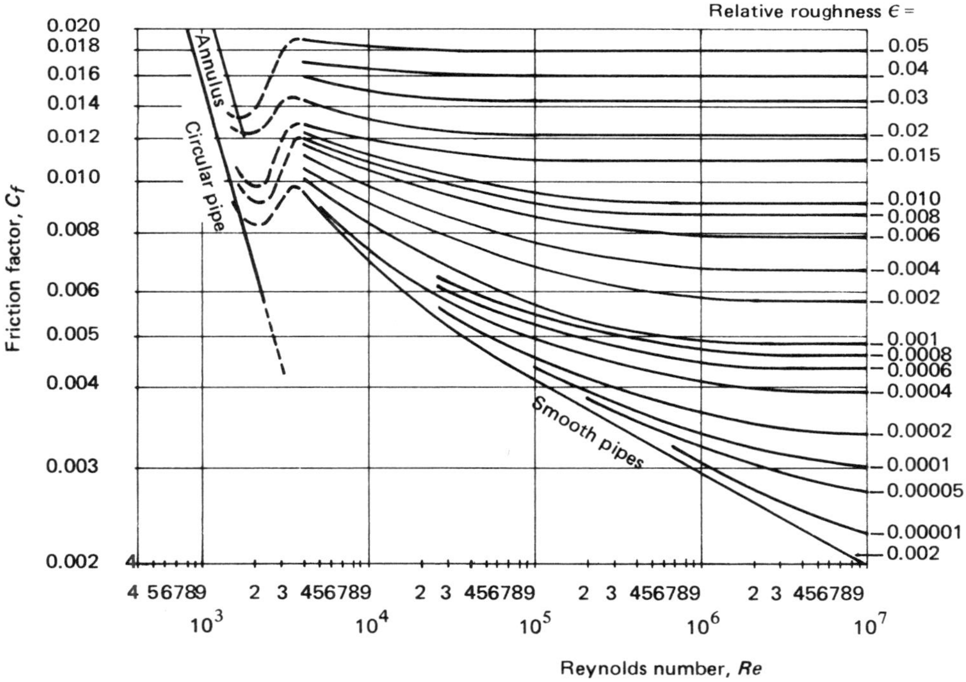


Fig. A.1 – Variation of friction factor with Reynolds number and surface roughness.

material. The relative roughness, $\epsilon = b/d$ where $d$ is the diameter of the pipe (mm) and values of $b$ for different materials are given in Table A.1.

**Table A.1 – '$b$' values for several materials.**

| Material | $b$ (mm) |
|---|---|
| Drawn tubing | 0.0015 |
| Commercial steel | 0.046 |
| Wrought iron | 0.046 |
| Galvanised iron | 0.15 |

## A.2 PRESSURE DROP ACROSS FITTINGS

The data in Table A.2 refer to **tubulent flow** conditions ($Re > 4000$). In the **transition flow** region ($2000 < Re < 4000$), the value of $q$ may be taken as twice the value from the preceding turbulent flow data. In the **laminar flow** region ($Re < 2000$), $q$ may vary between values lower than those quoted in the preceding data and about twice these values, the higher values being for Reynolds numbers approaching 2000.

Source: Rose, J. W. and Copper, J. R. (eds.) (1977), *Technical data on fuel*, British National Committee, World Energy Conference, 7th edition.

**Table A.2 – Pressure drop across fittings (tubulent flow of constant density fluid; $A$ = area).**

| Fitting | Value of $q$ |
|---|---|
| Abrupt enlargement 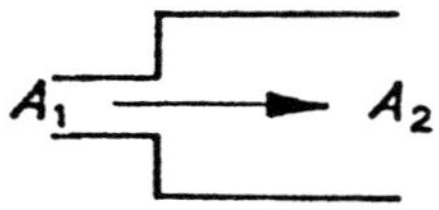 | $\left(\frac{A_2}{A_1} - 1\right)^2$ Note: for $A_2/A_1 \rightarrow \infty$ use $q = 1$ and *upstream* velocity in flow equation. |
| Abrupt contraction 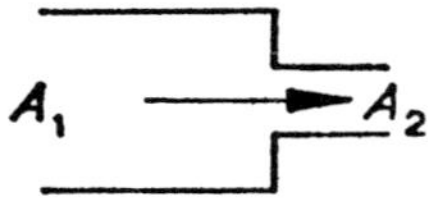 | $0.45\left(1 - \frac{A_2}{A_1}\right)$ |
| Sharp-edged orifice 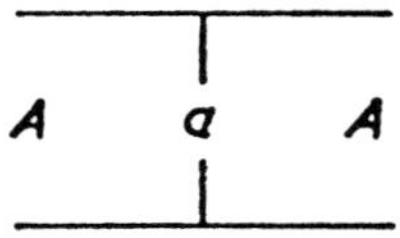 | $2.8\left(1 - \frac{a}{A}\right)\left[\left(\frac{A}{a}\right)^2 - 1\right]$ Note: *downstream* velocity is that based on area $A$ |

**Table A.2** – *Continued.*

| Fitting | Value of $q$ |
|---|---|

**Enlarger**

$$\beta\left(\frac{A_2}{A_1}-1\right)^2$$

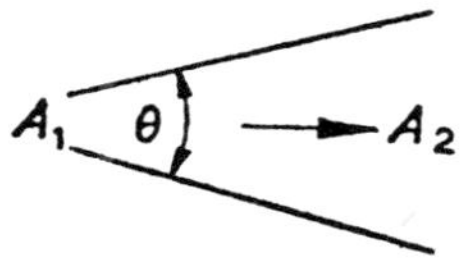


| $\theta$ | 3° | 5°–8° | 10° | 14° | 20° |
|---|---|---|---|---|---|
| $\beta$ | 0.18 | 0.14 | 0.16 | 0.25 | 0.45 |

| $\theta$ | 30° | 45° | 60° | >90° |
|---|---|---|---|---|
| $\beta$ | 0.7 | 0.95 | 1.1 | 1.0 |

**Reducer**

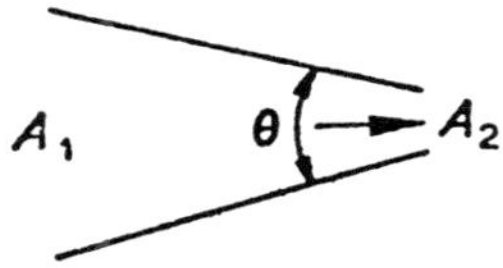


0.05 for $\theta < 30°$

**Mitred elbow**

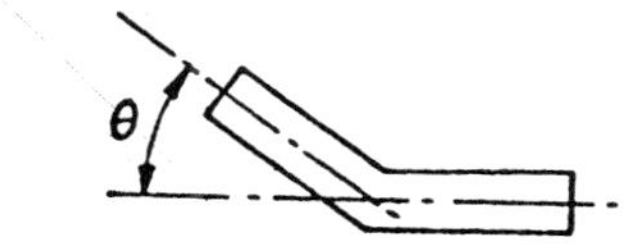


| $\theta$ | 90° | 60° | 45° | 30° | 15° |
|---|---|---|---|---|---|
| $q$ | 1.0 | 0.36 | 0.18 | 0.07 | 0.02 |

**90° bend**

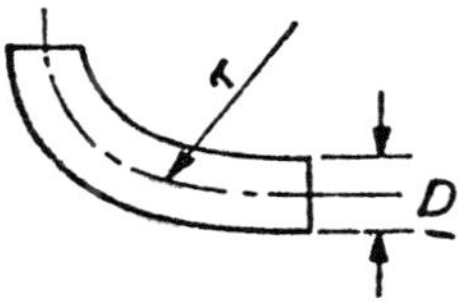


| $r/D$ | 0.5 | 0.75 | 1.0 | 1.5 | 2.0 | 4.1 |
|---|---|---|---|---|---|---|
| $q$ | 1.0 | 0.75 | 0.6 | 0.47 | 0.40 | 0.30 |

**Conical exit from a large tank**

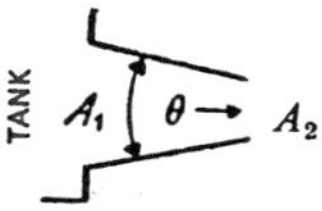


$$0.45\left(\frac{A_2}{A_1}\right)^2 + 0.05$$

for $\theta$ less than 30°

**Table A.2** – *Continued.*

| Fitting | Value of $q$ |
|---|---|
| Re-extrant (Borda) mouthpiece running full 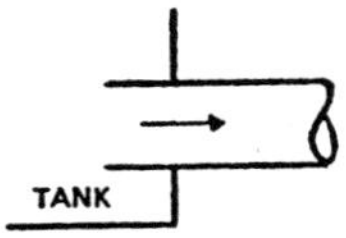 | 0.8 |
| Square exit from a large tank 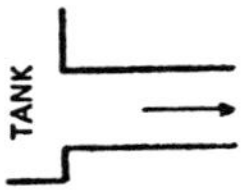 | 0.5 |

Miscellaneous fittings

| Fitting | | | $q$ Range | $q$ Mean |
|---|---|---|---|---|
| 90° elbows: | mitred, sharp intersections | | 1.0– 1.5 | 1.25 |
| | sharp bends, $r/D = 0.5$ | | 0.8– 1.2 | 1.0 |
| | castings, $r/D = 1.0$ | | 0.5– 1.0 | 0.75 |
| Standard tee, 90°, equal areas: | | from barrel to branch | 1.0– 1.4 | 1.2 |
| | | from branch to barrel | 1.2– 1.8 | 1.5 |
| | | through barrel | 0.2– 0.3 | 0.25 |
| (Third leg stagnant in each case) | | | | |
| Close return bend, 180° $r/D = 1.0$ | | | 1.5– 2.5 | 2.0 |
| Globe valve | fully open | | 5 –15 | 10 |
| Angle valve, 45° | ,, | | 3 – 7 | 5 |
| Flap non-return valve | ,, | | 1 – 3 | 2 |
| Gate valve | ,, | | 0.1– 0.2 | 0.15 |
| | $\frac{3}{4}$ open (by area) | | 0.6– 1.0 | 0.8 |
| | $\frac{1}{2}$ open ,, | | 4 – 5 | 4.5 |
| | $\frac{1}{4}$ open ,, | | 25 –30 | 28 |

## A.3 FLOW IN ENTRANCE TO A PIPE

The friction factor ($C_f$) in **laminar flow** in the entrance of a pipe is high at the entrance and decreases until the flow becomes fully-developed. Table A.3 gives the average friction factor between the entrance and a point a distance $l$ from the entrance, with a pipe diameter $d$.

**Table A.3 – Average friction factor for laminar flow in the entrance to a circular pipe.**

| $\frac{l/d}{Re}$ | $4C_f\left(\frac{l}{d}\right)$ |
|---|---|
| 0.000205 | 0.0530 |
| 0.000830 | 0.0965 |
| 0.001805 | 0.1413 |
| 0.003575 | 0.2075 |
| 0.00535 | 0.2605 |
| 0.00838 | 0.340 |
| 0.01373 | 0.461 |
| 0.01788 | 0.547 |
| 0.02368 | 0.659 |
| 0.0341 | 0.845 |
| 0.0449 | 1.028 |
| 0.0620 | 1.308 |
| 0.0760 | 1.538 |

In **turbulent flows** the flow is laminar immediately near the entrance but then it passes through the transition zone and becomes turbulent. Fig. A.2 illustrates the change in both the thickness of the boundary layer (----) and the friction factor (——) with distance from the entrance.

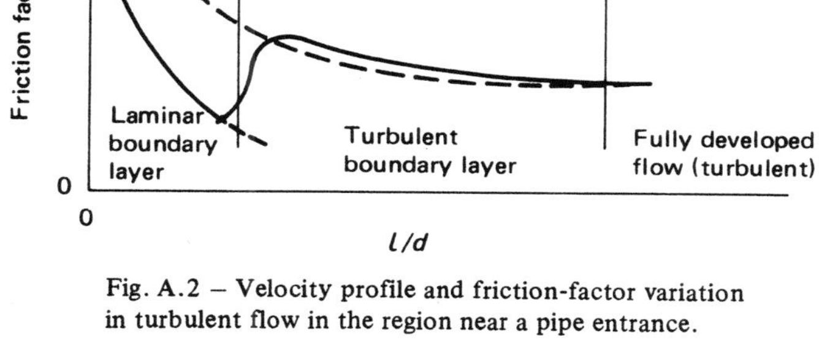


Fig. A.2 – Velocity profile and friction-factor variation in turbulent flow in the region near a pipe entrance.

Fig. A.3 illustrates the situation considered in Fig. A.2 where $(p_1 - p_l)$ is the pressure drop between the entrance $(p_1)$ and a distance $l$ $(p_l)$ from the entrance.

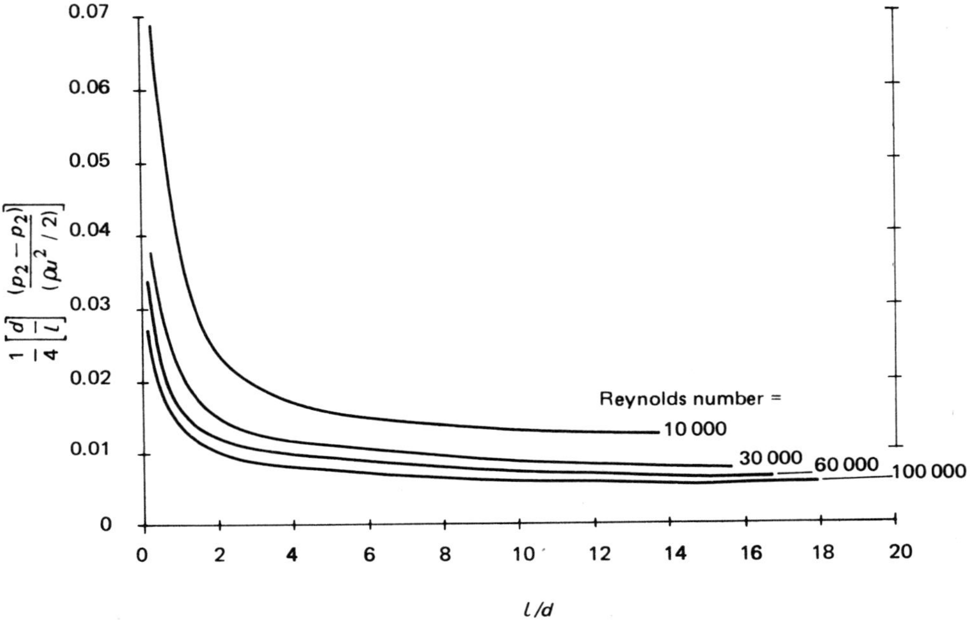


Fig. A.3 – Static pressure drop in the entrance to a smooth, horizontal, circular tube in turbulent flow.

Source: Welty, J. R., Wicks, C. E. and Wilson, R. E. (1968), Fundamentals of momentum, heat and mass transfer. John Wiley-Interscience.

APPENDIX B

# Unsteady-State Conduction Charts

Source: *Charts for Determining Centre, Surface and Mean Temperatures in Regular Geometric Solids During Heating or Cooling* by Dalgleish, N. and Ede, A. J. (1965), Ministry of Technology, National Engineering Laboratory, NEL Report No. 192.

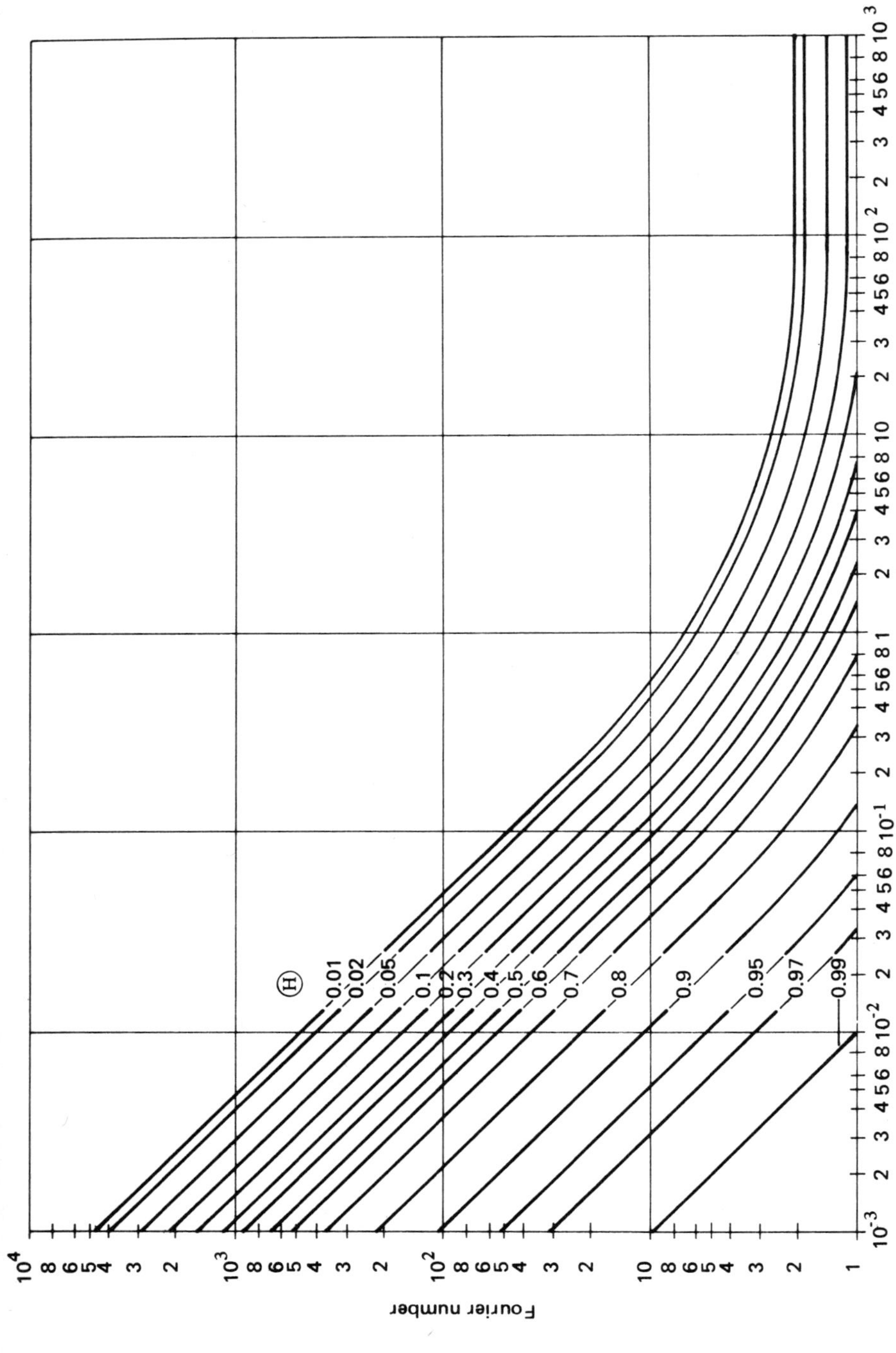


Fig. B.1(a) – Centre temperature of an infinite slab ($Fo > 1$).

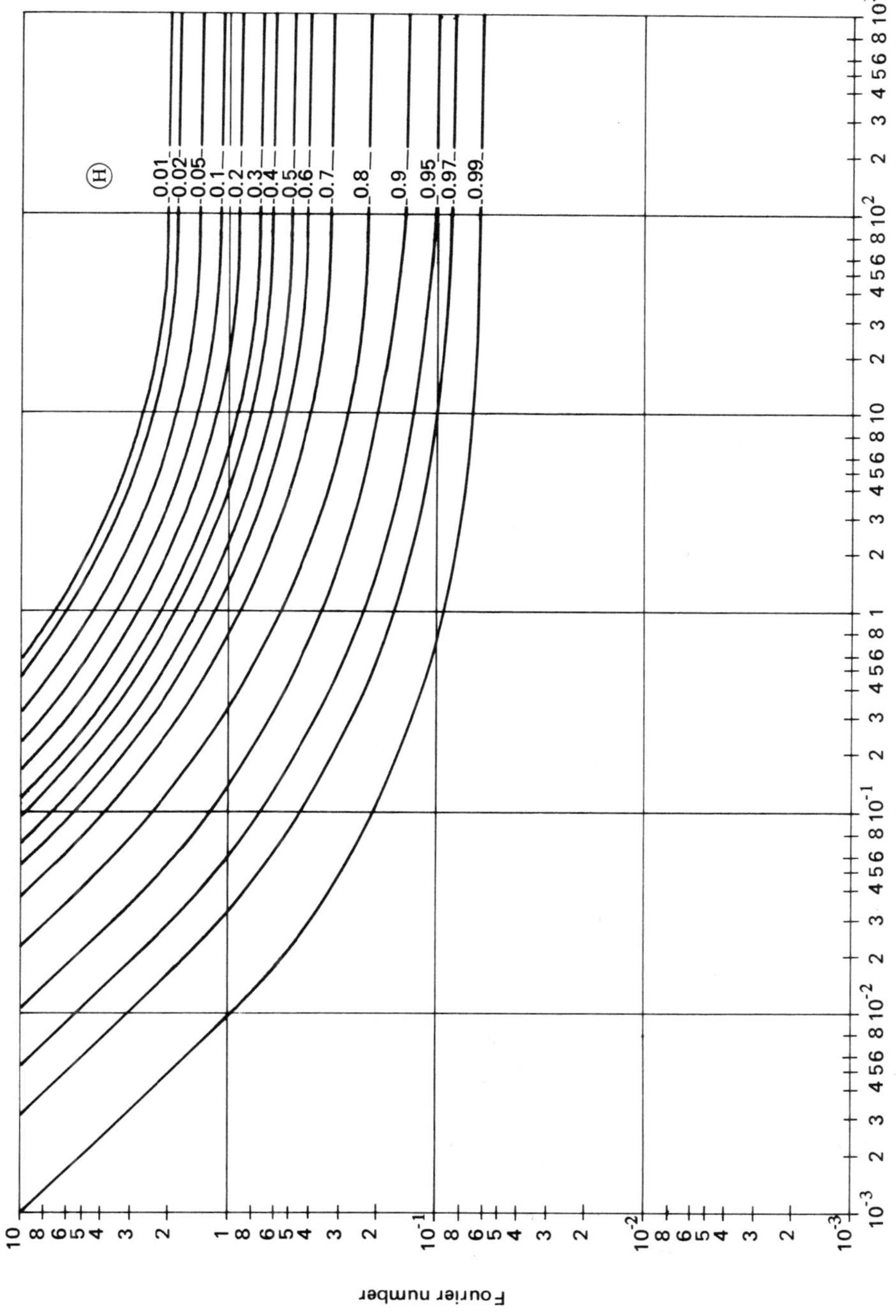


Fig. B.1(b) – Centre temperature of an infinite slab ($Fo < 10$).

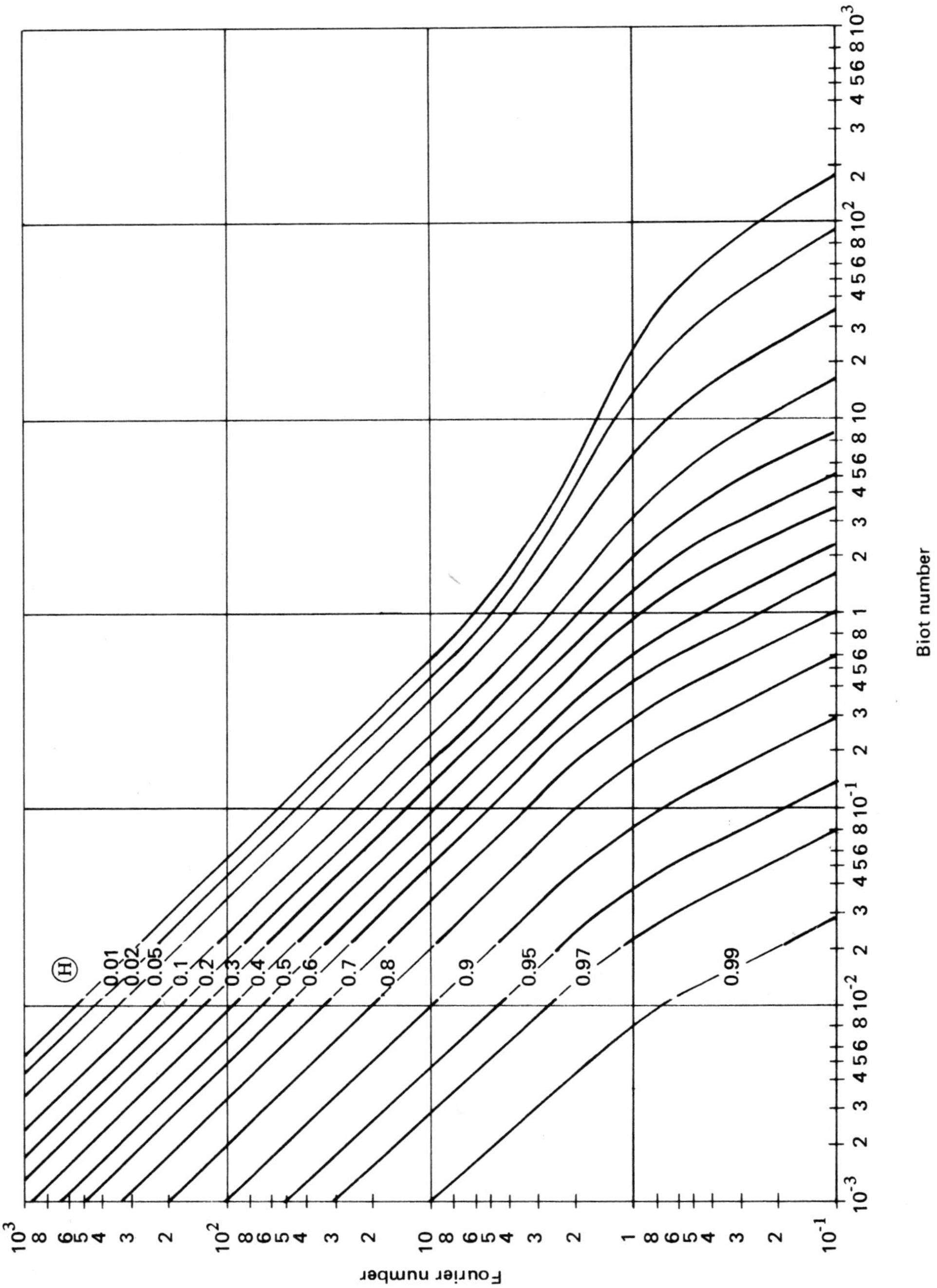


Fig. B.2 (a) — Surface temperature of an infinite slab ($Fo > 10^{-1}$).

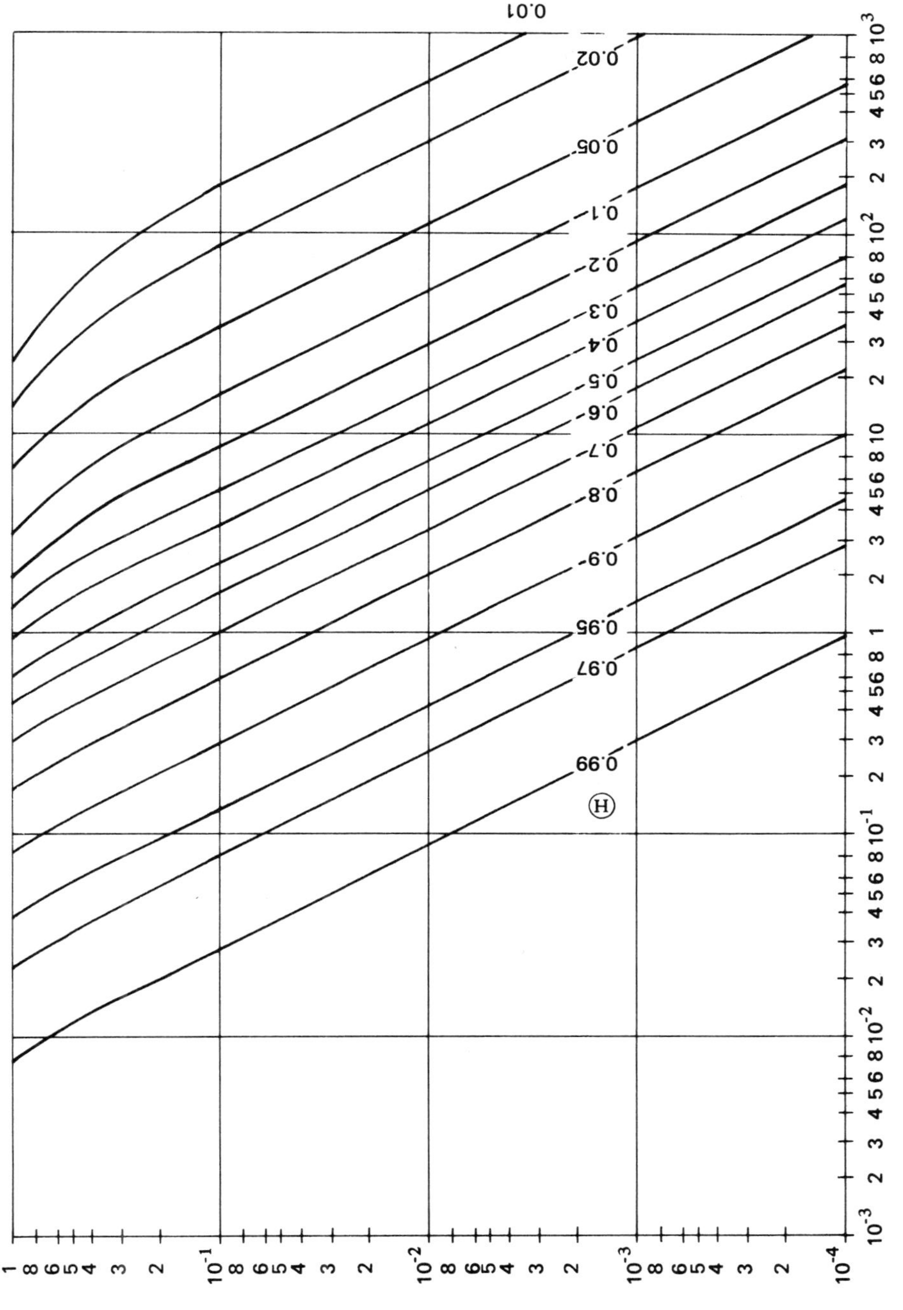


Fig. B.2(b) – Surface temperature of an infinite slab ($Fo < 1$).

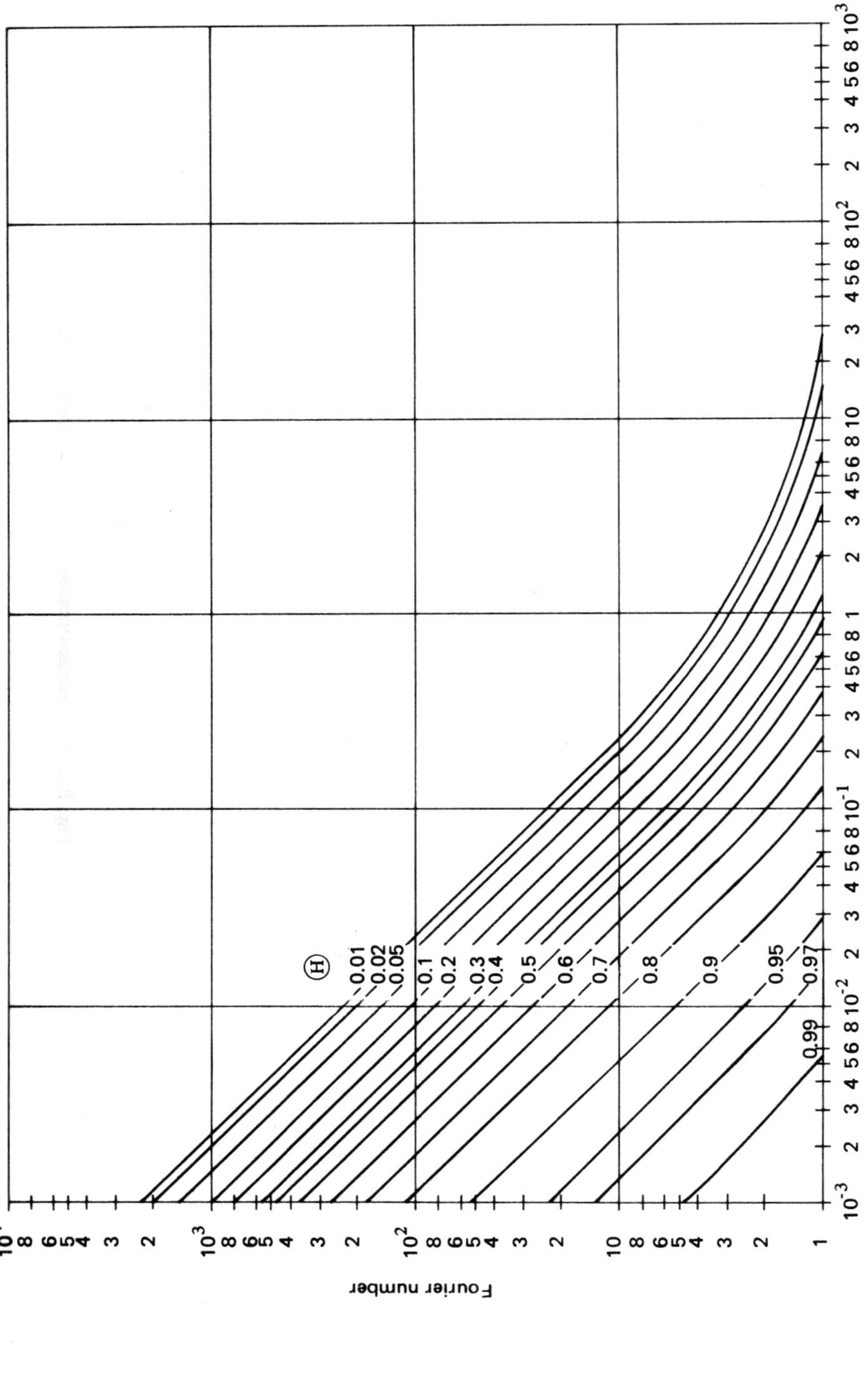


Fig. B.3 (a) – Centre temperature of an infinite circular cylinder ($Fo > 1$).

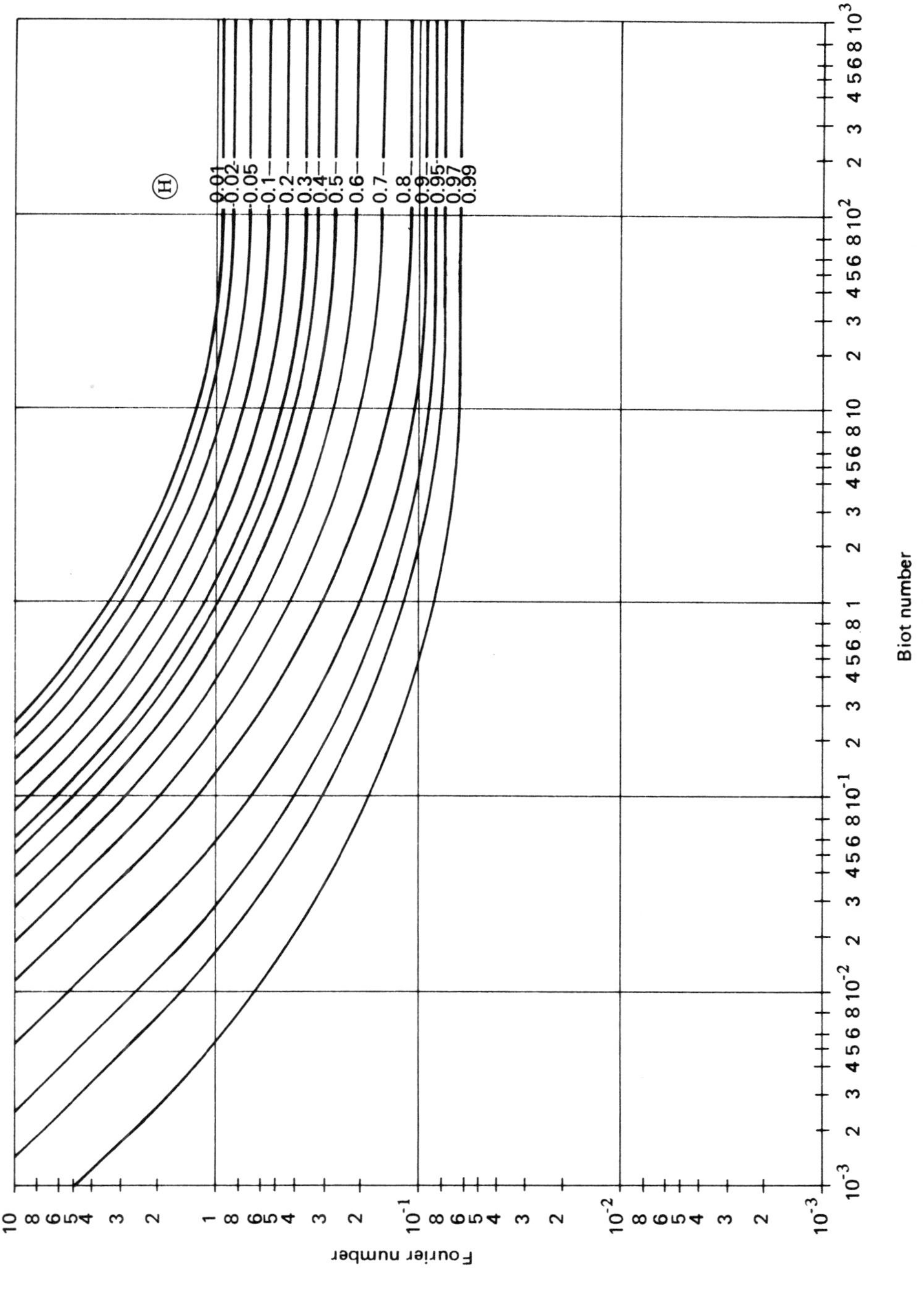


Fig. B.3 (b) – Centre temperature of an infinite circular cylinder ($Fo < 10$).

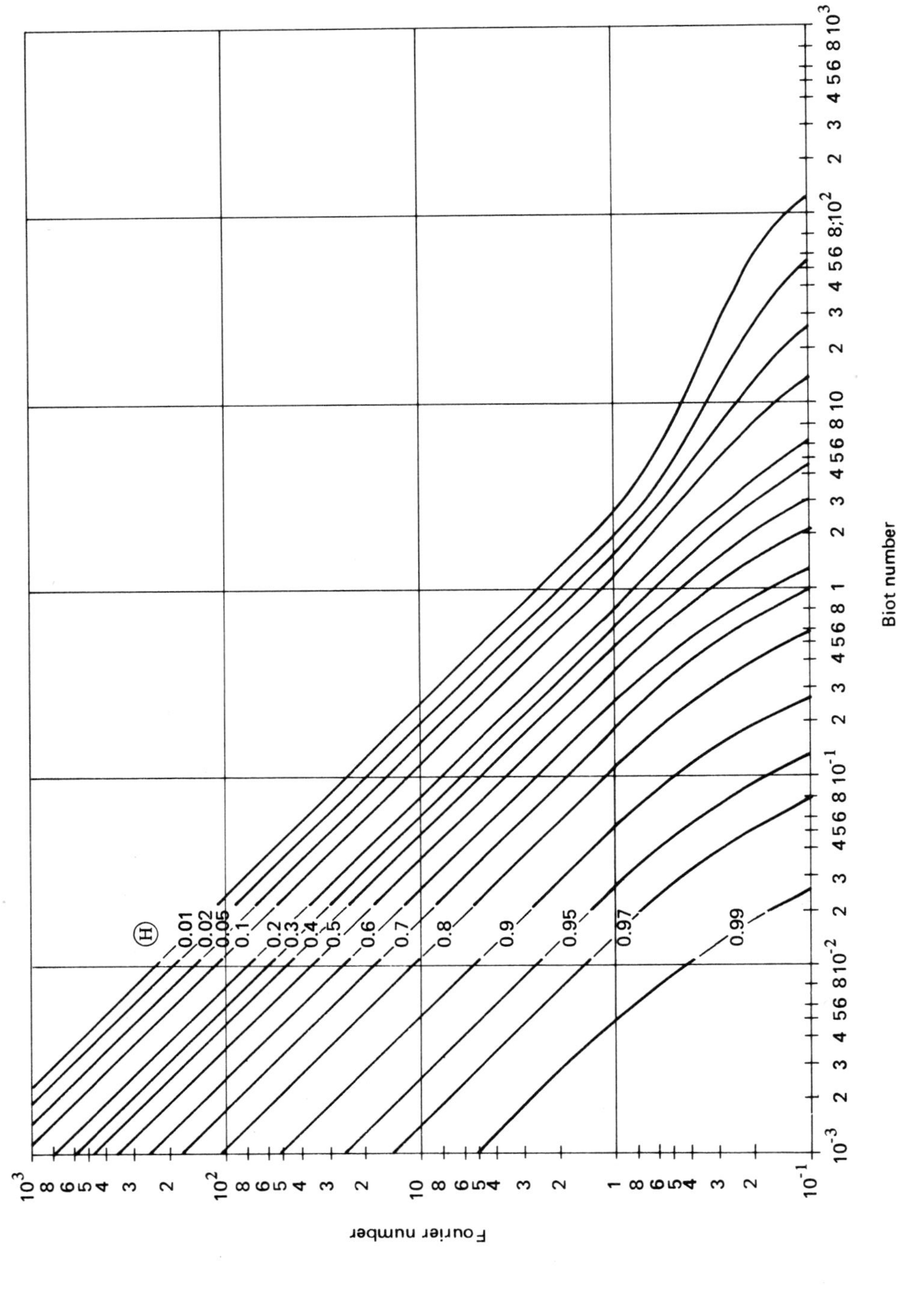


Fig. B.4(a) – Surface temperature of an infinite circular cylinder ($Fo > 10^{-1}$).

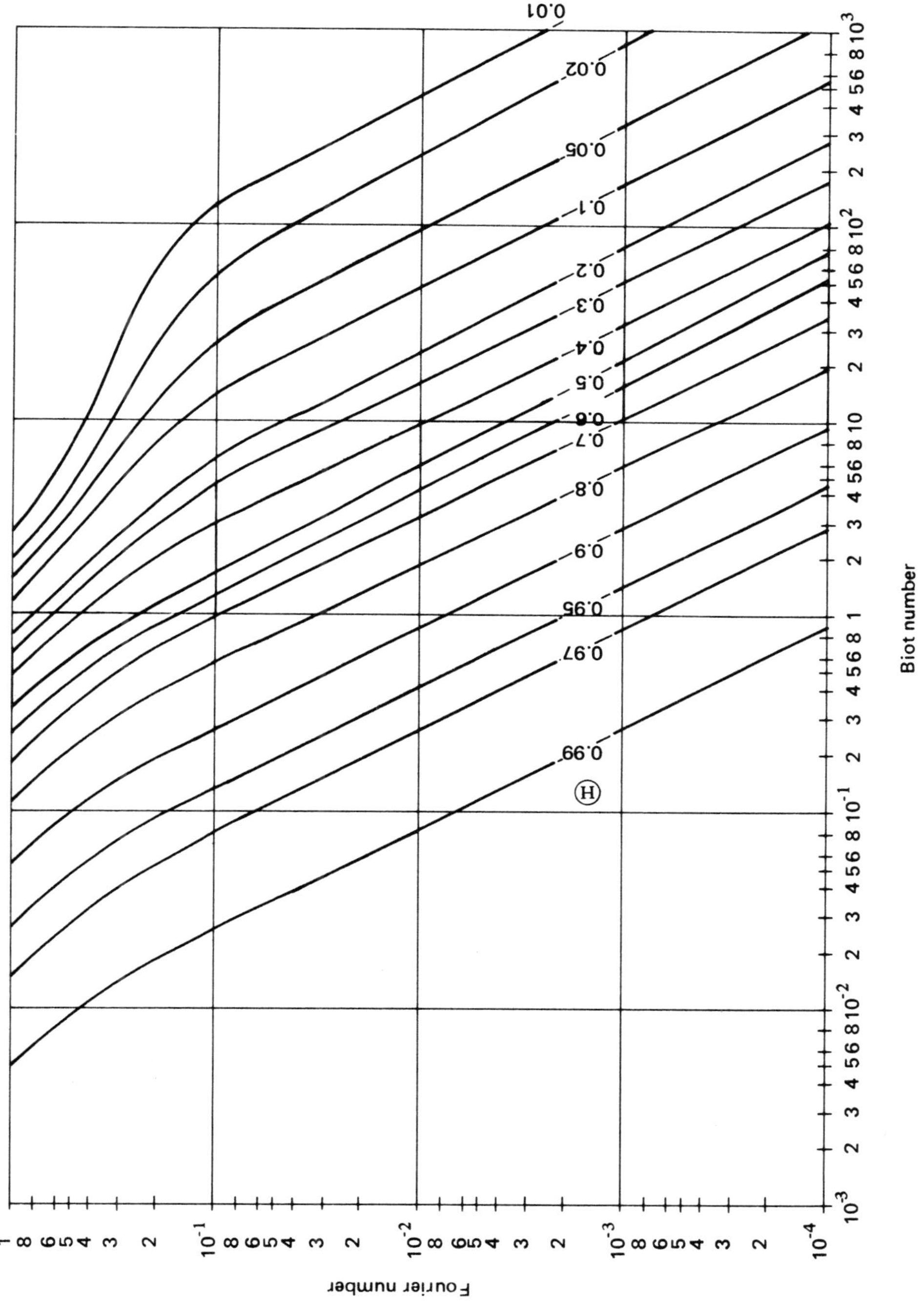


Fig. B.4 (b) – Surface temperature of an infinite circular cylinder ($Fo < 1$).

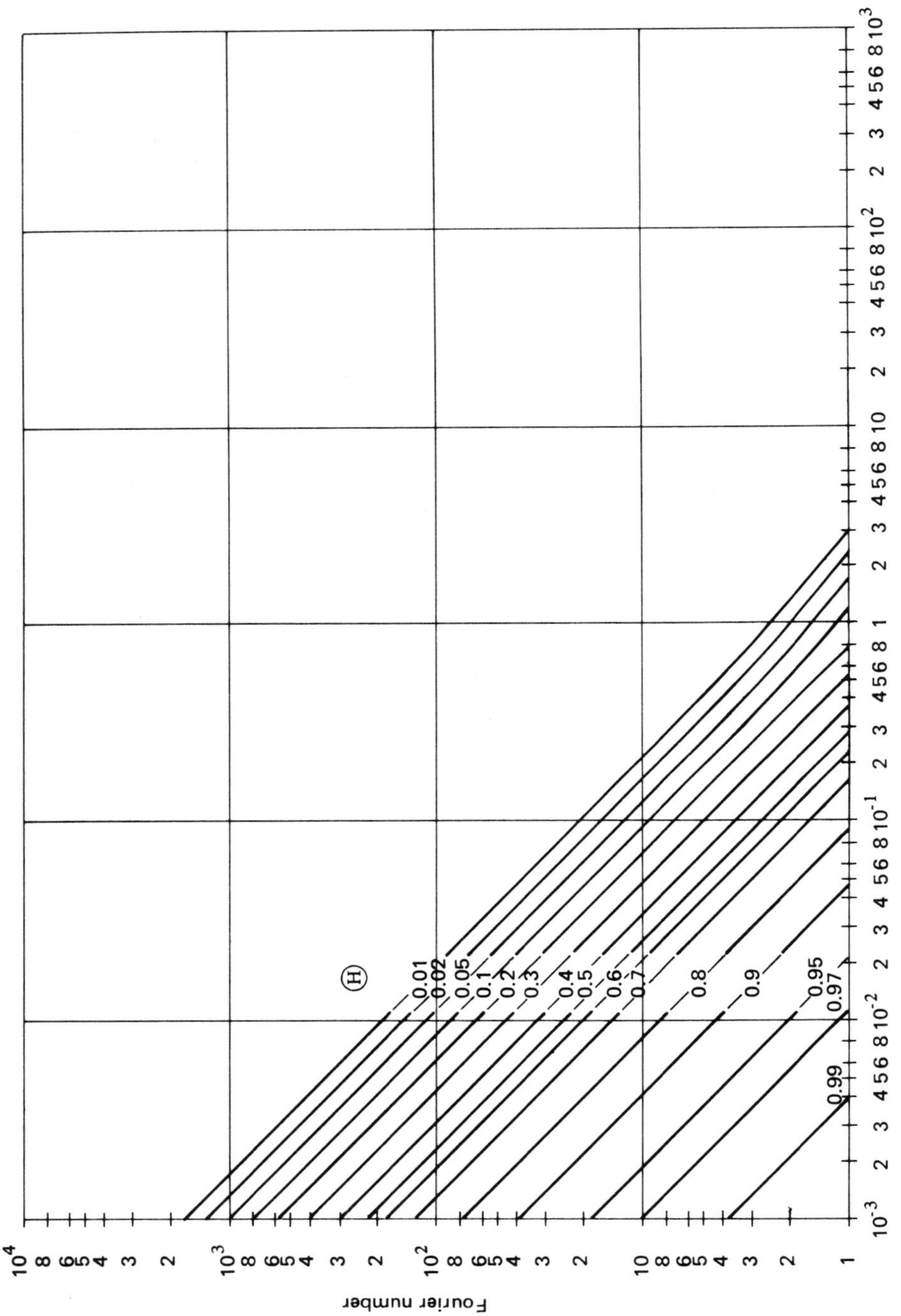


Fig. B.5 (a) – Centre temperature of a sphere ($Fo > 1$).

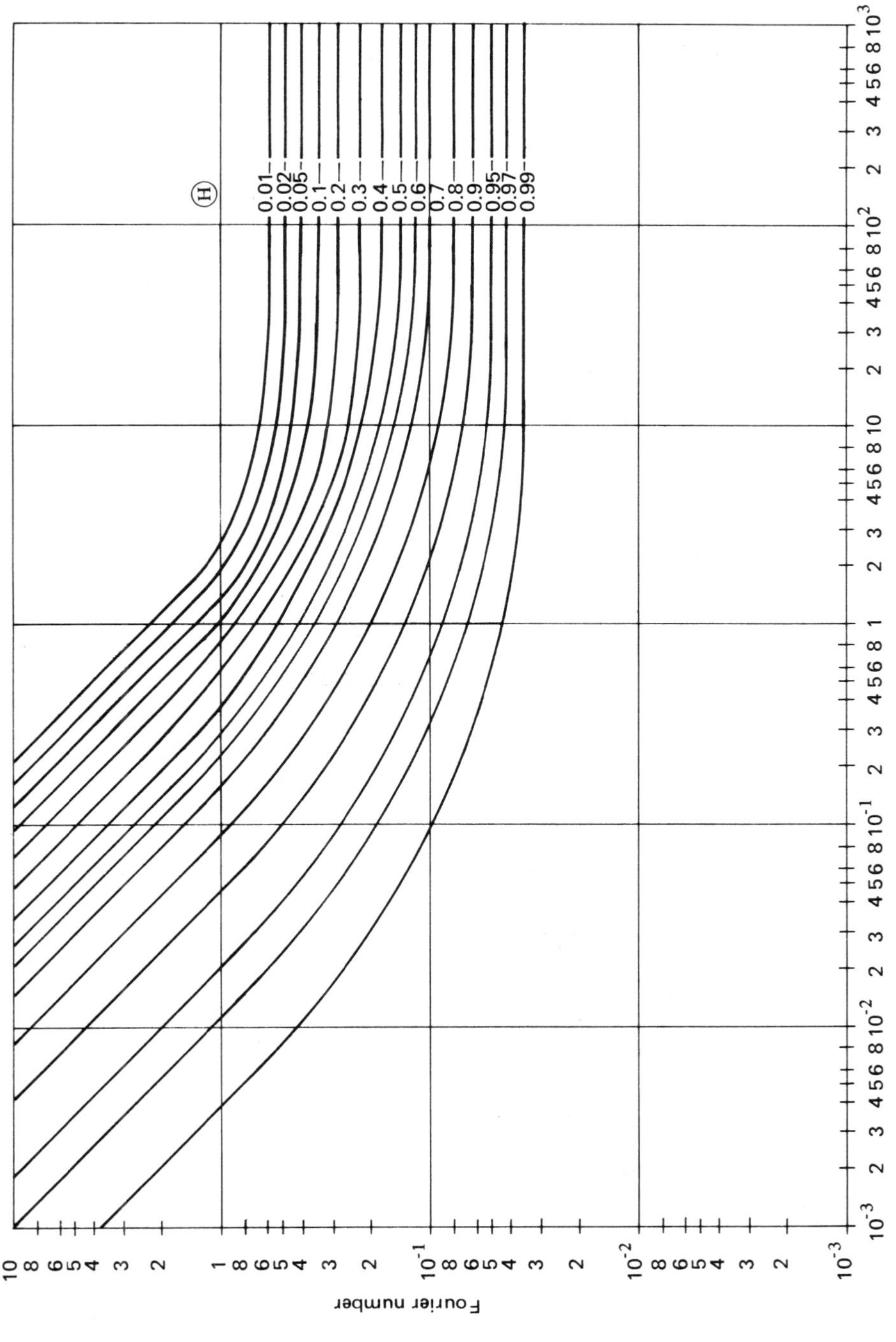


Fig. B.5(b) – Centre temperature of a sphere ($Fo < 10$).

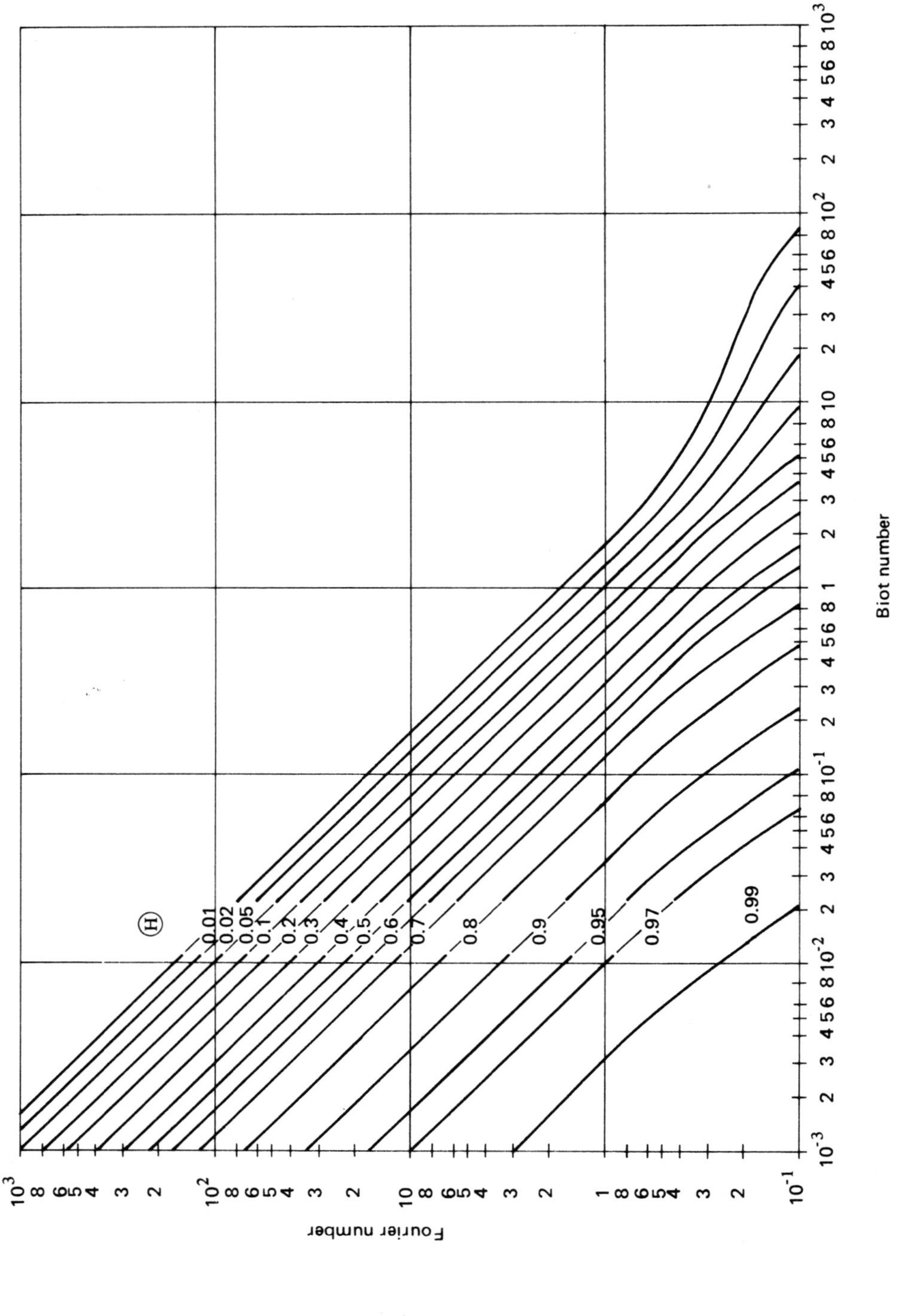


Fig. B.6(a) – Surface temperature of a sphere ($Fo > 10^{-1}$).

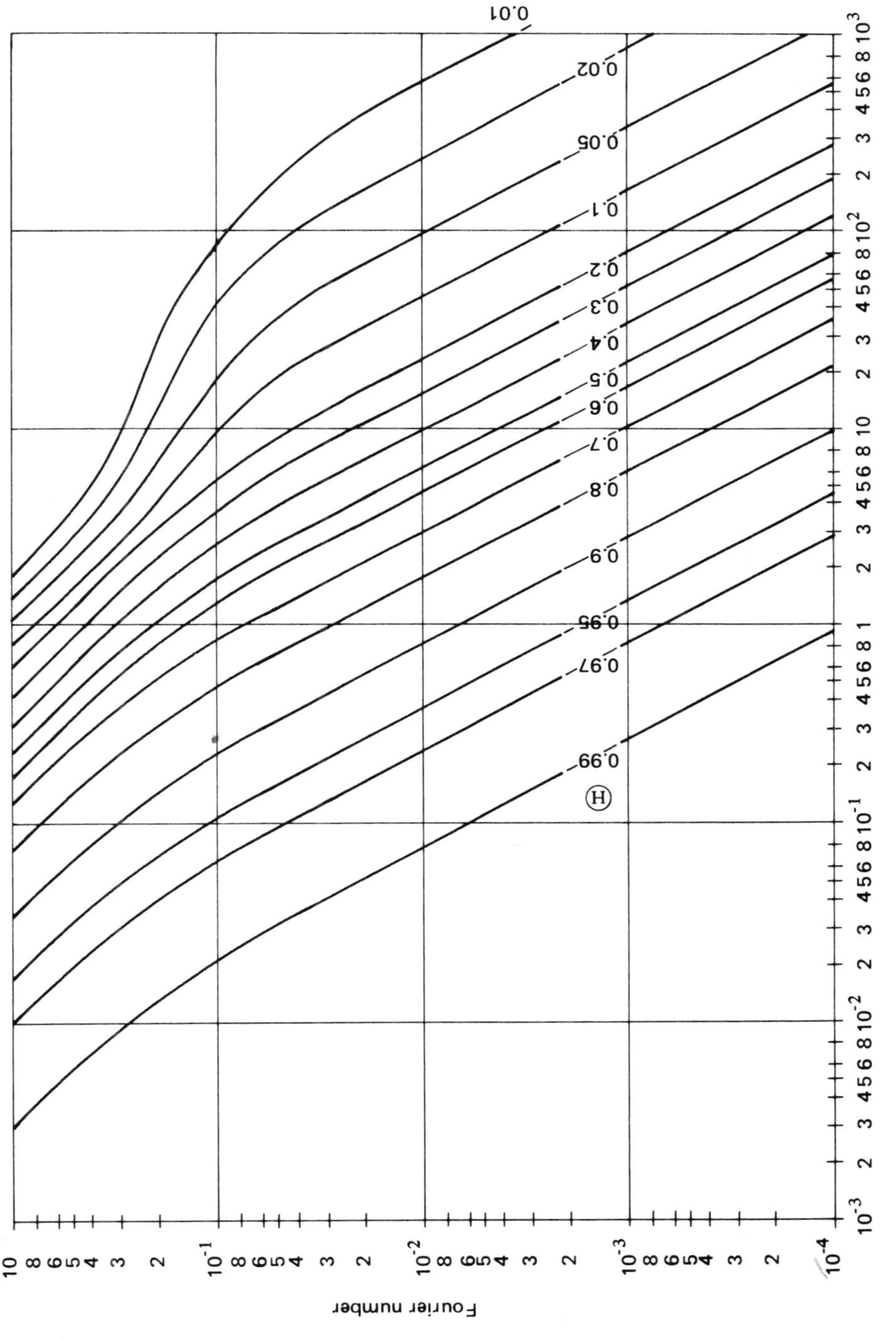


Fig. B.6 (b) – Surface temperature of a sphere ($Fo < 10^{-1}$).

APPENDIX C

# Convection Correlations

## C.1 INTRODUCTION

We present below a selection of correlation equations for convective heat-transfer, mainly taken from Welty, J. R., Wicks, C. E. and Wilson, R. E. (1969), *Fundamentals of momentum, heat and mass transfer*, John Wiley-Interscience, Chapter 20, and Ede, A. J. (1967), *An introduction to heat-transfer principles and calculations*, Pergamon Press. They in their turn have used books such as the following for their sources:

Jakob, M. (1949), *Heat transfer*, Vol. 1. John Wiley-Interscience.
McAdams, W. H. (1954), *Heat transmission*, 3rd ed. McGraw-Hill.
Fishenden, M. and Saunders, O. A. (1950), *An introduction to heat transfer*. Oxford: Clarendon Press.

Different sources often give slightly different equations for the same geometrical and flow situation. This is not surprising when one considers the scatter of the experiment data, collected from many investigators.

Two further publications which are useful data sources are:

Perry, J. H. (ed.), *Chemical engineers' handbook*, latest edition. McGraw-Hill.
Rose, J. W. and Cooper, J. R. (eds.) (1977), *Technical data on fuel*, British National Committee, World Energy Conference, 7th ed.

Note that **average** coefficients only are given unless specified otherwise. In addition film temperatures are used in estimating physical properties of the fluid unless specified otherwise. The film temperature is the arithmetic mean of the surface and main stream temperatures. The Prandtl number of combustion products may be taken as that for air. Calculated values given by empirical equations may be as much as 30% in error.

Newtonian fluids only are considered in Sections C.2-7; Section C.8 relates to non-Newtonian fluids.

## C.2 FORCED-CONVECTION CORRELATIONS

### C.2.1 Flow over flat plates

Laminar flow,

$$Nu_x = 0.664\, Re_x^{0.5}\, Pr^{0.33}$$

where $x$ = distance from leading edge.

Turbulent flow (starting at leading edge),

$$Nu_x = 0.037\, Re_x^{0.8}\, Pr^{0.33}$$

(free steam velocity is used in $Re$).

If it is assumed that flow is laminar up to $Re_x = 500\,000$ and thereafter it is turbulent:

$$Nu_x = 0.037\, Pr^{0.33}\, (Re_x^{0.8} - 23\,600).$$

### C.2.2 Flow inside tubes and ducts

In this case the boundary layer builds up from the entrance of the tube until it reaches the axis, beyond which point it cannot develop further (hence it is fully-developed at this stage).

**Laminar flow**

The starting length (length over which profile develops) is given by:

$$\frac{l_s}{d} = 0.05\, Re_d\, Pr$$

where $l_s$ = starting length.

Laminar flow:

$$Nu_d = 1.62 \left( Re_d\, Pr \frac{d}{l} \right)^{0.33} \qquad Re < 2000$$

where $l$ = distance from entrance and $d$ = internal pipe diameter. (This equation assumes fully-developed flow at the entrance to the pipe.)

In this equation the temperature difference for heat-transfer if taken as:

$$[(\theta_w - \theta_{in}) + (\theta_w - \theta_{out})]\,/2$$

where $\theta_w$ = wall temperature, $\theta_{in}$ = average temperature at inlet, and $\theta_{out}$ = average temperature at outlet.

**Turbulent flow** (fully developed)

For fluid in a pipe being heated:

$$Nu_d = 0.023\, Re_d^{0.8}\, Pr^{0.4}\,.$$

For fluid in a pipe being cooled:

$$Nu_d = 0.023\, Re_d^{0.8}\, Pr^{0.3}\,.$$

Physical properties and velocity are evaluated for 'bulk' fluid, that is, average temperature at any section. These equations are valid for $l/d > 60$ and for $Re >$ 2100 for gases and $Re > 5000$ for liquids.

For **fully-established** flow in ducts (rectangular cross-sections) the mean hydraulic diameter is used instead of the circular diameter of circular tubes (as in fluid flow). For the entrance region of tubes the following may be used:

$$\frac{h_l}{h} = 1 + \left(\frac{d}{l}\right)^{0.7} \qquad \text{for } 2 < l/d < 20$$

and

$$\frac{h_l}{h} = 1 + 6\frac{d}{l} \qquad \text{for } 20 < l/d < 60$$

where $h$ = coefficient for tube of infinite length (obtained from above equation for $l/d > 60$) and $h_l$ = average value of coefficient for finite length $l$.

The total heat transfer for a tube is given by:

$$q = h\pi dl\overline{\Delta\theta} \quad \text{where } \overline{\Delta\theta} = \frac{\Delta\theta_{in} - \Delta\theta_{out}}{\log_e\,(\Delta\theta_{in}/\Delta\theta_{out})}\,.$$

Here $\overline{\Delta\theta}$ is called the log mean temperature difference, $\theta_{out}$ = 'bulk' fluid temperature at outlet, $\theta_{in}$ = 'bulk' fluid temperature at inlet, $\theta_w$ = wall temperature, $\Delta\theta_{in} = \theta_{in} - \theta_w$, and $\Delta\theta_{out} = \theta_{out} - \theta_w$.

### C.2.3 Flow over various shapes

The data for a single object may be correlated by an equation of the form:

$$Nu_d = C\, Re_d^{\,m}\,.$$

The correlations are for a single object in an infinite flow of velocity $u$, which is used to define $Re_d$. Values of $C$ and $m$ are given (for air) in Table C.1.

**Table C.1 – Values of $C$ and $m$ for single objects.**

| Flow direction and profile | $Re_d$ | $C$ | $m$ |
|---|---|---|---|
| 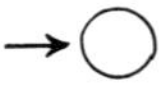 (cylinder) | 1-4 | 0.891 | 0.330 |
| | 4-40 | 0.821 | 0.385 |
| | 40-4000 | 0.615 | 0.466 |
| | 4000-40000 | 0.174 | 0.618 |
| | 40000-400000 | 0.024 | 0.805 |
| 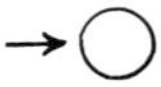 (sphere) | 20-150000 | 0.33 | 0.6 |
| 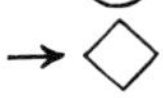 | 2500-7500 | 0.261 | 0.624 |
| | 5000-100000 | 0.222 | 0.588 |
|  | 5000-100000 | 0.138 | 0.638 |
| 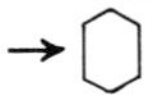 | 5000-19500 | 0.144 | 0.638 |
| | 19500-100000 | 0.035 | 0.782 |
| 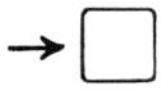 (square) | 2500-8000 | 0.160 | 0.699 |
| | 5000-100000 | 0.092 | 0.675 |
| 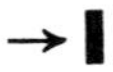 | 4000-15000 | 0.205 | 0.731 |
|  | 2500-15000 | 0.224 | 0.612 |
| 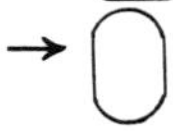 | 3000-15000 | 0.085 | 0.804 |

Note: $d$ is defined by the vertical heights of the various objects shown.

In the case of cylinders, if the air strikes at an angle of 45° (rather than perpendicular) the coefficient is reduced by about 25% and for flow parallel to axis by about 50%.

## C.3 NATURAL CONVECTION CORRELATIONS

Note that these correlations refer to freely-exposed surfaces.

### C.3.1 Flat surfaces: horizontal, vertical and inclined

For a vertical plate:

$$Nu = 0.62\,(Gr\,Pr)^{0.25} \quad \text{for } Gr < 10^{10},$$

$$Nu = 0.12\,(Gr\,Pr)^{0.33} \quad \text{for } Gr > 10^{10}$$

and the characteristic dimension is the height of plate.

For warm horizontal surface facing upwards and a cool horizontal surface facing downwards:

$$Nu = 0.54\,(Gr\,Pr)^{0.25} \quad \text{for } Gr\,Pr < 10^9\,,$$

$$Nu = 0.14\,(Gr\,Pr)^{0.33} \quad \text{for } 10^9 < Gr\,Pr$$

where the characteristic dimension is the mean length of sides.

For warm horizontal surface facing downwards and a cool horizontal surface facing upwards:

$$Nu = 0.25\,(Gr\,Pr)^{0.25} \quad \text{for all values of } Gr.$$

### C.3.2 Curved surfaces

A large diameter vertical cylinder may be treated as a vertical plate. For a small vertical cylinder and horizontal cylinder:

$$Nu = 0.52\,(Gr\,Pr)^{0.25} \quad \text{for } 10^4 < Gr\,Pr < 10^9\,,$$

$$Nu = 0.10\,(Gr\,Pr)^{0.33} \quad \text{for } 10^9 < Gr\,Pr$$

(constant = 0.17 for water).

## C.4 BOILING HEAT-TRANSFER COEFFICIENTS

The following relationships apply for water at atmospheric pressure. Temperature excess, $\Delta\theta = \theta_{\text{pan surface}} - \theta_{\text{water temperature}}$.

Horizontal surfaces:

For $\Delta\theta \leqslant 7.5°\text{C}$ and $q/A \leqslant 15\,000\ \text{W/m}^2$:

$$h = 1046(\Delta\theta)^{\frac{1}{3}} = 183\,(q/A)^{\frac{1}{4}}\ \text{W/m}^2.\text{K}\,.$$

For $7.5°\text{C} \leqslant \Delta\theta \leqslant 15°\text{C}$ or $15\,000 < q/A < 240\,000\ \text{W/m}^2$:

$$h = 5.56(\Delta\theta)^3 = 1.54\,(q/A)^{\frac{3}{4}}\,\text{W/m}^2.\text{K}\,.$$

Inside vertical tubes:

As above plus 25%.

Vertical surfaces:

For $\Delta\theta \leqslant 4.5°C$ or $q/A \leqslant 3200\ W/m^2$:

$$h = 542(\Delta\theta)^{\frac{1}{7}} = 247\ (q/A)^{\frac{1}{8}}\ W/m^2.K\ .$$

For $4.5 < \Delta\theta < 9.5°C$ or $3200 < (q/A) < 63\,000\ W/m^2$:

$$h = 7.95(\Delta\theta)^3 = 1.7\ (q/A)^{\frac{3}{4}}\ W/m^2.K\ .$$

## C.5 CONDENSING-STEAM HEAT-TRANSFER COEFFICIENTS

The following relationships apply to filmwise condensation.

Vertical surface:

For $Re < 1800$ (flow of condensate is laminar):

$$C_o = 1.88\ Re^{-\frac{1}{3}}$$

where

$$C_o = h\left(\frac{\mu^2}{\rho^2 g \lambda^3}\right)^{\frac{1}{3}}$$

and

$$Re = \frac{4(q/w)}{L\mu} = \frac{4G}{\mu}.$$

Where $C_o$ = condensation number, $h$ = the average coefficient, $q$ = heat flux, $G$ = mass flow rate per unit width of plate, $w$ = width of surface, $L$ = latent heat of water, and $\mu$ = viscosity of water. The physical properties are evaluated at $\theta_0 + \frac{1}{4}(\theta_1 - \theta_0)$, $\theta_1$ = temperature of water, that is condensation temperature, and $\theta_0$ = temperature of plate.

For $Re > 1800$ (flow of condensate turbulent):

$$C_o = 0.0077\ Re^{0.4}\ .$$

For a vertical tube:

Use the equations given above but if the lower part of the tube is flooded with condensate this may give the wrong result.

For horizontal surfaces:

This cannot be calculated from general expressions since so much depends upon the method of draining away the condensate.

## C.6 SCRAPED-SURFACE HEAT EXCHANGERS

Skelland, A. H. P., Oliver, D. R. and Tooke, S. (1962), *British Chemical Engineering*, 7(5), 346–353 derived the following relationship for scraped-surface heat exchangers:

$$\frac{hD_t}{\lambda} = \alpha\left(\frac{C_p\mu}{\lambda}\right)^{\beta}\left(\frac{(D_t - D_s)u\rho}{\mu}\right)^{1.00}\left(\frac{D_tN}{u}\right)^{0.62}\left(\frac{D_s}{D_t}\right)^{0.55} n_B^{0.53}$$

where $h$ = convection coefficient for scraped surface (W/m$^2$.K), $D_t$ = internal diameter of cylinder (m), $D_s$ = diameter of shaft (m), $u$ = average fluid velocity along tube (m/s), $\rho$ = density of fluid (kg/m$^3$), $\mu$ = viscosity of fluid (kg/m.s), $N$ = shaft speed (revolutions/s), $n_B$ = number of blades on shaft, $C_p$ = specific heat of fluid (J/kg.K), $\lambda$ = thermal conductivity of fluid (W/m.K), $\alpha = 0.014$ for thick, viscous liquids, 0.039 for thin, mobile liquids, and $\beta = 0.96$ for thick, viscous liquids, 0.70 for thin, mobile liquids.

The ratio, $L/D_t$ ($L$ = length of heat exchanger) was constant throughout the work and was equal to 6.0. Earlier work by Skelland, A. H. P. (1978), *Chem. Eng. Sci.*, 7, 166ff showed that the Nusselt group varied as $(D_t/L)^{0.37}$.

## C.7 IMPINGING JET FLAMES

See: Chigier, N. A. and Milson, A. (1973), *La Rivista dei Combustibili*, **27**(4–5), 137–146 for a review of this situation.

## C.8 NON-NEWTONIAN FLUIDS

Correlations for non-Newtonian fluids are more complex than those given above because of the need to include the flow constants. This area is also under-researched. The reader is advised to consult Skelland, A. H. P. (1967), *Non-Newtonian flow and heat transfer*, Wiley-Interscience for further information.

APPENDIX D

# Radiation Charts and Data

## D.1 VIEW FACTORS

There are two principal compilations of view factors: Love, T. J. (1968), *Radiative heat transfer*, Appendix II, C. E. Merrill; Sparrow, E. M. and Cess, R. D. (1970), *Radiation heat transfer*, Appendix A, Brook/Cole Publishing Co. although most books on heat transfer also contain data for the more common situations.

### D.1.1 Black body completely surrounded by black surfaces

$$F_{12} = 1 .$$

The area is that of the black body.

### D.1.2 Grey body completely surrounded by black surfaces

$$\text{The resistance to heat flow} = \frac{1}{A_G \epsilon_G}$$

where subscript $G$ refers to the grey body.

### D.1.3 Small grey body completely surrounded by grey surfaces

The equation given in D.1.2 is no longer valid but may be used, without significant loss of precision, if body is small compared to the surrounding surfaces.

### D.1.4 Parallel, directly opposed, plane circular discs

The radii of discs $= r_1$ and $r_2$. The distance between discs $= d$.

$$F = \frac{[x - (x^2 - 4r_2^2/r_1^2)^{0.5}]}{2}$$

where $x = 1 + d^2/r_1^2 + r_2^2/r_1^2$.

### D.1.5 Identical parallel rectangular plates

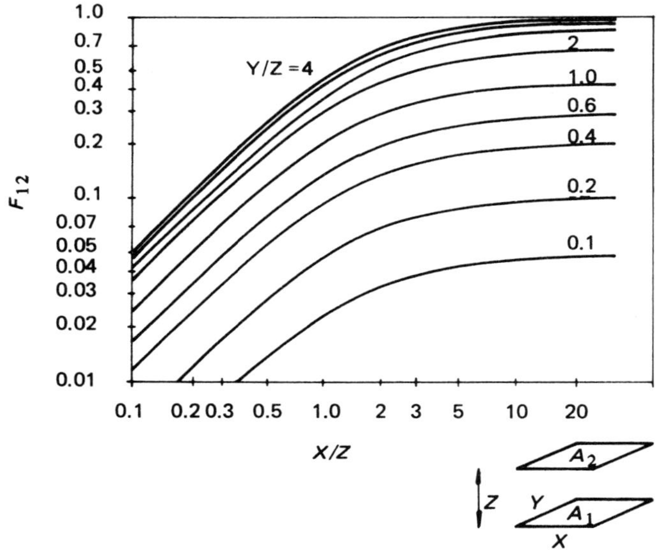


### D.1.6 Infinite parallel grey surfaces

$$\frac{1}{F} = \left[\frac{1}{\epsilon_1} + \frac{1}{\epsilon_2} - 1\right].$$

## D.1.7 Touching perpendicular planes

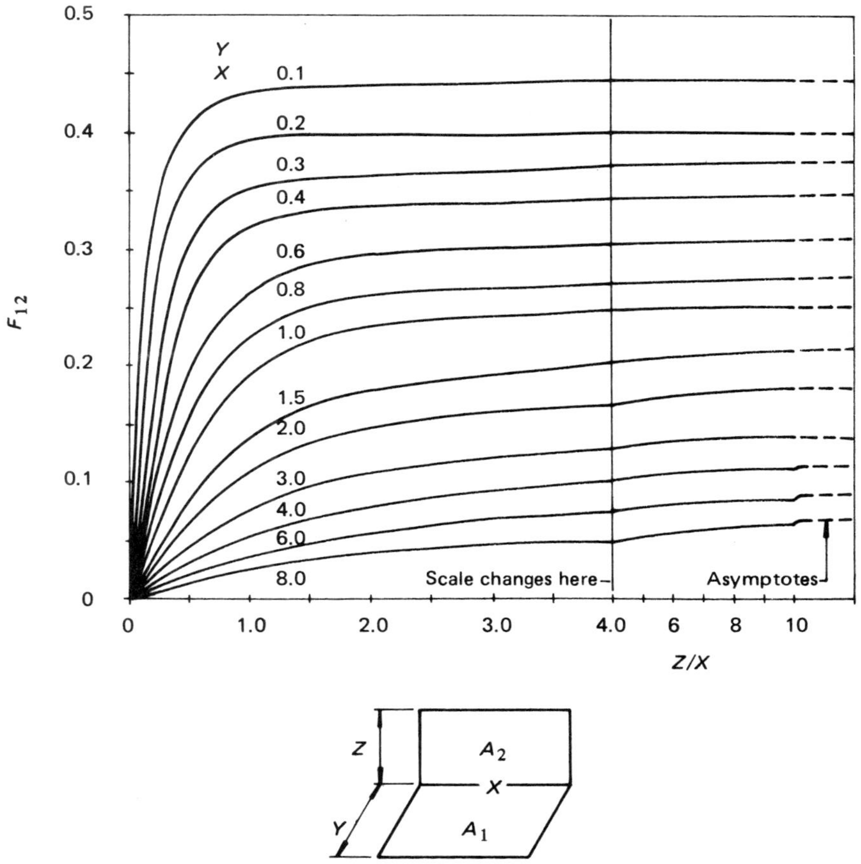

## D.1.8 Differential plate element orientated below a corner of a parallel finite rectangle

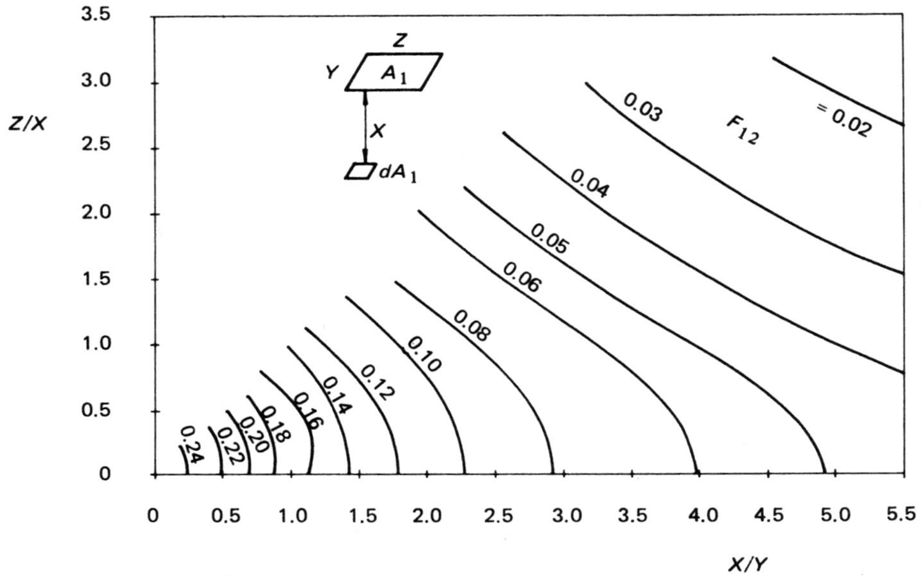

**D.1.9 Differential plate element orientated below a corner of a finite rectangle perpendicular to it**

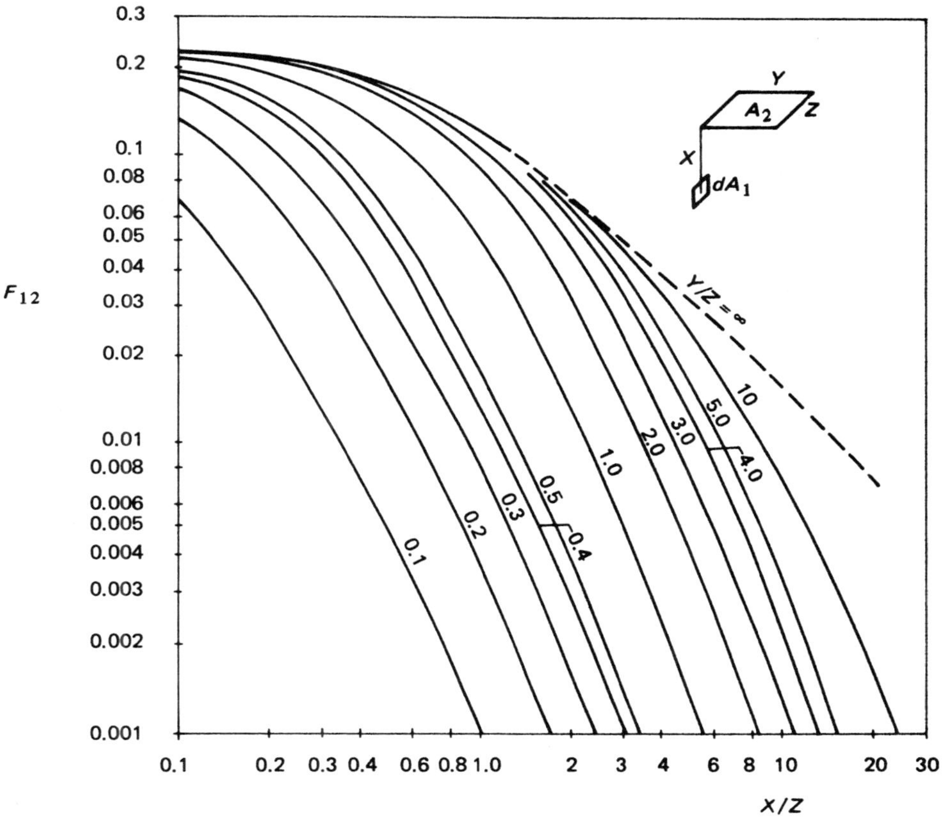


**D.1.10 Completely enclosed body, large compared to enclosing body**

As Section D.1.6.

## D.2 TECHNIQUES FOR CALCULATING VIEW($F$) FACTORS

The data already given are not sufficient for the calculation of all radiation problems and some additional techniques are given here.

### D.2.1 View factor between perpendicular plates which are not touching

The chart for perpendicular plates which are touching is given in Section D.1.7. This may be used for the case where the plates are not touching, for example, $F_{AC}$ in Fig. D.2.1.

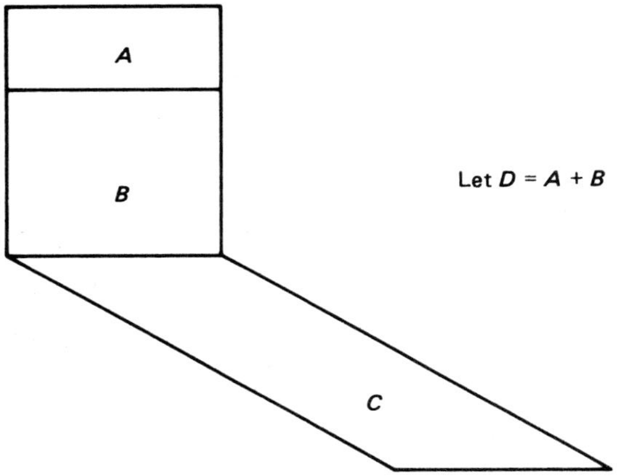


Fig. D.2.1

In this case $F_{BC}$ and $F_{DC}$ may both be obtained from the chart. Then, $F_{AC} = F_{DC} - F_{BC}$.

### D.2.2 View factor between a differential plate element not orientated below a corner of a parallel finite rectangle

The chart which is utilised is given in Section D.1.8.

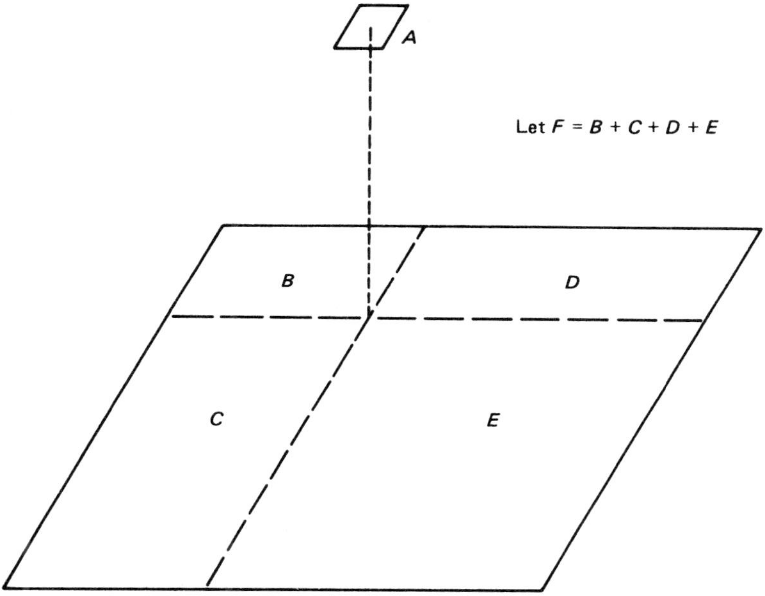


Fig. D.2.2

In this case $F_{AB}$, $F_{AC}$, $F_{AD}$ and $F_{AE}$ may be obtained from the chart. Then, $F_{AF} = F_{AB} + F_{AC} + F_{AD} + F_{AE}$.

### D.2.3 Special reciprocity

This principle helps in the deduction of view factors where symmetry or similarity exists. It is best illustrated by studying Fig. D.2.3.

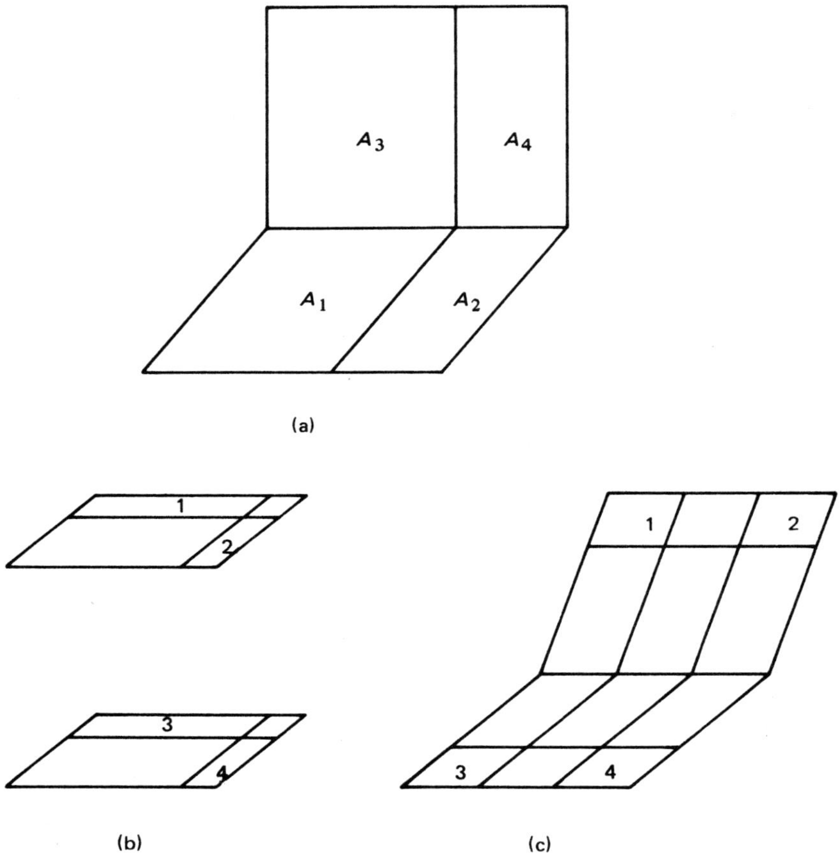


Fig. D.2.3

In each case the special reciprocity relationship is:

$$A_1 F_{14} = A_2 F_{23} .$$

It is possible to determine the view factor $F_{14}$ in the upper diagrams by using the relationship and the chart in Sections D.1.7 and D.1.5.

If

$$A_5 = A_1 + A_2 \text{ and } A_6 = A_3 + A_4,$$

then

$$A_5F_{56} = A_1F_{13} + A_1F_{14} + A_2F_{23} + A_2F_{24}$$

but

$$A_1F_{14} = A_2F_{23}$$

therefore

$$2A_1F_{14} = A_5F_{56} - A_1F_{13} - A_2F_{24}$$

and then

$$F_{14} = \frac{A_5F_{56} - A_1F_{13} - A_2F_{24}}{2A_1} \ .$$

The charts give $F_{56}$, $F_{13}$ and $F_{24}$ and the dimensions $A_5, A_1$ and $A_2$ are known.

## D.3 GAS RADIATION

The emissivities of water vapour and carbon dioxide may be obtained from Figs. D.3.1 and D.3.2. (The emissivities of nitrogen and oxygen, the constituents

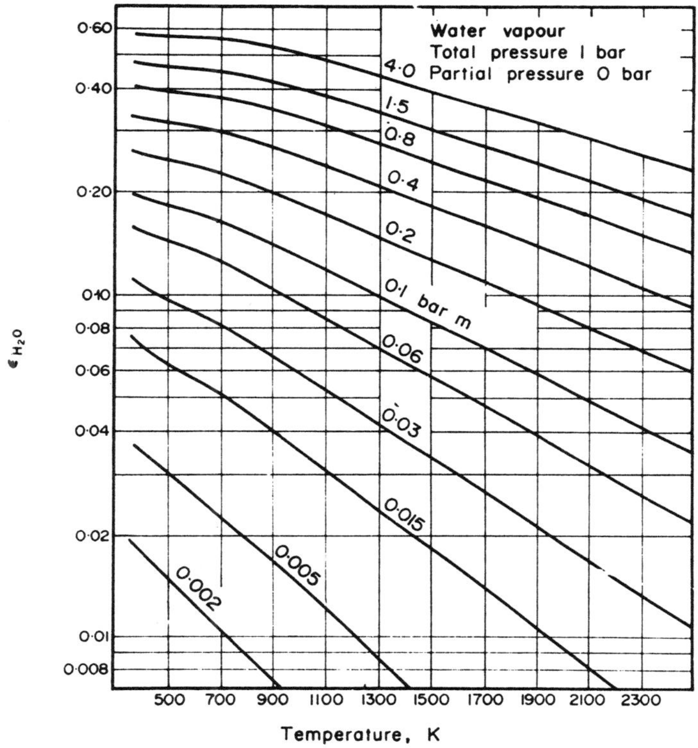


Fig. D.3.1 – Emissivity of water vapour.

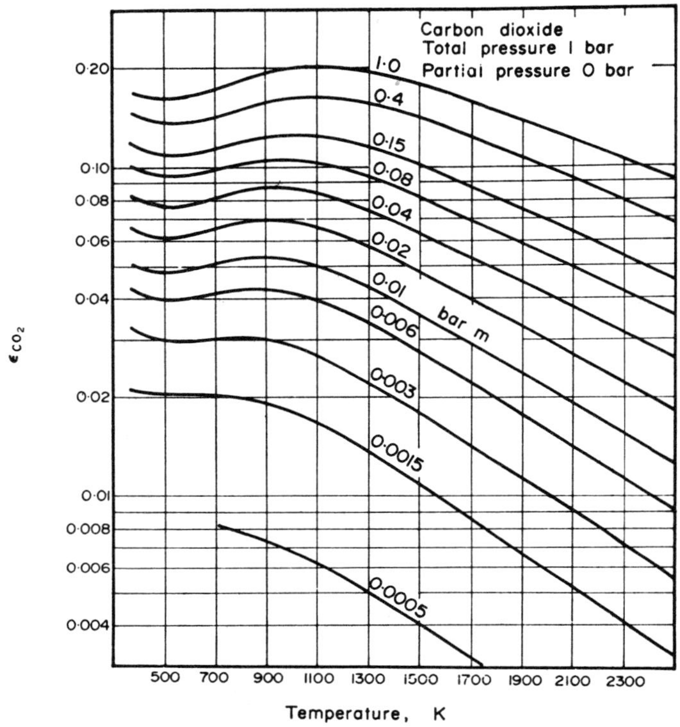


Fig. D.3.2 – Emissivity of carbon dioxide.

of air and the other constituents of combustion products are negligible.) The emissivity depends on the gas temperature and $pL$ where $p$ = partial pressure of the gas and $L$ = beam length. The partial pressure of the gas is the volume per cent of the gas in the mixture divided by 100. The beam length depends on the shape of the gas and may be obtained from Table D.3.1. For shapes not included in the table, the value of $L$ may be taken as:

$$L = 3.4 \times \frac{\text{Gas volume}}{\text{Area of retaining walls}} .$$

For combustion products, radiation from the carbon dioxide and the water vapour interact. Fig. D.3.3 enables the correction factor to be estimated at two different temperatures (415 K or 142°C and 800 K or 527°C). The emissivity of the gas mixture is given by:

$$\epsilon_{\text{gas mixture}} = \epsilon_{H_2O} + \epsilon_{CO_2} - \Delta\epsilon$$

where $\epsilon_{H_2O}$ is obtained from Fig. D.3.1, $\epsilon_{CO_2}$ is obtained from Fig. D.3.2 and $\Delta\epsilon$ is obtained from Fig. D.3.3.

**Table D.3.1 – Beam lengths for several shapes.**

| Shape | $L$ |
|---|---|
| Sphere radiating to surface | 0.63 X diameter |
| Cube radiating to each surface | 0.60 X side length |
| Cylinder, height = 0.5 X diameter | 0.45 X diameter |
| Cylinder, height = 2.0 X diameter | 0.73 X diameter |
| Infinite cylinder radiating to walls | 0.94 X diameter |
| Cylinder, height = diameter, radiating to whole surface | 0.60 X diameter |
| Space between infinite parallel planes, radiating to planes | 1.76 X distance between planes |
| Rectangular box, 1:1:4, radiating to each surface | 0.8 X shortest edge |
| Rectangular box, 1:2:6, radiating to each surface | 1.1 X shortest egde |

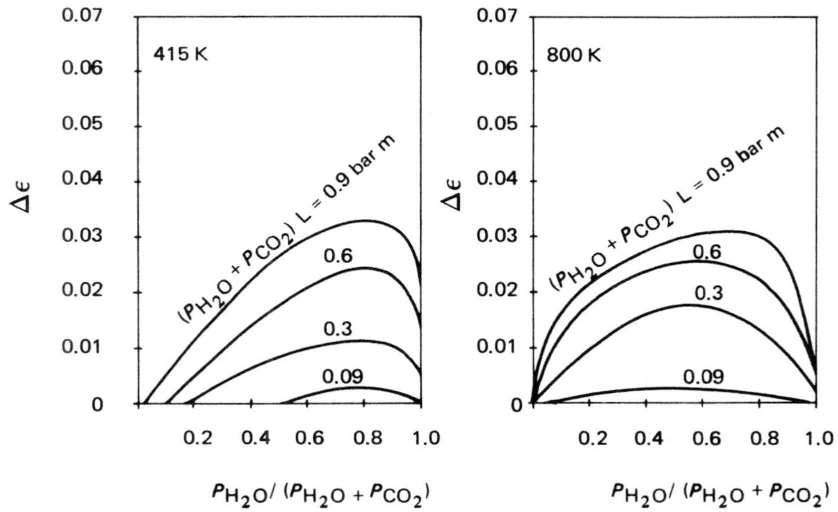


Fig. D.3.3 – Correction factor for interaction of $CO_2$ and $H_2O$.

If combustion products have the composition (% by volume): $CO_2$ 7.5%, $H_2O$ 15.0%, $O_2$ 5.0%, and $N_2$ 72.5% then partial pressures ($p$) are $CO_2$ 0.075, $H_2O$ 0.150, $O_2$ 0.050 and $N_2$ 0.725.

Then

$$\frac{p_{H_2O}}{p_{H_2O} + p_{CO_2}} = \frac{0.150}{0.150 + 0.075} = 0.67.$$

APPENDIX E

# Mass-Transfer Correlations

Average coefficients are given for liquids and gases unless otherwise stated. Fluid properties are evaluated at the average conditions between the surface and the bulk fluid.

**(1) Inside circular pipes**

$j_D = 0.023\ Re^{-0.17}$,

$Sh = 0.023\ Re^{0.83}\ Sc^{\frac{1}{3}}$

for $Re = 4000 - 60\,000$ and $Sc = 0.6 - 3000$.

**(2) Flow at right angles to a circular cylinder**

$Sh = 0.43 + 0.532\ Re^{0.5}\ Sc^{0.31}$

for $Re = 1 - 4000$.

**(3) Parallel to flat plates**

$j_D = 0.664\ Re^{-0.5}$

for $Re < 80\,000$ (this should hold to $Re = 350\,000$ if laminar flow is maintained.

$j_D = 0.036\ Re^{-0.2}$

for $Re > 500\,000$.

(The characteristic dimension is the distance from start of plate.)

For gases only $j_D = 0.037\ Re^{-0.2}$ is valid in the range $8000 < Re < 300\,000$.

**(4) Flow over single sphere**

$Sh_{av} = Sh_o + 0.347\ (Re\,Sc^{0.5})^{0.62}$

where $Sh_o = 2.0 + 0.569\ (Gr\,Sc)^{0.250}$ for $Gr\,Sc < 10^8$

and $Sh_o = 2.0 + 0.0254\ (Gr\,Sc)^{0.333}\ Sc^{0.244}$ for $Gr\,Sc > 10^8$.

Applicable for $Sc = 0.6 - 3200$ and $Re\,Sc^{0.5} = 1.8 - 600\,000$.

**(5a) Gases flowing through packed beds**

(i) $j_D = 2.06/\epsilon\,(Re)^{-0.575}$ for $Sc = 0.6$ and $Re = 90 - 4000$.

(ii) $j_D = 20.4/\epsilon\,(Re)^{-0.815}$ for $Re = 5000 - 10\,300$.

Note that $a = 6(1 - \epsilon)/d_p$ where $a$ = specific solid surface ($m^2/m^3$ of bed), $d_p$ = diameter of sphere or, for non-spherical particles, diameter of a sphere of the same surface area as particle (m).

**(5b) Packed beds (gases and liquids)**

$j_D = 1.17\,(Re)^{-0.415}$ for $10 < Re < 2500$

where velocity is the average over flow cross-section and the characteristic dimension is particle diameter.

Note that this correlation does not allow for changes in $\epsilon$ (the void fraction) which is typically 0.4 to 0.44 but may range from 0.3 to 0.5.

The above equation applies to non-spherical particles. Much lower values of $j_D$ than suggested by the above equation for very small granular particles ($d_p < 0.5$ mm) has been found.

**(6) Liquids flowering through fixed bed of pellets**

(i) $(F/G)Sc^{0.58} = 2.40(Re/\epsilon)^{-0.66}$

for $Sc = 159\text{–}13\,260$ and $Re/\epsilon = 0.08\text{–}125$.

(ii) $(F/G)Sc^{0.58} = 0.442(Re/\epsilon)^{-0.31}$

for $Re/\epsilon = 125\text{–}5000$

where $\epsilon$ is defined in (5) above and $F$ = mass-transfer coefficient expressed as $m/A$, and $G$ = mass velocity expressed as k.C.

Source: Treybal, R. E. (1968), *Mass-transfer operations*, McGraw-Hill, 2nd ed.

APPENDIX F

# Properties of Water

The properties of water are:

| | | | | |
|---|---|---|---|---|
| Temperature, °C | 0 | 20 | 50 | 100 |
| Density, kg/m$^3$ | 1000 | 998 | 988 | 958 |
| Viscosity, g/m.s† | 1.786 | 1.002 | 0.5478 | 0.2831 |
| Thermal conductivity, mW/m.K | 569 | 603 | 643 | 681 |
| Specific heat, kJ/kg.K | 4.223 | 4.182 | 4.181 | 4.218 |

† Multiply by $10^{-3}$ to convert to basis of m, kg and s.

The latent heat of vapourisation of water at 100°C and 1 atmosphere pressure is 2257 kJ/kg.

Source: Rose, J. W., and Cooper, J. R. (eds.) (1977), *Technical data on fuel*, British National Committee, World Energy Conference, 7th ed.

The emissivity of water is 0.95 and that of ice 0.97.

APPENDIX G

# Properties of Solid Foods

## G.1 INTRODUCTION

Sources of data on the physical properties of foods are often difficult to find and, even where available, these values have to be used with considerable care. In a proposed system for the classification of foods and their physical properties, Jowitt (1974) includes 77 properties of which 40 are mechanical, 10 thermal, 9 diffusional and 18 electromagnetic and electrostatic. The purpose of this Appendix is to provide data for the calculations given in the preceding chapters. We have included selected data on density, specific heat, latent heat, enthalpy, thermal conductivity, thermal diffusivity and transmission characteristics for infra red and microwave radiation. In some cases equations have been derived, based either on correlations of measured data or on theoretical thermodynamic considerations, which allow the prediction of properties from compositional data and/or temperature; where relevant these have been included. In addition references are given to review articles on physical properties which provide a more exhaustive source of data than this Appendix.

As mentioned above, it is often difficult to ascertain the applicability of literature values of physical properties to a particular situation. Most of these properties are temperature dependent and, in some cases, the temperature range over which measurement was made is not stated. Also, because foods are biological in origin, they vary in their chemical composition and in the relative proportions of differentiated tissue (for example the ratio of bone fat and lean meat in a joint of meat). These variations depend on a large number of factors including species, cultural conditions and time and conditions of storage. Most of the calculations given in preceding chapters assume perfect homogeneity, which is only found in a few foods. A further complication is that many foods on heating undergo a physical or chemical change, for example the hydration of starch, hydrolysis of gelatin and denaturation of protein: this can make the selection of physical properties for an unsteady-state heat-transfer calculation difficult.

Because of the factors given above, where accurate figures are required on a particular sample of food, it is often necessary to resort to measurement of the relevant data.

## G.2 DENSITY

The density of fresh foods varies with the state of maturity and the storage history of the sample. In the case of foods such as cakes and bread, the density changes during cooking, caused by the loss of water and by the generation of carbon dioxide. In the case of particulate foods the figure given is usually a bulk density or a packing density.

**Table G.1 – Density of solid foods.**

| | | |
|---|---|---|
| Fresh fruit – range of values | 865-1057 kg/m$^3$ | Dickerson (1969) |
| Fresh fruit 20°C range of values | 811-1064 kg/m$^3$ | Kirk: unpublished data |
| Fresh vegetables – range of values | 801-1095 kg/m$^3$ | Kirk: unpublished data |
| Frozen fruit −15 to −18°C | 625-801 kg/m$^3$ | Dickerson (1969) |
| Frozen vegetables −15 to −18°C | 561-977 kg/m$^3$ | Dickerson (1969) |
| Frozen whole egg −8°C | 977 kg/m$^3$ | Dickerson (1969) |
| Fish – above 0°C | 967 kg/m$^3$ | Jason *et al.* (1969) |
| Fish – below 0°C | 1056 kg/m$^3$ | Jason *et al.* (1969) |
| Meat | 1.07 (S.G.) | Morley (1972a) |
| Fats | 0.92 (S.G.) | Morley (1972a) |
| Lard 15.5°C | 0.863 (S.G.) | Pearson (1970) |
| Suet 15.5°C | 0.945 (S.G.) | Pearson (1970) |
| Oats (bulk density) | 513 kg/m$^3$ | Kramer *et al.* (1970) |
| Wheat (bulk density) | 785 kg/m$^3$ | Kramer *et al.* (1970) |
| Flour (bulk density) | 449 kg/m$^3$ | Kramer *et al.* (1970) |

## G.3 SPECIFIC HEAT, LATENT HEAT AND ENTHALPY

Particular care is needed when using literature values of thermal properties where a change of state in involved. Many of the early papers on the thermal properties of food divide the heat extracted on freezing into separate latent heat and specific heat components, assuming that there is a change of state at a constant temperature – the freezing point of the food. In practice, although foods do have a freezing point this is the temperature at which a change of phase starts and, as the temperature of the food is lowered, the proportion of unfrozen water decreases. However, even at very low temperatures a proportion of water remains unfrozen. As a further complication, the proportion of frozen water is time dependant, a fast rate of heat extraction leading to an initial supercooling followed by rapid ice crystal formation.

As an alternative to separate specific and latent heats, current practice is to give a specific heat value incorporating both latent and sensible heat; these values are obviously very temperature dependant. A similar approach is to give enthalpy values over the temperature range of interest. As an example of this, Table G.3 gives specific heat and enthalpy for beef and lamb. Table G.2 provides specific heats above and below freezing and latent heat; these are needed for freezing time calculations.

Several attempts have been made to predict values for the specific heat of foods (above freezing point) based on the specific heat of their components. Heldman (1975) compares a number of these equations with experimentally determined values of specific heat. The simplest of these equations uses the water content only, whereas others are based on a proximate analysis.

$$C_p = 0.84 + 3.35\,X_m \text{ kJ/kg.K},$$

$$C_p = 2.092\,X_f + 1.25\,X_s + 4.18\,X_m \text{ kJ/kg.K},$$

$$C_p = 1.142\,X_c + 1.54\,X_p + 1.67\,X_f + 0.84\,X_a + 4.18\,X_m \text{ kJ/kg.K}$$

**Table G.2 – Specific heat of solid foods.**

| | Specific heat below freezing kJ/kg.K | Specific heat above freezing kJ/kg.K | Latent heat kJ/kg |
|---|---|---|---|
| Vegetables | 1.92 ± 0.08† | 3.72 ± 0.25† | 287 ± 23† |
| Fruits | 1.92 ± 0.08† | 3.72 ± 0.17† | 286 ± 14† |
| Fish | 1.67 | 3.34 | 245 |
| Cheese | 1.30 | 2.09 | 126 |
| Butter | – | 1.38 | 54 |
| Beef | 1.59 to 1.80 | 2.92 to 3.51 | 207 to 256 |
| Pork | 1.26 to 1.34 | 2.01 to 2.26 | 117 to 140 |
| Lamb | 1.59 to 2.13 | 2.85 to 3.18 | 200 to 233 |

† 1 standard deviation.

Source: Adapted from Smith and Minor (1974). For more detailed specific heats see ASHRAE (1964) and, for meats, Morley (1972b).

**Table G.3 – Specific heat and enthalpy for meats.**

| | Temperature °C | | | | | | | | |
|---|---|---|---|---|---|---|---|---|---|
| | −40 | −34.4 | −28.9 | −23.3 | −17.8 | −12.2 | −6.7 | −1.1 | +4.4 |
| *Lean beef* (76% water, 1.1% fat) | | | | | | | | | |
| specific heat | 1.76 | 1.76 | 1.88 | 2.18 | 2.64 | 3.35 | 6.66 | 4.19 | 3.43 |
| enthalpy | 0.00 | 9.65 | 19.60 | 30.70 | 44.00 | 60.20 | 85.40 | 279.40 | 300.30 |
| *Lamb (loin)* (64.9% water, 11.7% fat) | | | | | | | | | |
| specific heat | 1.30 | – | 1.51 | 1.93 | 2.47 | 3.60 | 6.70 | 96.30 | 3.39(0°C) |
| enthalpy | 0.00 | – | 14.40 | 24.70 | 36.3 | 50.50 | 77.20 | 211.70 | 291.00(0°C) |

Enthalpy datum: −40°C.
Source: Morley (1972b).

where $X$ is the fraction of a component (for example 50% gives $X = 0.05$) and the subscript $m$ is for moisture, $f$ for fat, $s$ for solid, $c$ for carbohydrate, $p$ for protein and $a$ for ash.

Many of the literature values of latent heat (ASHRAE 1964) are calculated from the latent heat of the aqueous component:

$$L_f = 333.5\,X_m \text{ kJ/kg} .$$

In the case of frozen foods, the thermodynamic relationship between enthalpy, percentage of frozen water, water content of the unfrozen material and temperature have been investigated by Reidel (1957). He has developed charts for predicting the percentage of unfrozen water and enthalpy for a number of foods including meats (Dickerson 1969). Schwartzberg (1976) has described a method for calculating the specific heat and enthalpy of frozen foods using the freezing point depression.

## G.4 THERMAL CONDUCTIVITY

Thermal conductivity is dependant upon a number of factors including temperature and packing density. Smith *et al.* (1954) demonstrated the effect of packing density on the thermal conductivity of frozen carrots:

| | | |
|---|---|---|
| Whole carrots | 625 kg/m$^3$ | 0.623 W/m.K |
| Sliced carrots | 641 kg/m$^3$ | 0.675 W/m.K |
| Carrot puree | 897 kg/m$^3$ | 1.260 W/m.K . |

Reported values for the thermal conductivity of foods have been surveyed by Woodams and Nowrey (1968) and for meats by Morley (1972b). Selected values are given in Table G.4.

Several workers have looked at factors which affect thermal conductivity and these studies have lead to the generation of equations relating thermal conductivity of foods to temperature and/or water content. An alternative approach is to develop a theoretical model for the thermal conductivity of non-homogeneous materials; Heldman (1975) reviews equations for the prediction of thermal conductivity for a material composed of a spherical dispersed phase and also for the case of a meat-like model with a parallel fibre structure.

Sweat (1974) gives an equation for predicting the effect of water content on the thermal conductivity of fruits and vegetables between 25 and 30°C and with a water content above 60%:

$$\lambda = 0.148 + (0.00493\,W)\text{ W/m.K} \pm 15\%$$

where $W$ is the percentage water content on a wet basis. In a further paper

(Sweat 1975) he gives equations for meat. In the case of meats above 0°C and below 60°C it is considered unnecessary to include a temperature term.

$$\lambda = 0.080 + (0.0052\,W)\ \mathrm{W/m.K}$$

for the ranges: 0 to 60°C, 60 to 80% water, standard deviation 8.6%,

$$\lambda = 0.28 + (0.019\,W) - (0.0042\,T)$$

for the ranges: −40 to −5°C, 65 to 85% water, standard devaition 9.4%.

**Table G.4 – Thermal conductivity of solid foods (units: W/m.K).**

| | | |
|---|---|---|
| Fruits – frozen | 0.570 ± 0.322† | from Dickerson (1969) |
| Fruits – fresh | 0.381 to 0.502 | from Dickerson (1969) |
| Vegetables – frozen | 0.716 ± 0.300† | from Dickerson (1969) |
| Vegetables – fresh | 0.424 ± 0.093† | from Dickerson (1969) |
| Meats and poultry – above 0°C | 0.473 ± 0.038† | from Fennema (1973) |
| Meats and poultry – 0 to −10°C | 1.196 ± 0.146† | from Fennema (1973) |
| Meats and poultry – −10°C | 1.317 ± 0.151† | from Fennema (1973) |
| Meats – −20°C | 1.430 ± 0.160† | from Morley (1972b) |
| Meats – +20°C | 0.451 ± 0.050† | from Morley (1972b) |
| Fish – above 0°C | 0.502 ± 0.054† | from Fennema (1973) |
| Fish – 0 to −10°C | 1.259 ± 0.230† | from Fennema (1973) |
| Fish – −10°C | 1.422 ± 0.160† | from Fennema (1973) |

† 1 Standard deviation.

## G.5 THERMAL DIFFUSIVITY

Thermal diffusivity can be measured directly rather than by obtaining it from measured values of specific heat, thermal conductivity and density (Dickerson *et al.* 1968). These values, where available, are likely to be more accurate than calculated vaules based on each of the physical properties from different sources. Values of thermal diffusivity for several foods are given in Table G.5.

Reidel (1969) gives an equation for predicting the thermal diffusivity of high water content foods based on their water content ($X_m$) and the thermal diffusivity of water at the same temperature ($\alpha_w$).

$$\alpha\ (\mathrm{cm^2/min}) = 0.053 + (\alpha_w - 0.053)X_m.$$

This equation is valid for foods with water contents from 40 to 80% and over the temperature range 5 to 65°C.

**Table G.5 – Thermal diffusivity of solid foods.**

| | Temperature °C | | | |
|---|---|---|---|---|
| | −25 | 0 | +30 | |
| Beef (74.5% water) | $0.669 \times 10^{-6} m^2/s$ | $0.117 \times 10^{-6} m^2/s$ | $0.125 \times 10^{-6} m^2/s$ | Morley (1972b) |
| Pork (76.8% water) | $0.461 \times 10^{-6} m^2/s$ | $0.117 \times 10^{-6} m^2/s$ | $0.125 \times 10^{-6} m^2/s$ | Morley (1972b) |
| Pork sausage – 0 to 25°C | $0.116 \times 10^{-6} m^2/s$ | Morley (1972b) | | |
| Pork sausage – 20 to70°C | $0.149 \times 10^{-6} m^2/s$ | Morley (1972b) | | |
| Fruit – above 0°C | $0.121$ to $0.126 \times 10^{-6} m^2/s$ | Dickerson (1969) | | |
| Fish – above 0°C | $0.147 \times 10^{-6} m^2/s$ | Dickerson (1969) | | |
| Poultry – above 0°C | $0.108$ to $0.139 \times 10^{-6} m^2/s$ | Dickerson (1969) | | |

## G.6 TRANSMISSION CHARACTERISTICS: INFRA RED

Because foods usually contain large amounts of water their emissivities are similar to that of water. Seveik *et al.* (1962) give the following figures for beef:

| | | |
|---|---|---|
| Lean beef (69.5% water) | 20-30°C | $\epsilon = 0.74$ , |
| Beef fat | 20-35°C | $\epsilon = 0.78$ . |

Emissivities can be misleading as most foods do not behave as grey bodies but as spectral absorbers. IR spectra for several foods are given by De Werth (1962) and these indicate that foods have distinct absorption bands. This is most noticeable in the case of oils, fats and fatty foods. As an example, the transmission figures for pork do not vary significantly (over the range 1-15$\mu$m) whereas pork fat shows distinct absorption bands at 3.4, 5.7 and 8 to 9.5 $\mu$m. Penetration depths for meat products are given in Table 6.1.

## G.7 TRANSMISSION CHARACTERISTICS: MICROWAVE

Because the most common frequency for catering microwave ovens in 2.45 GHz, attention will be concentrated on literature values measured at around this frequency. The effect of temperature on the dielectric properties of foods are contained in papers by Ohlsson *et al.* (1975) and Bengtsson *et al.* (1971). Data from these sources are contained in Table G.6. A number of people have looked at the effect of chemical composition on dielectric properties; Roebuck *et al.* (1972) studied carbohydrate water mixtures and agar gels, and Ohlsson *et al* (1974) studied model meat emulsions. To *et al.* (1974) indicate that the dielectric properties of foods are dependent on:

(a) water content – the predominant effect, but its effect if reduced by irrotationally bound water;
(b) salts, because of the conductivity losses. Here again this is often lower than would be expected from the ash content because some of the salts are chemically complexed and not free to migrate in the electrical field.

Fig. G.1 shows the effect of water content on dielectric properties.

**Table G.6 – Dielectric properties of solid foods.**

| | | 3.0 GHz: Bengtsson *et al.* (1971) | | | | 2.8 GHz: Ohlsson *et al.* (1975) | | | |
|---|---|---|---|---|---|---|---|---|---|
| | | Temperature °C | | | | Temperature °C | | | |
| | | −20 | −10 | 0 | +20 | +40 | +60 | +80 | +100 |
| Beef (raw) | $\epsilon_r'$ | 4.8 | 6.2 | 47.8 | 47.7 | 45.2 | 44.4 | 42.6 | 41.2 |
| | $\epsilon_r''$ | 0.57 | 1.57 | 16.1 | 13.4 | 12.5 | 12.0 | 13.1 | 13.7 |
| | $\alpha$ | 6.66 | 16.05 | 58.92 | 49.29 | 47.24 | 45.78 | 50.90 | 54.02 |
| | $P\frac{1}{e}$ | 0.075 | 0.031 | 0.009 | 0.010 | 0.011 | 0.011 | 0.010 | 0.009 |
| Beef (cooked) | $\epsilon_r'$ | 5.3 | 6.1 | 35.4† | 35.4 | 43.0 | 41.8 | 40.7 | 39.6 |
| | $\epsilon_r''$ | 0.66 | 0.98 | 13.7† | 11.6 | 10.9 | 11.0 | 11.3 | 11.9 |
| | $\alpha$ | 7.34 | 10.14 | 58.02† | 49.37 | 42.30 | 43.27 | 45.01 | 47.98 |
| | $P\frac{1}{e}$ | 0.068 | 0.049 | 0.009† | 0.010 | 0.012 | 0.012 | 0.011 | 0.010 |
| Mashed potato | $\epsilon_r'$ | 4.6 | – | 66.7† | 64.6 | 60.6 | 56.3 | 54.2 | 52.9 |
| | $\epsilon_r''$ | 0.32 | – | 25.5† | 21.7 | 17.4 | 16.5 | 16.5 | 15.8 |
| | $\alpha$ | 3.82 | – | 78.71† | 68.52 | 56.76 | 55.82 | 56.85 | 55.12 |
| | $P\frac{1}{e}$ | 0.131 | – | 0.006† | 0.007 | 0.009 | 0.009 | 0.009 | 0.009 |
| Carrots | $\epsilon_r'$ | 4.2 | 5.5 | 75.6† | 71.5 | 70.1 | 65.5 | 61.8 | 58.0 |
| | $\epsilon_r''$ | 0.34 | 0.35 | 25.4† | 17.9 | 11.8 | 11.8 | 12.5 | 13.4 |
| | $\alpha$ | 4.24 | 3.83 | 73.92† | 53.88 | 36.02 | 37.25 | 40.58 | 44.84 |
| | $P\frac{1}{e}$ | 0.118 | 0.131 | 0.067† | 0.009 | 0.014 | 0.013 | 0.012 | 0.011 |

† −3°C

Values of attenuation coefficient ($\alpha$) and $P\frac{1}{e}$ calculated from data of Ohlsson and Bengtsson.

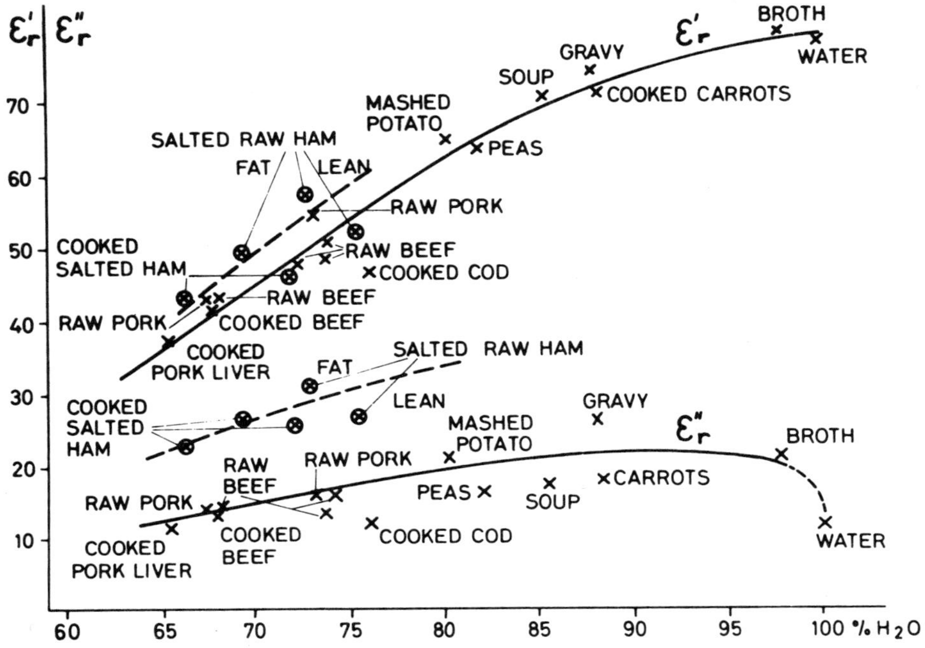


Fig. G.1 – Dielectric properties of foods at 20° C.

## G.8 REFERENCES


ASHRAE (1964), *Guide and data book*, Am. Soc. Heating Refrig. and Air Cond. Eng., New York.

Bengtsson, N. E. and Risman, P. O. (1971), *J. Microwave Power*, **6**(2), 107–123.

De Werth, D. W. (1962), AGA Research Bulletin 92.

Dickerson, R. W. and Read, R. B. (1968), *Food Technology*, **22**, 1533–1548.

Dickerson, R. W. (1969), Thermal properties of foods, in *Freezing preservation of foods*, 4th ed., Vol. 2, Tressler, D. K., Van Arsdel, W. B. and Copley, M. J. (eds.). Westport, Conn.: AVI.

Fennema, O. R., Pourie, W. D. and Maith, E. H. (1973), *Low temperature preservation of foods*. New York: Dekker.

Heldman, D. R. (1975), *Food process engineering*. Westport, Conn.: AVI.

Jason, A. C. and Jowitt, R. (1969), Physical properties of foodstuffs in relation to engineering design, *Dechema-Monographien*, Vol. 63.

Jowitt, R. (1974), *Lebensmittel-Wiss. und Technol.*, 7 (6), 358–371.

Kramer, A. and Twigg, B. A. (1970), *Quality control for the food industry*, Vol. II, p. 519. Westport, Conn.: AVI.

Morley, M. J. (1972a), *Thermal properties of meat, MRI Symposium No. 2.* Bristol: Meat Research Institute.

Morley, M. J. (1972b), *Thermal properties of meat.* Bristol: Meat Research Institute.

Ohlsson, T. and Bengtsson, N. E. (1975), *J. Microwave Power,* **10**(1), 93-108.

Pearson, D. (1970), *The chemical analysis of foods,* 6th ed. London: Churchill.

Reidel, L. (1957), *Kaltechnic,* **9**(2), 38-40.

Reidel, L. (1969), *Kaltechnic-Klimatierung,* **21**(11), 315-316.

Roebuck, B. D., Goldblith, S. A. and Westphal, W. B. (1972), *J. Food Science,* **37**, 199-204.

Schwartzberg, H. G. (1976), *J. Food Science,* **41**, 152-6.

Seveik, V. J. and Sunderland, J. E. (1962), *Food Technology,* **16**(9), 124-6.

Smith, L. L. W. and Minor. L. J. (1974), *Food Service Science,* Westport, Conn.: AVI.

Smith, J. C., Ede, A. J. and Gane, R. (1952), *Modern Refrigeration,* **55**(Aug.), 494-502.

Sweat, V. E. (1974), *J. Food Science,* **39**, 1080-3.

Sweat, V. E. (1975), *Trans. ASAE,* **18**(3), 564-8.

To, E. C., Mudgett, R. E., Wang, D. I. C., Goldblith, S. A. and Decareau, R. V. (1974), *J. Microwave Power,* **9**(4), 304-15.

Woodams, E. E. and Nowrey, J. E. (1968), *Food Technology,* **22**, 494-502.

APPENDIX H

# Properties of Liquid Foods

## H.1 INTRODUCTION

Many of the general comments relating to difficulties associated with the use of literature values of solid foods' physical properties (Appendix G) also apply to liquid foods. However liquid foods are often homogenous, which leads to greater sample uniformity. The properties covered in this Appendix are density, specific heat (and enthalpy), thermal conductivity, thermal diffusivity, viscosity and microwave dielectric properties.

## H.2 DENSITY

Densities and specific gravities are much easier to measure in the case of liquid foods and are often used for quality control and product characterisation: for these reasons literature values are easier to find for liquid foods than is the case for solid foods. A selection of densities and specific gravities for liquid foods is given in Table H.1.

## H.3 SPECIFIC HEAT, LATENT HEAT AND ENTHALPY

The general comments relating to the specific heat and latent heat of solid foods apply equally to liquid foods. Literature values of latent heat are normally calculated from the foods' water content; but it must be remembered that foods do not have a single freezing point at which all of the latent heat is extracted.

$$L_f = 333.5 X_m \,\text{kJ/kg} .$$

Specific heats for sugar solutions and other liquid foods are given in Table H.2 and enthalpies in Table H.3. Reidel (1951, 1957) has investigated the freezing of sugar solutions, fruit and vegetable juices and egg whites and yolks. Dickerson (1969) reproduces this data for fruit and vegetable juices in the form of a chart relating water content and temperature to % unfrozen water and enthalpy. For

**Table H.1 – Density of liquid foods.**

| | | | |
|---|---|---|---|
| Sugar solution, 20°C | 1 g/100 ml | 1002 kg/m$^3$ | Norish (1969) |
| | 10 g/100 ml | 1037 kg/m$^3$ | Norish (1969) |
| | 20 g/100 ml | 1075 kg/m$^3$ | Norish (1969) |
| | 40 g/100 ml | 1150 kg/m$^3$ | Norish (1969) |
| | 80 g/100 ml | 1300 kg/m$^3$ | Norish (1969) |
| Fruit juices | | 1040–1072 kg/m$^3$ | adapted from Dickerson (1969) |
| Vegetable oils | | 0.927 ± 0.017 (S.G.) | adapted from Pearson (1970) |
| Whole egg | | 1.036 (S.G.) | Pearson (1970) |
| Egg white | | 1.055 (S.G.) | Pearson (1970) |
| Egg yolk | | 1.026 (S.G.) | Pearson (1970) |
| Skim milk | | 1.035 (S.G.) | Pearson (1970) |
| Whole milk | | 1.032 (S.G.) | Pearson (1970) |
| Evaporated milk | | 1.085 (S.G.) | Pearson (1970) |
| Sweetened condensed milk | | 1.305 (S.G.) | Pearson (1970) |

friut juices with more than 50% water, Dickerson (1969) gives the following equation:

$$C_p = 1.67 + (2.51\,X_m)\,\mathrm{kJ/kg.K}\ .$$

**Table H.2 – Specific heats of liquid foods above their freezing point (Adapted from Heldman (1975)).**

| | | |
|---|---|---|
| Whole milk | | 3.85 kJ/kg.K |
| Skim milk | | 3.97 kJ/kg.K |
| Whole egg | | 3.09 kJ/kg.K |
| Sugar solution, 21°C | 5% | 4.06 kJ/kg.K |
| | 10% | 3.93 kJ/kg.K |
| | 15% | 3.85 kJ/kg.K |
| | 20% | 3.60 kJ/kg.K |

**Table H.3 – Specific heats and enthalpies of foods below their freezing point.**

| | Temperature °C | | | | | | |
|---|---|---|---|---|---|---|---|
| | −40 | −20 | −10 | −5 | −1 | 0 | |
| Sugar solutions | | | | | | | |
| specific heat (kJ/kg.K) 5% | 1.71 | 2.17 | – | – | 4.06 | – | adapted from Short *et al.* (1949) |
| 10% | 1.92 | 2.63 | – | – | 3.93 | – | adapted from Short *et al.* (1949) |
| 15% | 1.88 | 2.93 | – | – | 3.85 | – | adapted from Short *et al.* (1949) |
| 20% | 1.92 | 3.22 | – | – | 3.60 | – | adapted from Short *et al.* (1949) |
| Whole egg | | | | | | | |
| enthalpy (kJ/kg) | 0.00 | 36.4 | 60.7 | 80.3 | 174.6 | 280.5 | from Morley (1972) |
| Egg yolk | | | | | | | |
| enthalpy (kJ/kg) | 0.00 | 38.9 | 64.9 | 84.6 | 154.5 | 228.2 | from Morley (1972) |
| Fruit and veg. juice enthalpy (kJ/kg) | | | | | | | |
| 90% water | 0.00 | 39.5 | 86.0 | 123.0 | 332.0 | 367.0 | calculated from Dickerson (1969) |
| 80% water | 0.00 | 51.0 | 114.0 | 186.0 | 325.0 | 332.0 | calculated from Dickerson (1969) |
| 70% water | 0.00 | 116.0 | 144.0 | 248.0 | 288.0 | – | calculated from Dickerson (1969) |

Enthalpy datum: −40°C.

## H.4 THERMAL CONDUCTIVITY AND THERMAL DIFFUSIVITY

Thermal conductivities for a number of liquid foods and biological materials are contained in a survey of reported values by Woodams and Nowrey (1968). Lentz (1961) gives values of the thermal conductivity of gelatin gels and of mixtures of gelatin gels and butterfat between −25 and 0°C. The results are compared with theoretical models for materials with a dispersed spherical phase. Morley (1972) gives thermal conductivities for egg products. An equation for predicting the thermal conductivity of fruit juices is given by Dickerson (1969),

$$\lambda = (326 + 1.04\theta - 0.00304\theta^2)(0.80 + 0.01W) \times 10^{-3}\,\mathrm{W/m.K}$$

where $\theta$ is the temperature in °C and $W$ is the water content. Values of thermal conductivity for a range of liquid goods is given in Table H.4. Thermal diffusivities for several fruit juices are given by Dickerson (1969) and for eggs by Morley (1972). These figures are summarised in Table H.5.

**Table H.5 – Thermal diffusivity of some liquid foods.**

| | | | |
|---|---|---|---|
| Fruit juices | | $0.132 \times 10^{-6}$ to $0.137 \times 10^{-6}\,\mathrm{m^2/s}$ | calculated from Dickerson (1969) |
| Whole egg | 5°C | $0.127 \times 10^{-6}\,\mathrm{m^2/s}$ | from Morley (1972) |
| | 45°C | $0.142 \times 10^{-6}\,\mathrm{m^2/s}$ | from Morley (1972) |
| | 65°C | $0.145 \times 10^{-6}\,\mathrm{m^2/s}$ | from Morley (1972) |

## H.5 VISCOSITY OF FOODS

Values for the viscosity of Newtonian and non-Newtonian foods are given by Heldman (1975) and Muller (1973). Viscosity related to the sensory properties of foods is discussed by Sherman (1977). Values for the viscosity of Newtonian foods are given in Table H.6. Care must be used in the interpretation of literature values of viscosity because measurement is often made at a single rate of shear and may not be valid for other rates. Table H.7 gives flow properties for some non-Newtonian foods in the form of power law constants:

$$\tau = c + bD^s$$

where $c$ is the yield value, $b$ the consistency index and $s$ the flow behaviour index.

**Table H.4 – Thermal conductivities of some liquid foods.**

| | | | | |
|---|---|---|---|---|
| Fruit juices | 10–15% solids | 20°C | 0.559 W/m.K | calculated from Woodams *et al.* (1968) |
| | 10–15% solids | 80°C | 0.609 W/m.K | calculated from Woodams *et al.* (1968) |
| | 30–40% solids | 20°C | 0.493 W/m.K | calculated from Woodams *et al.* (1968) |
| | 30–40% solids | 80°C | 0.551 W/m.K | calculated from Woodams *et al.* (1968) |
| | 60–65% solids | 20°C | 0.396 W/m.K | calculated from Woodams *et al.* (1968) |
| | 60–65% solids | 80°C | 0.441 W/m.K | calculated from Woodams *et al.* (1968) |
| Whole milk | | 1°C | 0.554 W/m.K | calculated from Woodams *et al.* (1968) |
| | | 20°C | 0.550 W/m.K | calculated from Woodams *et al.* (1968) |
| | | 50°C | 0.587 W/m.K | calculated from Woodams *et al.* (1968) |
| | | 80°C | 0.614 W/m.K | calculated from Woodams *et al.* (1968) |
| Vegetable oils | | 0 to 25°C | 0.183 ± 0.008 W/m.K | calculated from Woodams *et al.* (1968) |
| Egg white | | 33 to 38°C | 0.557 W/m.K | from Morley (1972) |
| Egg yolk | | 24 to 38°C | 0.421 W/m.K | from Morley (1972) |

**Table H.6 – Viscosities of Newtonian fluids.**

| | | | | |
|---|---|---|---|---|
| Sucrose | 20°C | 20% | 0.002 N.s/m$^2$ | adapted from Norish (1967) |
| | 20°C | 40% | 0.006 N.s/m$^2$ | adapted from Norish (1967) |
| | 20°C | 60% | 0.058 N.s/m$^2$ | adapted from Norish (1967) |
| | 20°C | 70% | 0.48 N.s/m$^2$ | adapted from Norish (1967) |
| Milk | 20°C | | 0.00146 N.s/m$^2$ | from Muller (1973) |
| Salad oils | 20°C | | 0.07 to 0.16 N.s/m$^2$ | from Muller (1973) |
| Frying oil | 25°C | | 0.073 N.s/m$^2$ | Kirk (unpublished data) |
| | 80°C | | 0.030 N.s/m$^2$ | Kirk (unpublished data) |
| Apple juice | 27°C | 20°Brix | 0.0021 N.s/m$^2$ | adapted from Heldman (1975) |
| | 27°C | 60°Brix | 0.03 N.s/m$^2$ | adapted from Heldman (1975) |
| Cream | 3°C | 20% fat | 0.0062 N.s/m$^2$ | adapted from Heldman (1975) |

**Table H.7 – Flow properties of non-Newtonian fluids.**

| | | | $c (N/m^2)$ | $b (N.s/m^2)$ | s | |
|---|---|---|---|---|---|---|
| Mayonnaise | 25°C | | 67 | 14 | 0.37 | Kirk (unpublished data) |
| Custard | 80°C | | – | 7.24 | 0.36 | Kirk (unpublished data) |
| Gravy | 80°C | | – | 2.88 | 0.39 | Kirk (unpublished data) |
| Apple puree | 25°C | | 55 | 50 | 0.23 | Kirk (unpublished data) |
| Sausage meat | – | | – | 586 | 0.36 | adapted from Hawkins (1971) |
| Pear puree | 27°C | 18.3% solids | – | 2.25 | 0.486 | adapted from Heldman (1975) |
| | 27°C | 45.8% solids | – | 35.5 | 0.479 | adapted from Heldman (1975) |

## H.6 DIELECTRIC PROPERTIES OF LIQUID FOODS

In addition to the references given in Appendix G (Section G.7), investigations into the dielectric properties of solutions of non-fat dried milk (Mudgett *et al.* 1974a) and oil and water mixtures (Mudgett *et al.* 1974b) have been described. The dielectric properties of vegetable soup and gravy are given in Table H.8.

## H.7 REFERENCES

Bengtsson, N. E. and Risman, P. O. (1971), *J. Microwave Power*, **6**(2), 107-123.

Dickerson, R. W. (1969), Thermal properties of foods, in *Freezing preservation of foods*, 4th ed., Vol. 2, Tressler, D. K., Van Arsdel, W. B. and Copley, M. J. (eds.). Westport, Conn.: AVI.

Hawkins, A. E. (1971), *The Chemical Engineer*, No. 245, 19-23.

Heldman, D. R. (1975), *Food Process Engineering*. Westport, Conn.: AVI.

Lentz, C. P. (1961), *Food Technology*, **15**(5), 243-7.

Mudgett, R. E., Smith, A. C., Wang, D. I. C. and Goldblith, S. A. (1974a), *J. Food Science*, **39**, 52-4.

Mudgett, R. E., Wang, D. I. C. and Goldblith, S. A. (1974b), *J. Food Science*, **39**, 632-5.

Morley, M. J. (1972), *Thermal properties of meat*. Bristol: Meat Research Institute.

Muller, H. G. (1973), *An introduction to food rheology*. London: Heinemann.

Ohlsson, T. and Bengtsson, N. E. (1975), *J. Microwave Power*, **10**(1), 93-108.

Norish, R. S. (1967), Selected tables of physical properties of sugar solutions, *Scientific and Technical Survey No. 51*. Leatherhead: BFMIRA.

Pearson, D. (1970), *The chemical analysis of foods*, 6th ed. London: Churchill.

Reidel, L. (1951), *Refrigeration Engineer*, **59**, 670-3.

Reidel, L. (1957), *Kaltechnik*, **9**(11), 342-5.

Woodhams, E. E. and Nowrey, J. E. (1968), *Food Technology*, **22**, 150-8.

**Table H.8 – Dielectric properties of liquid foods.**

| | | 3.0 GHz: Bengtsson *et al.* (1971) | | | | 2.8 GHz: Ohlsson *et al.* (1975) | | | |
|---|---|---|---|---|---|---|---|---|---|
| | | Temperature °C | | | | Temperature °C | | | |
| | | −20 | −10 | 0 | +20 | +40 | +60 | +80 | +100 |
| Gravy | $\epsilon_r'$ | 5.2 | 9.2 | 76.2 | 73.4 | 76.4 | 75.5 | 73.6 | 70.4 |
| | $\epsilon_r''$ | 1.3 | 4.3 | 27.7 | 26.4 | 24.1 | 26.6 | 26.2 | 26.3 |
| | $\alpha$ | 14.5 | 35.45 | 80.12 | 77.83 | 69.88 | 77.37 | 77.16 | 79.08 |
| | $P_{\frac{1}{e}}$ | 0.040 | 0.014 | 0.006 | 0.007 | 0.007 | 0.007 | 0.007 | 0.006 |
| Vegetable soup | $\epsilon_r'$ | 3.9 | 5.1 | 74.7 | 70.2 | 65.8† | 63.6† | | |
| | $\epsilon_r''$ | 0.29 | 0.9 | 23.9 | 17.5 | 16.9† | 18.2† | | |
| | $\alpha$ | 3.76 | 10.18 | 70.06 | 53.17 | 53.01† | 57.96† | | |
| | $P_{\frac{1}{e}}$ | 0.133 | 0.05 | 0.008 | 0.009 | 0.009† | 0.009† | | |

† 3.0 GHz.
Values of attenuation coefficient ($\alpha$) and $P_{\frac{1}{e}}$ are calculated from the data of Ohlsson and Bengtsson.

APPENDIX I

# Properties of Fuel Gases, Air and Combustion Products

**I.1 NET CALORIFIC VALUES** (at 15°C and atmospheric pressure):

| | |
|---|---|
| Methane | 33.9 $MJ/m^3$ |
| Propane | 86.42 $MJ/m^3$ |
| Butane | 112.41 $MJ/m^3$ |

**I.2 GROSS CALORIFIC VALUES** (at 15°C and atmospheric pressure):

| | |
|---|---|
| Methane | 37.71 $MJ/m^3$ |
| Propane | 93.94 $MJ/m^3$ |
| Butane | 121.80 $MJ/m^3$ |

**I.3 SPECIFIC GRAVITIES** (air = 1):

| | |
|---|---|
| Methane | 0.55 |
| Propane | 1.52 |
| Butane | 2.01 |
| Oxygen | 1.10 |
| Nitrogen | 0.97 |
| Carbon dioxide | 1.52 |
| Water vapour | 0.62 |

**I.4 DENSITIES** (at 15°C and atmospheric pressure):

| | |
|---|---|
| Oxygen | 1.353 $kg/m^3$ |
| Nitrogen | 1.185 $kg/m^3$ |
| Air | 1.225 $kg/m^3$ |
| Carbon dioxide | 1.861 $kg/m^3$ |
| Water vapour | 0.762 $kg/m^3$ |

Densities at $t$°C may be obtained by multiplying the data above by $(283/(t + 273))$.

Source of data in Sections I.1, I.2, I.3 and I.4: Rose, J. W. and Cooper, J. R. (ed.) (1977), *Technical data of fuel*, British National Committee, World Energy Conference, 7th ed., Table 5.14.

Note that the above properties are additive, that is the property of a mixture may be calculated from the properties of its constituents in relation to the proportions present.

## I.5 MEAN SPECIFIC HEATS OF GASES AT CONSTANT PRESSURE BETWEEN 0°C and *T*°C

Units: kJ/m³.K where the volume of the gas is expressed as at 0°C and 760mm Hg.

| Temperature °C | Oxygen | Nitrogen | Air | Carbon dioxide | Water vapour |
|---|---|---|---|---|---|
| 100 | 1.32 | 1.30 | 1.30 | 1.71 | 1.50 |
| 200 | 1.34 | 1.31 | 1.31 | 1.80 | 1.52 |
| 300 | 1.36 | 1.31 | 1.32 | 1.88 | 1.54 |
| 400 | 1.38 | 1.32 | 1.33 | 1.94 | 1.56 |
| 500 | 1.40 | 1.34 | 1.34 | 2.01 | 1.58 |
| | $CH_4$ | $C_3H_6$ | $C_4H_{10}$ | | |
| 0 | 1.49 | 2.96 | 3.71 | | |
| 100 | 1.69 | 3.36 | 4.23 | | |

Adapted from: Spiers, H. M. (ed.) (1962), *Technical data on fuel*, British National Committee, World Power Conference, London, 6th ed.

Note that mean specific heat may be regarded as an additive property.

## I.6 VISCOSITIES (dynamic)

Units: mg/m.s (that is multiply number by $10^{-6}$ to change value to basis of kg, m and s).

| Temperature °C | Oxygen | Nitrogen | Air | Carbon dioxide | Water vapour |
|---|---|---|---|---|---|
| 100 | 24.5 | 21.2 | 22.0 | 18.6 | 12.2 |
| 150 | 27.0 | 23.3 | 24.2 | 20.9 | 14.2 |
| 200 | 29.4 | 25.2 | 26.2 | 23.1 | 16.3 |
| 250 | 31.6 | 27.0 | 28.1 | 25.1 | 18.4 |
| 300 | 33.7 | 28.7 | 29.9 | 27.1 | 20.5 |
| 350 | 35.7 | 30.3 | 31.7 | 29.0 | 22.6 |
| 400 | 37.6 | 31.9 | 33.3 | 30.8 | 24.6 |
| 450 | 39.5 | 33.4 | 34.8 | 32.5 | 26.6 |
| 500 | 41.3 | 34.9 | 36.4 | 34.1 | 28.6 |

Derived from: Rose, J. W. and Cooper, J. R., loc. cit., Table 4.10.

For a method of calculating the viscosities of gas mixtures, see Wilke, C. R. (1950), *Journal of Chemical Physics*, **18**(4), 517-519.

## I.7 THERMAL CONDUCTIVITIES

Units: mW/m.K.

| Temperature °C | Oxygen | Nitrogen | Air | Carbon dioxide | Water vapour |
|---|---|---|---|---|---|
| 100 | 31.9 | 31.1 | 31.8 | 22.4 | 25.0 |
| 150 | 35.4 | 34.0 | 35.3 | 26.3 | 28.9 |
| 200 | 38.8 | 36.8 | 38.6 | 30.3 | 34.2 |
| 250 | 42.1 | 39.6 | 41.7 | 34.1 | 39.2 |
| 300 | 45.3 | 42.3 | 44.6 | 37.9 | 44.3 |
| 350 | 48.5 | 44.9 | 47.4 | 41.5 | 48.9 |
| 400 | 51.7 | 47.5 | 50.2 | 45.1 | 53.8 |
| 450 | 54.9 | 50.0 | 53.0 | 48.7 | 59.4 |
| 500 | 58.1 | 52.3 | 55.7 | 52.1 | 64.8 |

Derived from: Rose, J. W. and Cooper, J. R., loc. cit., Table 4.12.

For a method of calculating the thermal conductivity of gas mixtures see Mason, E. A. and Saxena, S. C. (1958), *The Physics of fluids*, **1**(5), 361-369.

## I.8 SPECIFIC HEATS

Units: $kJ/m^3.K$.

| Temperature °C | Oxygen | Nitrogen | Air | Carbon dioxide | Water vapour |
|---|---|---|---|---|---|
| 100 | 0.98 | 0.95 | 0.95 | 1.31 | 1.11 |
| 150 | 0.87 | 0.83 | 0.84 | 1.21 | 0.99 |
| 200 | 0.79 | 0.75 | 0.76 | 1.13 | 0.90 |
| 250 | 0.73 | 0.69 | 0.70 | 1.06 | 0.83 |
| 300 | 0.67 | 0.63 | 0.64 | 0.99 | 0.77 |
| 350 | 0.62 | 0.59 | 0.60 | 0.93 | 0.71 |
| 400 | 0.58 | 0.55 | 0.56 | 0.88 | 0.67 |
| 450 | 0.55 | 0.52 | 0.52 | 0.84 | 0.63 |
| 500 | 0.52 | 0.49 | 0.50 | 0.80 | 0.60 |

Derived from: Rose, J. W. and Cooper, J. R., loc. cit., Table 4.6.

## I.9 PHYSICAL PROPERTIES OF DRY AIR AT ATMOSPHERIC PRESSURE

| Temperature | Density | Specific heat | Dynamic viscosity | Kinematic viscosity | Thermal conductivity | Prandtl number |
|---|---|---|---|---|---|---|
| $\theta$ | $\rho$ | $C_{(p)}$ | $\eta$ | $\nu \equiv \eta/\rho$ | $\lambda$ | $Pr$ |
| °C | kg/m$^3$ | $10^3$ J/kg.K | $10^{-5}$ kg/m.s | $10^{-5}$ m$^2$/s | $10^{-2}$ W/m.K | – |
| −40 | 1.51 | 1.005 | 1.51 | 1.00 | 2.09 | 0.73 |
| −30 | 1.45 | 1.005 | 1.56 | 1.08 | 2.17 | 0.72 |
| −20 | 1.40 | 1.005 | 1.61 | 1.16 | 2.25 | 0.72 |
| −10 | 1.34 | 1.006 | 1.67 | 1.24 | 2.33 | 0.72 |
| 0 | 1.30 | 1.006 | 1.72 | 1.33 | 2.41 | 0.72 |
| 10 | 1.25 | 1.006 | 1.76 | 1.42 | 2.49 | 0.71 |
| 20 | 1.20 | 1.007 | 1.81 | 1.51 | 2.57 | 0.71 |
| 30 | 1.16 | 1.007 | 1.86 | 1.60 | 2.65 | 0.71 |
| 40 | 1.13 | 1.008 | 1.90 | 1.69 | 2.72 | 0.71 |
| 50 | 1.09 | 1.008 | 1.95 | 1.78 | 2.80 | 0.70 |
| 60 | 1.06 | 1.009 | 2.00 | 1.88 | 2.88 | 0.70 |
| 70 | 1.03 | 1.010 | 2.04 | 1.98 | 2.95 | 0.70 |
| 80 | 1.00 | 1.010 | 2.08 | 2.08 | 3.02 | 0.70 |
| 90 | 0.97 | 1.011 | 2.13 | 2.19 | 3.10 | 0.70 |
| 100 | 0.94 | 1.012 | 2.17 | 2.30 | 3.18 | 0.69 |
| 110 | 0.92 | 1.013 | 2.21 | 2.40 | 3.25 | 0.69 |
| 120 | 0.90 | 1.014 | 2.26 | 2.51 | 3.32 | 0.69 |
| 300 | 0.62 | 1.050 | 2.94 | 4.75 | 4.5 | 0.69 |

## I.10 DIFFUSION COEFFICIENT, *Sc* AND *Pr* NUMBERS OF STEAM-AIR AT ATMOSPHERIC PRESSURE

| Temperature | $D$ | $Sc$ | $Pr$ |
|---|---|---|---|
| °C | $10^{-5}$ m$^2$/s | | |
| 0 | 2.2 | 0.60 | 0.72 |
| 20 | 2.5 | 0.60 | 0.71 |
| 40 | 2.8 | 0.60 | 0.71 |
| 60 | 3.1 | 0.61 | 0.70 |
| 80 | 3.4 | 0.61 | 0.70 |
| 100 | 3.8 | 0.61 | 0.69 |

## I.11 PROPERTIES OF AIR SATURATED WITH WATER VAPOUR AT ATMOSPHERIC PRESSURE

| Temperature | Vapour pressure | Humidity | Enthalpy |
|---|---|---|---|
| °C | $10^3$ kg/m.s$^2$ | kg water per kg dry air | kJ per kg dry air |
| 0 | 0.611 | 0.00382 | 9.59 |
| 10 | 1.23 | 0.00773 | 29.5 |
| 20 | 2.33 | 0.0149 | 57.9 |
| 30 | 4.24 | 0.0276 | 101 |
| 40 | 7.37 | 0.0495 | 168 |
| 50 | 12.3 | 0.0875 | 277 |
| 60 | 19.9 | 0.155 | 464 |
| 70 | 31.2 | 0.282 | 810 |
| 80 | 47.4 | 0.559 | 1560 |
| 90 | 70.1 | 1.46 | 3980 |
| 100 | 101.3 | ∞ | ∞ |

Source of data in Sections I.9, I.10, I.11: Ede, A. J. (1967), *An introduction to heat-transfer principles and calculations*. Pergamon Press.

# Properties of Construction Materials

## J.1 INSULATING MATERIALS

For the thermal conductivity of these, see Section 4.4. Some data are also given in Rose, J. W. and Cooper, J. R. (eds.) (1977), *Technical data on fuel*, British National Committee, World Energy Conference, 7th ed. This property is usually dependent on the bulk density of the material, which can be altered by the manufacturer in most cases. It is therefore advisable to check with him regarding the thermal properties of his products.

## J.2 METALS

The thermal properties of three relevant pure metals are given in Table J.1.

The properties of an 18/8 chrome-nickel stainless steel are as follows:

| | | |
|---|---|---|
| Linear thermal expansion coefficient: | 0–100°C | $14.82 \times 10^{-6}\ K^{-1}$ |
| | 0–200°C | $16.47 \times 10^{-6}\ K^{-1}$ |
| | 0–300°C | $17.11 \times 10^{-6}\ K^{-1}$ |
| | 0–400°C | $17.61 \times 10^{-6}\ K^{-1}$ |
| | 0–500°C | $18.04 \times 10^{-6}\ K^{-1}$ |

| | | |
|---|---|---|
| Mean specific heat: | 0–100°C | 0.502 kJ/kg.K |
| | 100–200°C | 0.528 kJ/kg.K |
| | 200–300°C | 0.544 kJ/kg.K |
| | 300–400°C | 0.565 kJ/kg.K |
| | 400–500°C | 0.590 kJ/kg.K |

| | | |
|---|---|---|
| Thermal conductivity: | 0°C | 15.1 W/m.K |
| | 100°C | 15.9 W/m.K |
| | 200°C | 17.2 W/m.K |
| | 300°C | 18.4 W/m.K |
| | 400°C | 20.1 W/m.K |
| | 500°C | 21.8 W/m.K |

Source: Rose, J. W. and Cooper, J. R. (eds.), loc. cit.

**Table J.1**

| Metal | Melting point (°C) | Coefficient of linear thermal expansion at 20°C ($K^{-1}$) | Thermal conductivity (W/m.K) | | | | Specific enthalpy (kJ/kg) above 15°C | | |
|---|---|---|---|---|---|---|---|---|---|
| | | | 0°C | 100°C | 300°C | 500°C | 100°C | 300°C | 500°C |
| Aluminium | 660 | $23.5 \times 10^{-6}$ | 236 | 240 | 235 | 221 | 91.3 | 291.2 | 511.2 |
| Copper | 1085 | $17.0 \times 10^{-6}$ | 401 | 393 | 384 | 372 | 38.8 | 120.7 | 205.3 |
| Iron | 1540 | $12.1 \times 10^{-6}$ | 84 | 72 | 57 | 45 | 47.0 | 152.0 | 276.0 |

## J.3 EMISSIVITIES OF MATERIALS

Emissivity values depend very much on the surface conditions and values given in reference books are only approximate. Most non-metallic surfaces have high emissivities and are usually considered grey. Polished metals have lower emissivities than non-metallic surfaces and their emissivity can vary considerably with wavelength. Overall emissivities are however usually listed. Dirt on the surface and oxidation of the surface makes for considerable increases in the surface's emissivities. Emissivities do not vary greatly over small temperature ranges. However, if a high-temperature source is radiating to a much cooler surface, it is usually necessary to use an emissivity corresponding to the lower temperature and an absorptivity corresponding to the higher temperature.

| *Material* | | *Emissivity* |
|---|---|---|
| Aluminium, | polished | 0.095 |
| | oxidised | 0.10-0.20 |
| Brass, | polished | 0.03 |
| | dull plate | 0.60 |
| Copper, | polished | 0.02 |
| | oxidised | 0.80 |
| Iron, | polished | 0.20 |
| | oxidised | 0.70 |
| Steel, | polished | 0.07 |
| | oxidised | 0.80 |
| Stainless steels | | 0.2 to 0.7 |
| Glass | | 0.90-0.95 |
| White paint | | 0.80-0.95 |
| Black paint | | 0.80-0.95 |

Compilations of data are given in:

Rose, J. W. and Cooper, J. R. (eds.) (1977), *Technical data on fuel*, British National Committee, World Power Conference, 7th ed.
Sparrow, E. M. and Cess, R. D. (1970), *Radiation heat transfer*. Brooks/Cole.
Singham, J. R. (1962), *Int. jnl. heat mass-transfer*, **5**, 67-76.

# Solutions to Problems

**CHAPTER 2: PROBLEM (1)**

**(a)**

**50% excess air**

Excess $O_2$: $2 \times \frac{50}{100} = 1.00$ .

Excess $N_2$: $7.52 \times \frac{50}{100} = 3.76$ .

Volume of combustion air/$m^3$ gas burnt $= 9.52 + 1.00 + 3.76 = 14.28\ m^3$ .

Volume of combustion products/$m^3$ gas burnt $= 10.52 + 1.00 + 3.76 = 15.28 m^3$ .

Composition of wet combustion products:

| | | | | |
|---|---|---|---|---|
| $CO_2$ | 1 | = | 1 $m^3$ | 6.5% |
| $H_2O$ | 2 | = | 2 $m^3$ | 13.1% |
| $N_2$ | 7.52 + 3.76 | = | 11.28 $m^3$ | 73.9% |
| $O_2$ | 1.00 | = | 1.00 $m^3$ | 6.5% |
| | | | 15.28 $m^3$ | 100.0% |

Composition of dry combustion products:

| | | |
|---|---|---|
| $CO_2$ | 1 $m^3$ | 7.5% |
| $N_2$ | 11.28 $m^3$ | 85.0% |
| $O_2$ | 1.00 $m^3$ | 7.5% |
| | 13.28 $m^3$ | 100.0% |

Heat lost in combustion products =

$$(1 \times 1.80 \times 10^3 \times 180) + (2 \times 1.52 \times 10^3 \times 180) + (11.28 \times 1.31 \times 10^3 \times 180) + (1.00 \times 1.34 \times 10^3 \times 180)\,\text{J} = 3.77\,\text{MJ}\,.$$

This heat loss as a proportion of net calorific value =

$$\frac{3.77}{35.77} \times 100\% = 10.5\%\,.$$

Volume of combustion products at 200°C =

$$15.28 \times \frac{(273 + 200)}{(273 + 20)} = 24.67\,\text{m}^3/\text{m}^3 \text{ gas burnt}\,.$$

**100% excess air**

Excess $O_2$: $2 \times \frac{100}{100} = 2.00\,.$

Excess $N_2$: $7.52 \times \frac{100}{100} = 7.52\,.$

Volume of combustion air/m$^3$ gas burnt $= 9.52 + 2.00 + 7.52 = 19.04\,\text{m}^3$ .

Volume of combustion products/m$^3$ gas burnt $= 10.52 + 2.00 + 7.52 = 20.04\,\text{m}^3$.

Composition of wet combustion products:

| | | | | |
|---|---|---|---|---|
| $CO_2$ | 1 | = | 1 m$^3$ | 5.0% |
| $H_2O$ | 2 | = | 2 m$^3$ | 10.0% |
| $N_2$ | 7.52 + 7.52 | = | 15.04 m$^3$ | 75.0% |
| $O_2$ | 2.00 | = | 2.00 m$^3$ | 10.0% |
| | | | 20.04 m$^3$ | 100.0% |

Composition of dry combustion products:

| | | |
|---|---|---|
| $CO_2$ | 1 m$^3$ | 5.5% |
| $N_2$ | 15.04 m$^3$ | 83.4% |
| $O_2$ | 2.00 m$^3$ | 11.1% |
| | 18.04 m$^3$ | 100.0% |

Heat lost in combustion products =

$$(1 \times 1.80 \times 10^3 \times 180) + (2 \times 1.52 \times 10^3 \times 180) + (15.04 \times 1.31 \times 10^3 \times 180) + (2.00 \times 1.34 \times 10^3 \times 180)\,\text{J} = 4.90\,\text{MJ}\,.$$

The heat loss as a proportion of net calorific value =

$$\frac{4.90}{35.77} \times 100\% = 13.7\% .$$

Volume of combustion products at 200°C =

$$20.04 \times \frac{(273 + 200)}{(273 + 20)} = 32.35 \text{ m}^3/\text{m}^3 \text{ gas burnt.}$$

*Summary of results of calculations:*

| | | | | |
|---|---|---|---|---|
| Excess air (%) | 0 | 25 | 50 | 100 |
| Volume combustion air ($m^3/m^3$) | 9.52 | 11.90 | 14.28 | 19.04 |
| Volume combustion products ($m^3/m^3$) | 10.52 | 12.90 | 15.28 | 20.04 |
| $CO_2$ in dry products (%) | 11.7 | 9.2 | 7.5 | 5.5 |
| $O_2$ in dry products (%) | 0.0 | 4.6 | 7.5 | 11.1 |
| $H_2O$ in wet products (%) | 19.0 | 15.5 | 13.1 | 10.0 |
| Heat in combustion products ($MJ/m^3$) | 2.64 | 3.21 | 3.77 | 4.90 |
| Heat in combustion products (%) | 7.4 | 8.9 | 10.5 | 13.7 |
| Volume of combustion products at 200°C ($m^3/m^3$) | 16.98 | 20.82 | 24.67 | 32.35 |

**(b)**

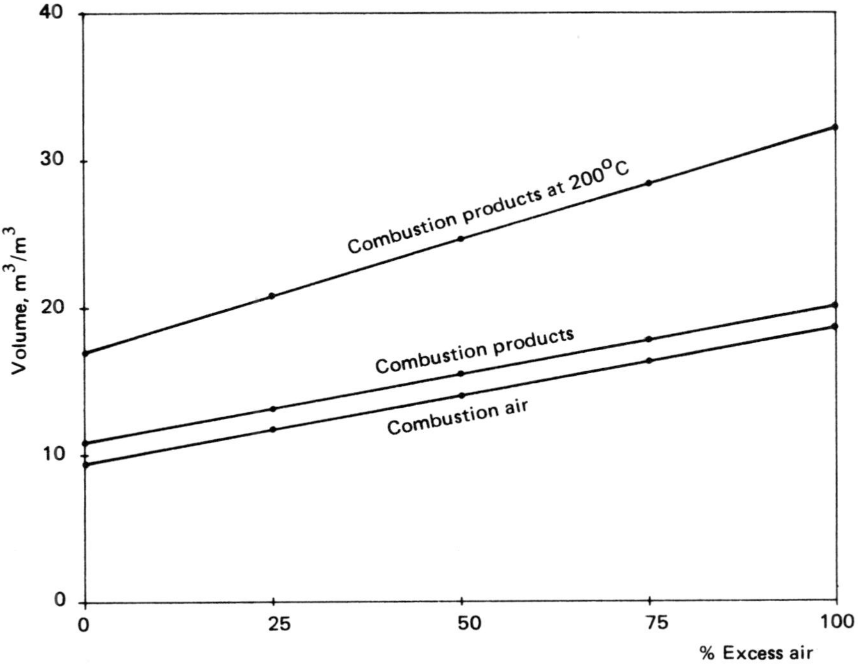

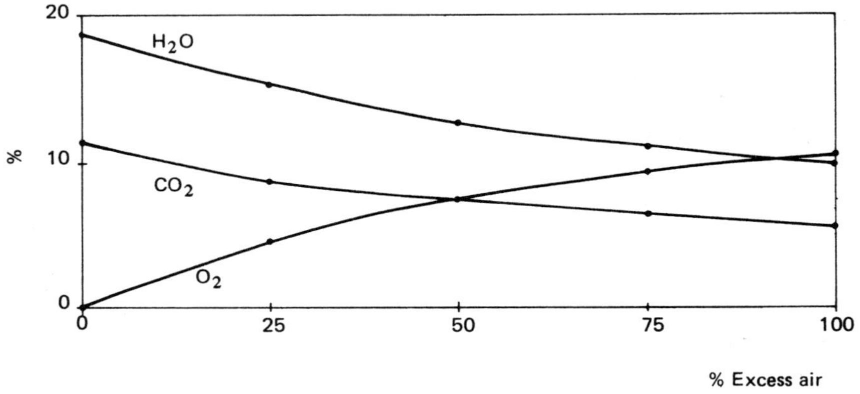


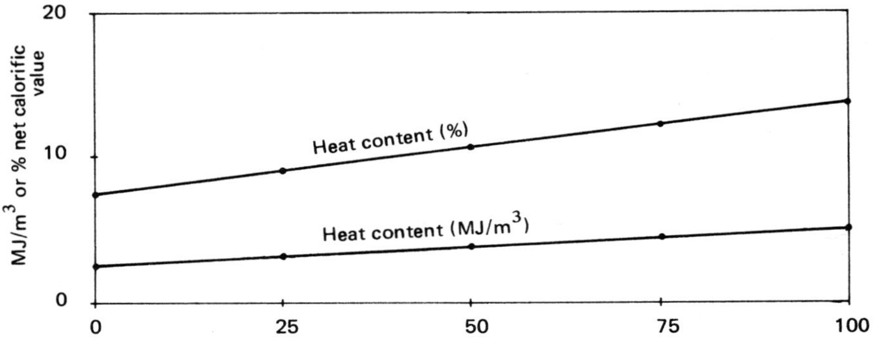


**(c)**

The significance of the calculations is dealt with in Sections 2.4, 2.5 and 2.6.

## CHAPTER 2: PROBLEM (2)

(a)

Theoretical energy consumption/kg of carrots =

specific heat of carrots + latent heat lost in water vapour
$= [1 \times 3720 \times (100 - 0)] + [0.036 \times 2\,257\,000]$ J
$= (372\,000 + 81\,252)$ J
$= 453.3$ kJ .

(b)

Energy used to raise steam in generator $= (0.85 \times 3.6)$ MJ
$= 3.06$ MJ .

Total energy used =

energy used in generator + energy used in cooking compartment.

This gives energy consumptions as follows:

| | Batch (kg) | Cooking energy (kWh) | Cooking energy (MJ) | Energy used in steam generator (MJ) | Total energy used (MJ) | Total energy used (MJ/kg) | Efficiency (%) |
|---|---|---|---|---|---|---|---|
| (i) | 1 X 0.5 | 0.2 | 0.72 | 3.06 | 3.78 | 7.56 | 6.0 |
| (ii) | 4 X 0.5 | 0.8 | 2.88 | 3.06 | 5.94 | 2.97 | 15.3 |
| (iii) | 1 X 1.0 | 0.3 | 1.08 | 3.06 | 4.14 | 4.14 | 10.9 |
| (iv) | 2 X 1.0 | 0.6 | 2.16 | 3.06 | 5.22 | 2.61 | 17.4 |
| (v) | 1 X 2.0 | 0.4 | 1.44 | 3.06 | 4.50 | 2.25 | 20.1 |
| (vi) | 4 X 2.0 | 1.6 | 5.76 | 3.06 | 8.82 | 1.10 | 41.2 |
| (vii) | 8 X 2.0 | 3.2 | 11.52 | 3.06 | 14.58 | 0.91 | 49.8 |

$$\text{Efficiency} = \frac{\text{theoretical energy consumption}}{\text{total energy used}} \times 100\% .$$

(c)

Comparison of the efficiencies for (i), (iii) and (v) shows that the larger the batch size, the greater the efficiency.

Comparison of the efficiencies for (i) and (ii), (iii) and (iv), and (v), (vi) and (vii) shows that the larger the number of batches the greater the efficiency.

See also the discussion of Section 2.8.

**CHAPTER 3: PROBLEM (1)**

For a power law fluid in laminar flow (Section 3.7):

$$Q = \frac{s\pi r^3}{3s+1}\left(\frac{r\Delta p}{2lb}\right)^{1/s}$$

$$= \frac{0.5 \times 3.142 \times 0.02^3}{1.5+1}\left(\frac{0.02 \times 10\,000}{2 \times 5 \times 0.6}\right)^{1/0.5} \frac{m^3}{s}$$

($s = 0.5, r = 2\,\text{cm} = 0.02\,\text{m}, \Delta p = 10\,000\,\text{N/m}^2, l = 5\,\text{m}$ and $b = 0.6$.)

$$= 0.000\,005 \times 1111\,\text{m}^3/\text{s}$$

$$= 0.0055\,\text{m}^3/\text{s}$$

$$= 0.0055 \times 3600\,\text{m}^3/\text{h}$$

$$= 19.8\,\text{m}^3/\text{h}.$$

**CHAPTER 3: PROBLEM (2)**

The pressure drop is calculated from the equation which makes allowances for both straight pipe and fittings (Section 3.7):

$$\Delta p = \left(4C_f\frac{l}{d} + q\right)\frac{\rho u^2}{2}.$$

**Pressure drop in straight pipe**

$C_f$ is obtained from Fig. A.1 but Reynolds number ($Re$) and relative roughness factor ($\epsilon$) are required.

$$Re = \frac{du\rho}{\mu}.$$

$$d = 2.5\,\text{cm} = 0.025\,\text{m}.$$

$$u = \frac{Q}{A}$$

$$Q = 0.7\,\text{m}^3/\text{h} = \frac{0.7}{3600}\,\text{m}^3/\text{s}.$$

$$A = \frac{\pi d^2}{4} = \frac{\pi 0.025^2}{4}$$

therefore

$$u = \frac{0.7}{3600} \times \frac{4}{\pi 0.025^2} \frac{\text{m}}{\text{s}} = 0.40 \frac{\text{m}}{\text{s}}$$

$$\rho = 1000\,\text{kg/m}^3$$

$$\mu = 1\,\text{g/m.s} = 0.001\,\text{kg/m.s}$$

therefore

$$Re = \frac{0.025 \times \text{m} \times 0.40 \times \text{m} \times 1000 \times \ \text{kg} \ \times \text{m} \times \text{s}}{\text{s} \times \ \text{m}^3 \ \times 0.001 \times \ \text{kg}}$$

$$= \frac{0.025 \times 0.40 \times 1000}{0.001} \times \frac{\text{m} \times \text{m} \times \text{kg} \times \text{m} \times \text{s}}{\text{s} \times \text{m}^3 \times \text{kg}}$$

$$= 10\,000 \text{ (dimensionless).}$$

Relative roughness = 0.046/25 = 0.0018 (using $\epsilon = b/d$ where $d = 25$ mm and $b = 0.046$ (Table A.1)).

From Fig. A.1 (plot of $C_f$ against $Re$), $C_f = 0.0081$ ($Re = 10^4$, $\epsilon = 0.0018$).

**Pressure drop in fittings**

Abrupt contraction (Table A.2):

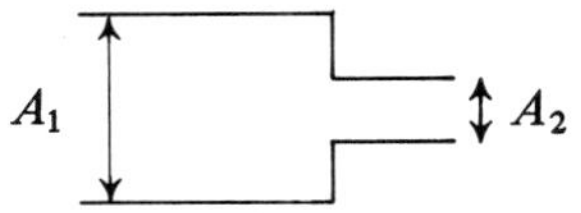


$$q = 0.45\,(1 - A_2/A_1) \text{ from Table A.2}$$

$$A_1 = \frac{\pi d_1^2}{4} \quad \text{and} \quad A_2 = \frac{\pi d_2^2}{4}$$

therefore

$$\frac{A_2}{A_1} = \frac{d_2^2}{d_1^2} = \frac{1^2}{2^2} = 0.25$$

therefore

$$q = 0.45\left(1 - \frac{A_2}{A_1}\right) = 0.45\,(1 - 0.25)$$

$$= 0.338\,.$$

From Table A.2: 90° bends: $r = 2$ in, $D = 1$ in, $r/D = 2.0$ and therefore $q = 0.40$.

$$\text{Adding } q \text{ for all fittings} = 0.338 + 0.40 + 0.40 + 0.40 + 0.40$$

$$= 1.938\,.$$

**Pressure drop for straight pipe and fittings**

$$\Delta p = \left(4C_f\frac{l}{d} + q\right)\frac{\rho u^2}{2}$$

$$= \left(4 \times 0.0081 \times \frac{20}{0.025} + 1.938\right)\frac{1000 \times 0.40^2}{2}\,\frac{\text{N}}{\text{m}^2}$$

$$= (25.92 + 1.938)\frac{1000 \times 0.40^2}{2}\,\frac{\text{N}}{\text{m}^2}$$

$$= 2229\ \text{N/m}^2$$

$$= 2.2\ \text{kN/m}^2.$$

**CHAPTER 3: PROBLEM (3)**

The equation to use (from Section 3.4) is:

$$t = \frac{A_s.2}{C_d A_N \sqrt{2g}}\left(\sqrt{h_2} - \sqrt{h_1}\right).$$

$$As = 2\,\text{m} \times 0.5\,\text{m} = 1.0\,\text{m}^2$$

$$A_N = \frac{\pi 0.05^2}{4}\text{m}^2 = 0.002\,\text{m}^2$$

therefore

$$\frac{A_s.2}{C_d A_N \sqrt{2g}} = \frac{1.0 \times 2}{0.8 \times 0.002\sqrt{2 \times 9.81}} = 282.2.$$

Thus:

$$282.2\left(\sqrt{h_2} - \sqrt{h_1}\right) = t.$$

(i) $h_2 = 0.60\,\text{m}$, $h_1 = 0.45\,\text{m}$

therefore $t = 282.2(\sqrt{0.60} - \sqrt{0.45})\ \text{s}$

$= 29.3\ \text{s}.$

(ii) $h_2 = 0.45\,\text{m}$, $h_1 = 0.30\,\text{m}$

therefore $t = 282.2(\sqrt{0.45} - \sqrt{0.30})\ \text{s}$

$= 34.6\ \text{s}.$

(iii) $h_2 = 0.30\,\text{m}$, $h_1 = 0.15\,\text{m}$

therefore $t = 282.2(\sqrt{0.30} - \sqrt{0.15})\ \text{s}$

$= 45.2\ \text{s}.$

(iv) $h_2 = 0.15\,\text{m},\ h_1 = 0\,\text{m}$

$$\text{therefore } t = 282.2\,(\sqrt{0.15} - \sqrt{0})\,\text{s} = 109.5\,\text{s}.$$

Check for total time:

$$t = 282.2\,(\sqrt{0.60} - \sqrt{0})\,\text{s} = 218.6\,\text{s}\ (= 29.3 + 34.6 + 45.2 + 109.5).$$

Note that about $\frac{3}{4}$ of water empties in $\frac{1}{2}$ the total time.

## CHAPTER 3: PROBLEM (4)

Refer to Section 3.7 for method.

$$[u]^{\alpha}[l]^{\beta}[\rho]^{\gamma}[\mu]^{\delta}[\sigma]^{\epsilon}[g]^{\theta}[F]^{1} = [1]^{0}$$

therefore

$$[Lt^{-1}]^{\alpha}[L]^{\beta}[ML^{-3}]^{\gamma}[ML^{-1}t^{-1}]^{\delta}[Mt^{-2}]^{\epsilon}[Lt^{-2}]^{\theta}[MLt^{-2}]^{1} = [1]^{0}.$$

Condition on $M$: $\gamma + \delta + \epsilon + 1 = 0$,

$L$: $\alpha + \beta - 3\gamma - \delta + \theta + 1 = 0$,

$t$: $-\alpha - \delta - 2\epsilon - 2\theta - 2 = 0$.

Solving in terms of $\delta$, $\epsilon$ and $\theta$:

From $t$: $\alpha = -\delta - 2\epsilon - 2\theta - 2$,

$M$: $\gamma = -\delta - \epsilon - 1$,

$L$: $\beta = -\alpha + 3\gamma + \delta - \theta - 1$
$= -(-\delta - 2\epsilon - 2\theta - 2) + 3(-\delta - \epsilon - 1) + \delta - \theta - 1$
$= \delta + 2\epsilon + 2\theta + 2 - 3\delta - 3\epsilon - 3 + \delta - \theta - 1$
$= -\delta - \epsilon + \theta - 2$.

Hence:

$$[u]^{-\delta - 2\epsilon - 2\theta - 2}[l]^{-\delta - \epsilon + \theta - 2}[\rho]^{-\delta - \epsilon - 1}[\mu]^{\delta}[\sigma]^{\epsilon}[g]^{\theta}[F]^{1} = [1]^{0}.$$

or

$$\left[\frac{ul\rho}{\mu}\right]^{-\delta}\left[\frac{\rho u^2 l}{\sigma}\right]^{-\epsilon}\left[\frac{u^2}{lg}\right]^{-\theta}\left[\frac{F}{\rho u^2 l^2}\right]^{1} = [1]^{0}$$

or

$$\frac{F}{\rho u^2 l^2} = \phi\left(\frac{ul\rho}{\mu}, \quad \frac{\rho u^2 l}{\sigma}, \quad \frac{u^2}{lg}\right)$$

Euler number — Reynolds number — Weber number — Froude number

$$\text{Euler number} = \frac{\Delta p}{\rho u^2} \quad \text{since} \quad \frac{F}{l^2} = \Delta p.$$

(7 variables, 3 primary quantities, therefore 4 dimensionless groups.)

Note this quick method:

| Variable | $u$ | $l$ | $\rho$ | $\mu$ | $\sigma$ | $g$ | $F$ |
|---|---|---|---|---|---|---|---|
| Index | $\alpha$ | $\beta$ | $\gamma$ | $\delta$ | $\epsilon$ | $\theta$ | 1 |
| Condition on $M$ | | | 1 | 1 | 1 | | 1 |
| $L$ | 1 | 1 | −3 | −1 | | 1 | 1 |
| $t$ | −1 | | | −1 | −2 | 2 | −2 |

Then one can write down directly from the Table the three simultaneous equations (conditions on $M$, $L$, and $t$).

## CHAPTER 4: PROBLEM (1)

See Section 4.3.

**(a)**

$$R = \left(\frac{l}{\lambda . A}\right) \quad \text{and} \quad R = R_{s.s.} + R_{m.w.} + R_{s.s.}$$

(i) **25 mm thick**

$$R = \frac{0.000\,05}{26.4 \times 50} + \frac{0.025}{0.0258 \times 50} + \frac{0.000\,05}{26.4 \times 50}$$

inner layer of steel — mineral wool — outer layer of steel

$$= \text{negligible} + 0.019 + \text{negligible}$$

$$= 0.019$$

therefore

$$q = \frac{(\theta_1 - \theta_2)}{R} = \frac{(22 - (-18))}{0.019} \text{ Watts}$$

$$= 2105 \text{ W}$$

(ii) **50 mm thick**

Since resistance of steel to heat flow is negligible, it may be ignored. Mineral wool is twice as thick as in (i) and therefore heat loss is halved to 1052 W.

(iii) **100 mm thick**

Same argument as (ii), $q = 526$ W.

**(b)**

Balance capital cost against energy by some suitable accounting procedure.

## CHAPTER 4: PROBLEM (2)

See Section 4.3.

**(a)**

As in the previous example, the resistance of the steel casing will be negligible. If thickness of insulation is doubled, the heat saving will be halved (50%).

**(b)**

$$\text{Heat loss (25 mm case)} = \frac{(\theta_1 - \theta_2)}{R}$$

$$R = \left(\frac{l}{\lambda . A}\right) = \frac{0.025}{0.025 \times 4} = 0.25$$

$$q = \frac{(170 - 20)}{0.25} = 600 \text{ W.}$$

Heat saving is therefore 600 W (since for $l = 12.5$ mm, $q = 1200$ W).

Heat saving is 600 W or 600 J/s.

In one year this heat saving is $\underset{\text{J/s}}{600} \times \underbrace{(4 + 6) \times 6 \times 50 \times 3600)}_{\text{seconds in year operated}}$

$$= 6480 \text{ MJ.}$$

The value of this saving is £6480 $\times \dfrac{0.5}{100}$ per annum

$$= £32.40 \text{ per annum.}$$

**(c)**

$$\text{Pay-back period} = \frac{100}{32.40} = 3.1 \text{ years.}$$

As life of equipment is likely to be more than 3.1 years, it would seem to be a good investment.

## CHAPTER 4: PROBLEM (3)

See Section 4.5.

**For flat plate:**

$$\begin{aligned} q &= hA(\theta_g - \theta_1) \\ &= 15 \times 1 \times (1250 - 150)\,\mathrm{W} \\ &= 16.5\,\mathrm{kW}. \end{aligned}$$

**For one stud:**

$$\begin{aligned} m &= \sqrt{\frac{hC_s}{\lambda A_s}} = \sqrt{\frac{2h}{\lambda r}} = \sqrt{\frac{2 \times 15}{25 \times 0.0025}} \qquad (r = 2.5\,\mathrm{mm}) \\ &= \sqrt{480} \\ &= 21.9. \end{aligned}$$

$$\begin{aligned} q_s &= m\lambda A_s \tanh ml\,(\theta_g - \theta_1) \\ &= 21.9 \times 25 \times (\pi \times 0.0025^2) \times \tanh(21.9 \times 0.01) \times (1250 - 150)\,\mathrm{W} \\ &= 21.9 \times 25 \times \pi \times 0.0025^2 \times 0.20 \times 1100\,\mathrm{W} \\ &= 2.36\,\mathrm{W} \end{aligned}$$

(since $\tanh 0.219 = 0.219/0.5 \times 0.4621 = 0.20$).

**For 2500 studs:**

$$\begin{aligned} q_s &= 2500 \times 2.36\,\mathrm{W} \\ &= 5.9\,\mathrm{kW}. \end{aligned}$$

Plane area when studs are present $= (1 - (2500 \times \pi \times 0.0025^2))\,\mathrm{m}^2$

$= (1 - 0.05)\,\mathrm{m}^2$

$= 0.95\,\mathrm{m}^2$.

Heat flow through this area $= 16.5 \times 0.95\,\mathrm{kW}$

$= 15.7\,\mathrm{kW}$.

Total heat flow (with studs) $= (15.7 + 5.9)\,\mathrm{kW} = 21.6\,\mathrm{kW}$.

Effect of using studs is to increase heat-transfer rate by $(21.6 - 16.5)/16.5 \times 100\% = 31\%$.

## CHAPTER 4: PROBLEM (4)

See Section 4.7.

$$\alpha = \frac{\lambda}{C_p.\rho} = \frac{1.06}{3760 \times 960} = 0.29 \times 10^{-6}\,\mathrm{m}^2/\mathrm{s}.$$

Example of tabulated solution (24 mm thick, 240°C on surface, centre to reach 80°C)

| | *G* | *F* | *E* | *D* | *C* | *B* | *A* |
|---|---|---|---|---|---|---|---|
| $t = 0$ | 240 | 0 | 0 | 0 | 0 | 0 | 240 |
| $t = 1$ | – | 120 | – | 0 | – | 120 | – |
| $t = 2$ | 240 | – | 60 | – | 60 | – | 240 |
| $t = 3$ | – | 150 | – | 60 | – | 150 | – |
| $t = 4$ | 240 | – | 105 | – | 105 | – | 240 |
| $t = 5$ | – | 172 | – | 105 | – | 172 | – |

Say, $t = 4.5$:

$$\Delta x = 24/6 = 4\,\mathrm{mm} = 4 \times 10^{-3}\,\mathrm{m},$$

$$\Delta t = \frac{(\Delta x)^2}{2\alpha} = \frac{4 \times 4 \times 10^{-3} \times 10^{-3}}{2 \times 0.29 \times 10^{-6}}$$

$$= 27.6\,\mathrm{s}.$$

therefore

$$\text{Time required} = 4.5 \times 27.6\,\mathrm{s}$$

$$= 124\,\mathrm{s}.$$

There are 3 centre temperatures, 4 thicknesses and 3 surface temperatures and therefore 3 × 4 × 3 = 36 possible combinations of these variables. To carry out all these calculations in the form of tables (12 tables are required as the same table will enable all three centre-temperature calculations to be carried out) is very time-consuming and use could be made of a computer to do this. Alternatively one graph with different axes will enable all the calculations to be carried out because the number of slices taken is the same in all cases. See the diagram.

**Number of time intervals (estimated from graph by interpolation):**

| Surface temperature | | 240°C | 220°C | 200°C |
|---|---|---|---|---|
| Number of | 40°C | 2.33 | 2.43 | 2.60 |
| time intervals | 60°C | 3.00 | 3.24 | 3.53 |
| to reach | 80°C | 3.89 | 4.22 | 4.60 |

**Duration of time intervals** ($\Delta t = (\Delta x)^2/2\alpha$):

Thickness:

24 mm, $\Delta x = 24/6 = 4$ mm, $\Delta t = (4 \times 10^{-3})^2/2 \times 0.29 \times 10^{-6}$ $= 27.6$ s.

18 mm, $\Delta x = 18/6 = 3$ mm, $\Delta t = 9/0.58 = 15.5$ s .

12 mm, $\Delta x = 12/6 = 2$ mm, $\Delta t = 4/0.58 = 6.9$ s .

6 mm, $\Delta x = 6/6 = 1$ mm, $\Delta t = 1/0.58 = 1.7$ s .

**Cooking time** (= number of time intervals × $\Delta t$):

| Surface temperature (°C) | | 240 | | | | 220 | | | | 200 | | | |
|---|---|---|---|---|---|---|---|---|---|---|---|---|---|
| Thickness (mm) | | 24 | 18 | 12 | 6 | 24 | 18 | 12 | 6 | 24 | 18 | 12 | 6 |
| Centre temperature | 40°C | 64 | 36 | 16 | 4 | 67 | 38 | 17 | 4 | 72 | 40 | 19 | 4 |
| | 60°C | 83 | 46 | 20 | 5 | 90 | 50 | 22 | 6 | 98 | 54 | 24 | 6 |
| | 80°C | 108 | 60 | 27 | 6 | 116 | 66 | 29 | 7 | 127 | 72 | 32 | 8 |

Graph shows that the major parameters affecting cooking time are hamburger thickness and centre-temperature, both of which are related to customer satisfaction. Griddle surface temperature is the least important.

The assumptions made in this method of calculation are discussed in Sections 4.6 and 4.7.

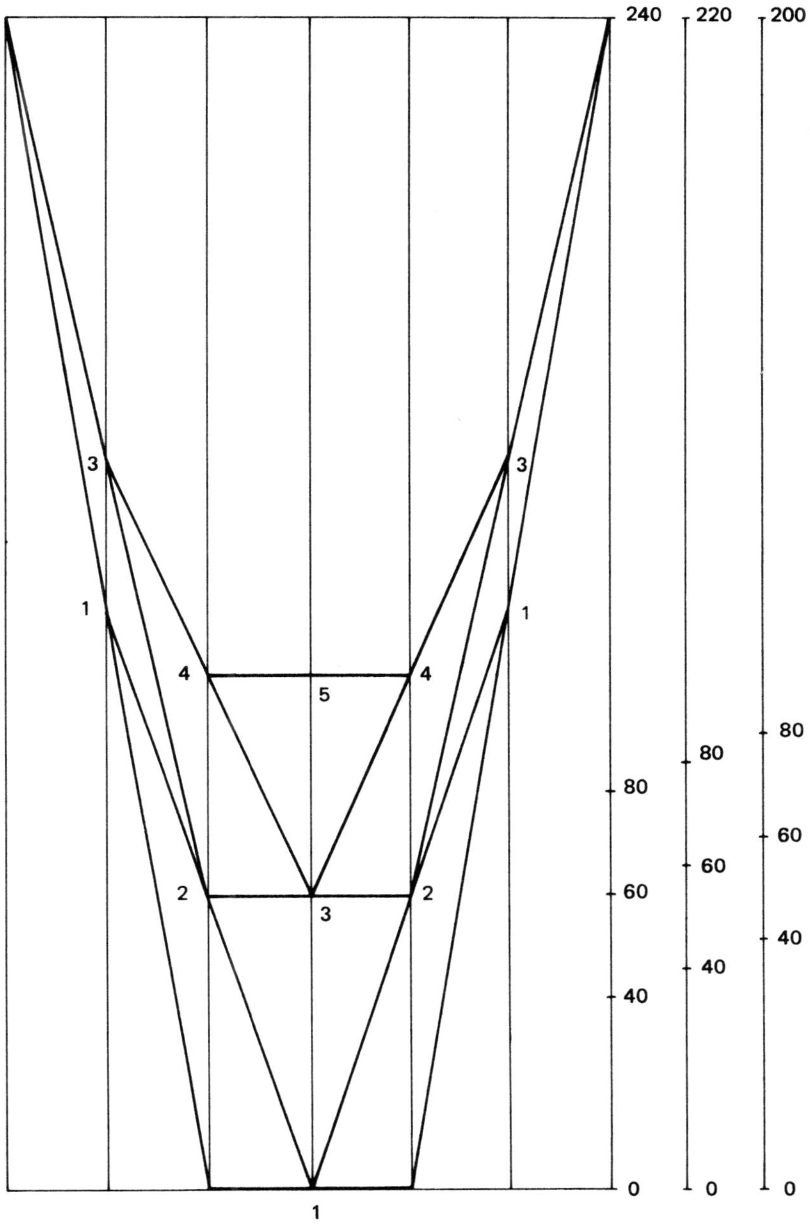
240
220
200
3
3
1
1
4
4
5
80
80
80
60
60
60
2
2
3
40
40
40
0
0
0
1

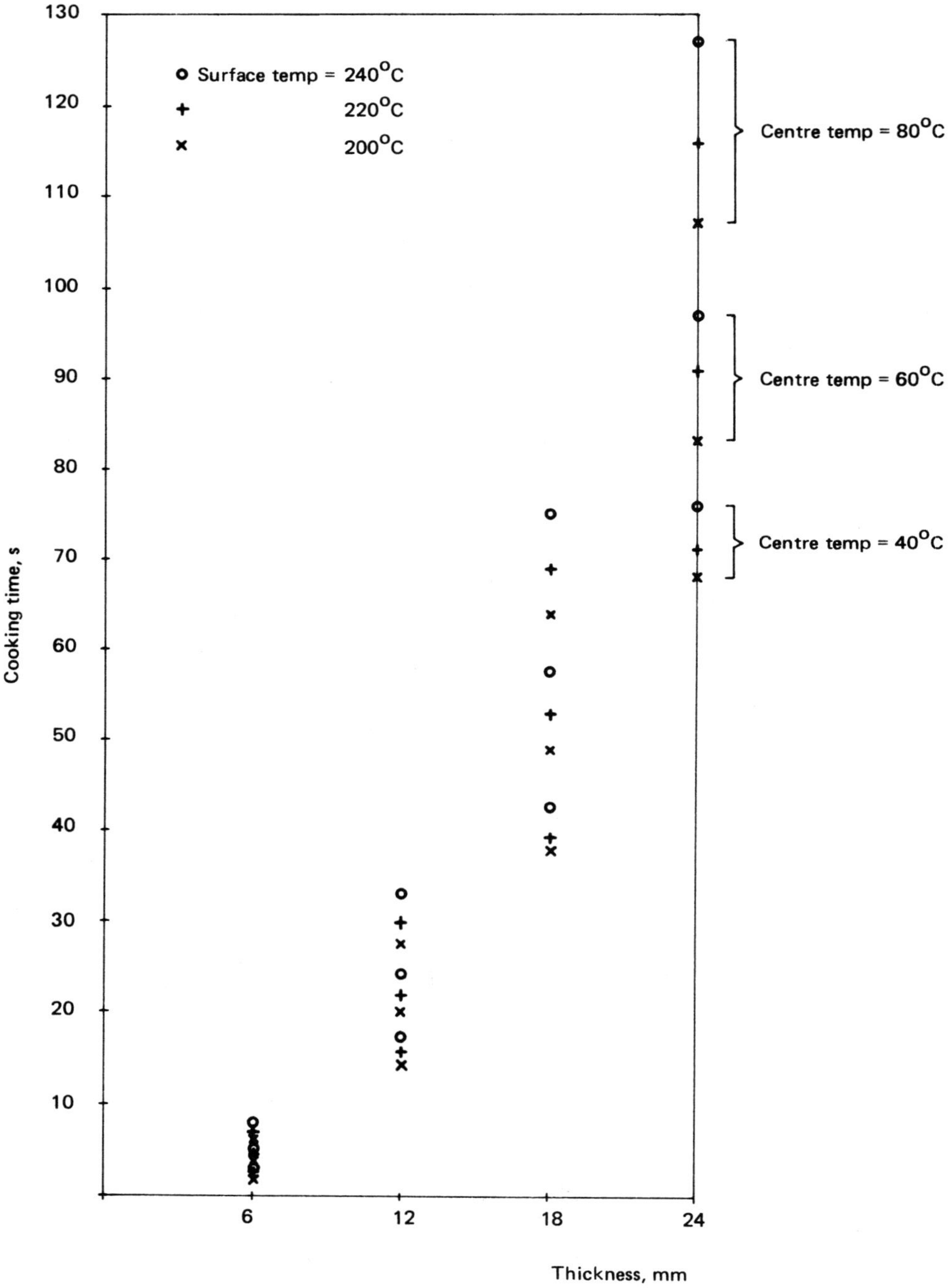

Surface temp = 240°C
220°C
200°C
Centre temp = 80°C
Centre temp = 60°C
Centre temp = 40°C
Cooking time, s
Thickness, mm
130
120
110
100
90
80
70
60
50
40
30
20
10
6
12
18
24

## CHAPTER 4: PROBLEM (5)

See Section 4.8.

It is necessary to calculate the radius of the sphere by converting the mass to a volume using its density and then calculating the radius from this volume.

$$V = \frac{m}{\rho} = \frac{2}{1000} = 0.002\,\text{m}^3$$

$$V = \frac{4}{3}\pi r_m^3 \quad \text{or} \quad r_m = \sqrt[3]{\frac{3 \times V}{4 \times \pi}}$$

$$= \sqrt[3]{\frac{3 \times 0.002}{4 \times \pi}}$$

$$= \sqrt[3]{0.0005}$$

$$= 0.08\,\text{m}\,.$$

$$\alpha = \frac{\lambda}{C_p\,\rho} = \frac{1}{4000 \times 1000} = 0.25 \times 10^{-6}\,\text{m}^2/\text{s}\,.$$

$$Bi = \frac{r_m\,h}{\lambda} = \frac{0.08 \times 15}{1} = 1.2\,.$$

$$Fo = \frac{\alpha t}{r_m^2} = \frac{0.25 \times 10^{-6} \times t}{0.08 \times 0.08}\,.$$

$$\text{Ⓗ} = \frac{\theta_s - \theta}{\theta_s - \theta_o} = \frac{175 - 95}{175 - 15} = \frac{80}{160} = 0.5\,.$$

From Fig. B.5, $Fo = 0.35$ (when $Bi = 1.2$ and Ⓗ $= 0.5$).

Therefore

$$0.35 = \frac{0.25 \times 10^{-6} \times t}{0.08 \times 0.08}$$

therefore time to reach centre temperature of 95°C,

$$t = 8960\,\text{s}$$

$$= 2.5\,\text{h}\,.$$

**Surface temperature when centre is at 95°C**

In this case, $Bi = 1.2$ and $Fo = 0.35$. From chart B.6:

$$Ⓗ = 0.3 = \frac{175 - \theta_{sur}}{175 - 15}$$

where $\theta_{sur}$ = surface temperature

therefore

$$0.3(175 - 15) = 175 - \theta_{sur}$$

therefore

$$48 = 175 - \theta_{sur}$$

therefore

$$\theta_{sur} = (175 - 48)°\text{C} = 127°\text{C}\,.$$

In practice, the surface temperature would not reach 127°C. It would begin to dry out at 100°C and would maintain this temperature until there was insufficient moisture present to allow surface drying. The temperature of 127°C is predicted by this method because the charts used assume that no phase changes take place.

**CHAPTER 4: PROBLEM (6)**

See Section 4.8.

This finite slab may be considered as the intersection of three finite slabs.

(i) Assume tray is a 0.1 m thick infinite slab, then:

$$Fo = \frac{\alpha t}{r_m^2} = \frac{0.15 \times 10^{-6} \times 5 \times 3600}{(0.1/2)^2} = 1.08\,.$$

$$Bi = \frac{r_m h}{\lambda} = \frac{0.05 \times 6}{0.52} = 0.58\,.$$

From Fig. B.1, Ⓗ = 0.64 .

(ii) Assume tray is a 0.2 m thick infinite slab:

$$Fo = \frac{0.15 \times 10^{-6} \times 5 \times 3600}{(0.2/2)^2} = 0.27\,.$$

$$Bi = \frac{0.1 \times 6}{0.52} = 1.15\,.$$

From Fig. B.1, Ⓗ = 0.9.

(iii) Assume tray is a 0.3 m thick infinite slab:

$$Fo = \frac{0.15 \times 10^{-6} \times 5 \times 3600}{(0.3/2)^2} = 0.12\,.$$

$$Bi = \frac{0.15 \times 6}{0.52} = 1.73.$$

From Fig. B.1, Ⓗ = 0.974.

Combining the three infinite slab solutions, gives 0.64 × 0.9 × 0.974 = 0.56.

Then:

$$0.56 = \frac{\theta_s - \theta}{\theta_s - \theta_o} = \frac{4 - \theta}{4 - 60}$$

therefore

$$4 - \theta = 0.56 \times (-56) = -31$$

therefore

$$\theta = 4 + 31 = 35°\text{C}\,.$$

The centre-temperature has only been reduced to 35°C after 5 hours of cooling. It has therefore been in the critical zone for microbiological growth for 5 hours and is still not into the safe region, say below 10°C. Note also that the heat abstracted from the food in the refrigerator will have warmed up the other food, possibly taking it into the microbial growth temperature zone. Faster cooling of the food may be achieved using a blast chiller.

## CHAPTER 5: PROBLEM (1)

See Section 4.8.

$$\text{Biot number, } Bi = \frac{r_m h}{\lambda}$$

$$= \frac{0.012 \times h}{1.06}.$$

$$\text{Reduced temperature, Ⓗ} = \frac{\theta_s - \theta}{\theta_s - \theta_o}.$$

$$= \frac{\theta_s - 80}{\theta_s - 0}$$

$$\text{Fourier number, } Fo = \frac{\lambda t}{\rho C_p r_m^2}$$

$$= \frac{1.06 \times t}{960 \times 3.76 \times 10^3 \times 0.012^2}$$

$$= 0.002\, t$$

therefore

$$t = \frac{Fo}{0.002} \text{(seconds)}.$$

**(a)**

In all cases: $\theta_s = 180°\text{C}$,

hence

$$\text{Ⓗ} = \frac{180 - 80}{180 - 0} = 0.56 .$$

When

$$h = 5, \quad Bi = \frac{0.012 \times 5}{1.06} = 0.06 ,$$

$$h = 15, \quad Bi = \frac{0.012 \times 15}{1.06} = 0.17 ,$$

$$h = 30, \quad Bi = \frac{0.012 \times 30}{1.06} = 0.34 .$$

Using Fig. B.1:

$$Bi = 0.06, \text{ Ⓗ} = 0.56, Fo = 10 \text{ therefore } t = \frac{10}{0.002} = 5000\,\text{s},$$

$$Bi = 0.17, \text{ Ⓗ} = 0.56, Fo = 3.8 \text{ therefore } t = \frac{3.8}{0.002} = 1900\,\text{s},$$

$$Bi = 0.34, \text{ Ⓗ} = 0.56, Fo = 2.1 \text{ therefore } t = \frac{2.1}{0.002} = 1050\,\text{s},$$

**(b)**

In all cases: $h = 15$, hence $Bi = 0.17$.

Using Fig. B.1 to obtain $Fo$ (and hence $t$) from Ⓗ and $Bi$:

| $\theta_s$ | Ⓗ | *Bi* | *Fo* | *t* |
|---|---|---|---|---|
| 150°C | 0.47 | 0.17 | 5.0 | 2500 s |
| 180°C | 0.56 | 0.17 | 3.8 | 1900 s |
| 210°C | 0.62 | 0.17 | 3.1 | 1550 s |
| 240°C | 0.67 | 0.17 | 2.7 | 1350 s |

**(c)**

With *Bi* and *Fo* from **(a)** and **(b)**, Ⓗ for the surface temperature ($\theta_{sur}$) may be obtained using Fig. B.2:

| $\theta_s$ | *h* | *Bi* | *Fo* | Ⓗ | $\theta_{sur}$ |
|---|---|---|---|---|---|
| 180°C | 5 | 0.06 | 10.0 | 0.54 | 83°C |
| 180°C | 15 | 0.17 | 3.8 | 0.51 | 88°C |
| 180°C | 30 | 0.34 | 2.1 | 0.47 | 95°C |
| 150°C | 15 | 0.17 | 5.0 | 0.42 | 87°C |
| 180°C | 15 | 0.17 | 3.8 | 0.51 | 88°C |
| 210°C | 15 | 0.17 | 3.1 | 0.57 | 90°C |
| 240°C | 15 | 0.17 | 2.7 | 0.60 | 96°C |

The results are plotted on the graphs below.

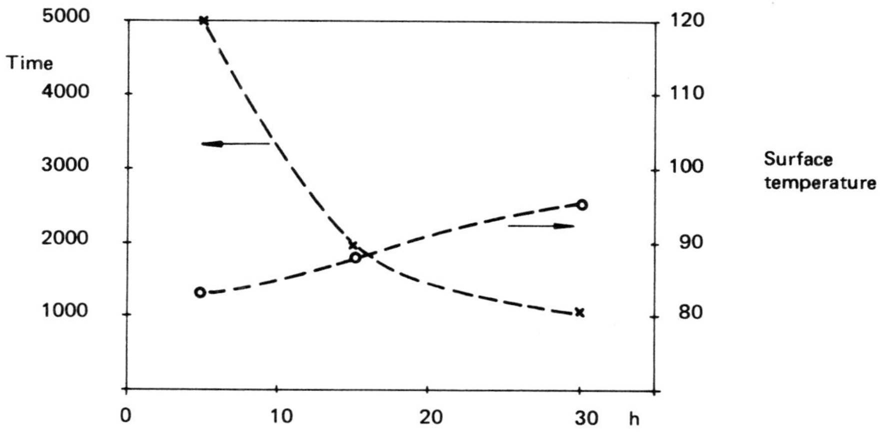

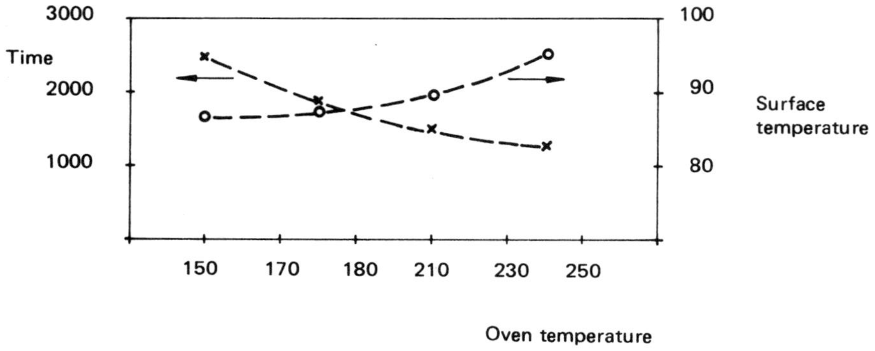


**(d)**

The effect of heat-transfer coefficient and air temperature on heating time are non-linear for the reasons explained in Section 5.2.1. There is a marked decrease in heating time for an increase in $h$ from 5 to 15, but a much lower increase if $h$ increases from 15 to 30.

The surface temperature may become excessive if the convective heat-transfer coefficient and the oven temperature are increased too far.

## CHAPTER 5: PROBLEM (2)

See Section 5.4.1.

**(a)**

$$Re_x = \frac{du\rho}{\mu} = \frac{0.2 \times 1 \times 0.6}{30 \times 10^{-6}} = 4000 .$$

Flow is laminar, that is, $Re_x < 500\,000$.

Then,

$$Nu_x = 0.664\ (4000)^{0.5}\ (0.68)^{0.33}$$

$$= 0.664 \times 63.25 \times 0.88$$

$$= 36.96 .$$

$$\frac{hd}{\lambda} = \frac{h \times 0.2}{0.04} = 36.96$$

$$h = \frac{0.04 \times 36.96}{0.2}$$

$$= 7.4\ \mathrm{W/m^2.K} .$$

(b)

$$Re_x = \frac{0.2 \times 4 \times 0.6}{30 \times 10^{-6}} = 16\,000 .$$

Flow is still laminar.

Then,

$$Nu_x = 0.664\,(16\,000)^{0.5}\,(0.68)^{0.33}$$
$$= 0.664 \times 126.5 \times 0.88$$
$$= 73.9 .$$
$$h = \frac{0.04 \times 73.9}{0.2}$$
$$= 14.8\ \mathrm{W/m^2.K} .$$

(c)

$$Re_x = \frac{0.2 \times 8 \times 0.6}{30 \times 10^{-6}} = 32\,000 .$$

Flow is still laminar.

Then,

$$Nu_x = 0.664\,(32\,000)^{0.5}\,(0.68)^{0.33}$$
$$= 0.664 \times 178.9 \times 0.88$$
$$= 104.5 .$$
$$h = \frac{0.04 \times 104.5}{0.2}$$
$$= 20.9\ \mathrm{W/m^2.K} .$$

(d)

$$Re_x = \frac{0.2 \times 12 \times 0.6}{30 \times 10^{-6}} = 48\,000 .$$

Flow is still laminar.

Then,

$$Nu_x = 0.664\,(48\,000)^{0.5}\,(0.68)^{0.33}$$
$$= 0.664 \times 219.1 \times 0.88$$
$$= 128.0 .$$
$$h = \frac{0.04 \times 128.0}{0.2}$$
$$= 25.6\ \mathrm{W/m^2.K} .$$

(e)

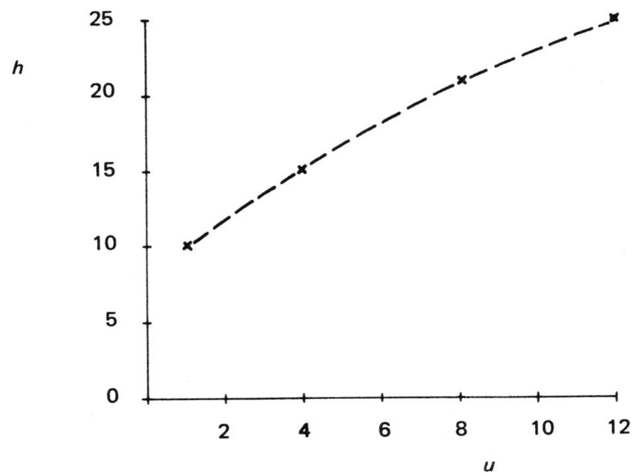


There is a progressive rise in heat-transfer rate from $u = 1$ to $u = 12$ m/s. Extrapolation back to $u = 0$, suggests that if there is no forced convection, $h \approx 8$ W/m$^2$.K.

## CHAPTER 5: PROBLEM (3)

See Section 5.2.2.

$$R = R_1 + R_2 + R_3 + ----$$

Resistance of metal casing will be negligible.

$$A = 0.5 \times 0.5 \times 2\,\text{m}^2 \qquad \text{base and top}$$
$$+ 0.5 \times 1.5 \times 4\,\text{m}^3 \qquad \text{sides (4)}$$
$$= 3.5\,\text{m}^2\,.$$

Resistance of inner surface ($R_1$) =

$$\frac{1}{hA} = \frac{1}{2 \times 3.5} = 0.143\,.$$

Resistance of glass fibre ($R_2$) =

$$\frac{l}{\lambda A} = \frac{0.075}{0.025 \times 3.5} = 0.857\,.$$

Resistance of outer surface ($R_3$) =

$$\frac{1}{hA} = \frac{1}{5 \times 3.5} = 0.057\,.$$

Then,

$$R = R_1 + R_2 + R_3$$
$$= 0.143 + 0.857 + 0.057$$
$$= 1.057\,.$$

Heat to be extracted,

$$q = \frac{(\theta_1 - \theta_2)}{R}$$
$$= \frac{(20 - 4)}{1.057}$$
$$= 15.1\,\mathrm{W}\,.$$

For film on outer surface:

$$q = \frac{(\theta_1 - \theta_s)}{R_3}$$ where $\theta_s$ = temperature of outer surface

therefore

$$15.1 = \frac{(20 - \theta_s)}{0.057}$$

therefore

$$20 - \theta_s = 15.1 \times 0.057 = 0.86$$

therefore

$$\theta_s = 20 - 0.86$$
$$\approx 19.1°\mathrm{C}\,.$$

Importance of surface films:

Total resistance = 1.057

Glass fibre resistance = 0.857.

Therefore, surface films increase resistance to heat-flow by:

$$\frac{(1.057 - 0.857)}{1.057} \times 100\%$$
$$= 18.9\%\,.$$

## CHAPTER 5: PROBLEM (4)

(a)

$$\mu_s = \mu \quad \text{therefore} \left(\frac{\mu_s}{\mu}\right)^{0.14} = 1\,.$$

$$\frac{h \times 0.1}{0.27}\left(\frac{2400 \times 1.34}{0.27}\right)^{-\frac{1}{3}} = 0.37\left(\frac{1.0^2 \times N \times 1262}{1.34}\right)^{\frac{2}{3}},$$

$$h \times \frac{0.37}{(11911)^{\frac{1}{3}}} = 0.37\,N^{\frac{2}{3}}\,(942)^{\frac{2}{3}}\,,$$

$$h \times \frac{0.37}{22.8} = 0.37 \times N^{\frac{2}{3}} \times 96.1\,,$$

$$h = 0.37 \times 96.1 \times \frac{22.8}{0.37} \times N^{\frac{2}{3}},$$

$$h = 2191 N^{\frac{2}{3}}\,.$$

(b)

$N = 1 \quad h = 2191 \times 1 = 2191\,\mathrm{W/m^2.K}$,

$N = 0.1 \quad h = 2191 \times 0.216 = 473\,\mathrm{W/m^2.K}$,

$N = 0.01 \quad h = 2191 \times 0.047 = 103\,\mathrm{W/m^2.K}$.

(c)

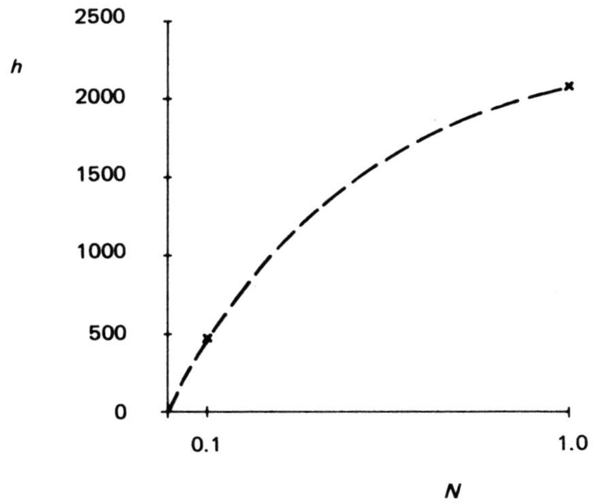

(d)

Advantages: increased heat-transfer rate gives increased throughput and smaller equipment with smaller floor space (lower costs) and reduced labour cf. hand stirring; there is a fuel saving since heat balance should improve.

Disadvantages: extra capital costs, is capital available? extra power cost (for stirring), operative training and acceptance, extra cleaning cost.

Possible effect of stirring on structure of food.

See Section 5.6.2.2.

(e)

If the steam side coefficient is 250 W/m².K, there would be little point in having $h$ (food side) more than 500 W/m².K since in this case (if $A = 0.3\ \mathrm{m}^2$):

$$R(\text{steam}) = \frac{1}{250 \times 0.3} = 0.013,$$

$$R(\text{food}) = \frac{1}{500 \times 0.3} = 0.0065$$

and steam side is controlling resitance.
The speed for which $h = 500\ \mathrm{W/m^2.K}$, is about 0.1 revolutions per second.

## CHAPTER 6: PROBLEM (1)

From Appendix D.1.2, equation to use is:

$$q = \frac{(e_1 - e_2)}{\left(\dfrac{1}{A_G \epsilon_G}\right)}$$

$$e_1 = \sigma\theta_1^4 = 56.7 \times 10^{-9} \times (175 + 273)^4 = 2284\ \mathrm{W/m^2}.$$

$$e_2 = \sigma\theta_2^4 = 56.7 \times 10^{-9} \times (100 + 273)^4 = 1098\ \mathrm{W/m^2}.$$

$$\frac{1}{A_G\epsilon_G} = \frac{1}{0.05 \times 0.85} = 23.5\ \mathrm{m^2}.$$

$$q = \frac{(2284 - 1098)}{23.4}$$

$$= 50.7\ \mathrm{W}.$$

## CHAPTER 6: PROBLEM (2)

The solution follows the theory presented in Section 6.4.

$$e_1 = \sigma\theta_1^4 = 5.67 \times 10^{-8} \times 1000^4 = 56\,700\ \mathrm{W/m^2}.$$

$$e_2 = \sigma\theta_2^4 = 5.67 \times 10^{-8} \times 300^4 = 459\ \mathrm{W/m^2}.$$

$$(e_1 - e_2) = (56\,700 - 459)\mathrm{W/m^2} = 56\,241\mathrm{W/m^2}.$$

Area of surface 1 = area of surface 2 = $1.0 \times 0.5 = 0.5\ \mathrm{m^2}$.

Area of surface $s$ = $0.5 \times 0.2 = 0.1\ \mathrm{m^2}$.

**(a)**

For infinite parallel plates,

$$\frac{1}{A_1F_{12}} = \frac{1}{0.5 \times 1} = 2,$$

$F_{12} = 1$ since, for infinite parallel plates, all radiant heat leaving surface 1 will be intercepted by surface 2.

$$q = \frac{(e_1 - e_2)}{\left(\dfrac{1}{A_1F_{12}}\right)} = \frac{56\,241}{2} = 28.1\,\mathrm{kW}.$$

**(b)**

In this case (Section D.1.6):

$$\frac{1}{F} = \left(\frac{1}{\epsilon_1} + \frac{1}{\epsilon_2} - 1\right)$$

$$= \left(\frac{1}{0.9} + \frac{1}{0.8} - 1\right)$$

$$= 1.36$$

$$F = 1/1.36 = 0.74.$$

$$\frac{1}{AF} = \frac{1}{0.5 \times 0.74} = 2.70.$$

$$q = \frac{56\,241}{2.70} = 20.8\,\mathrm{kW}.$$

**(c)**

In this case Section 6.4.2 applies

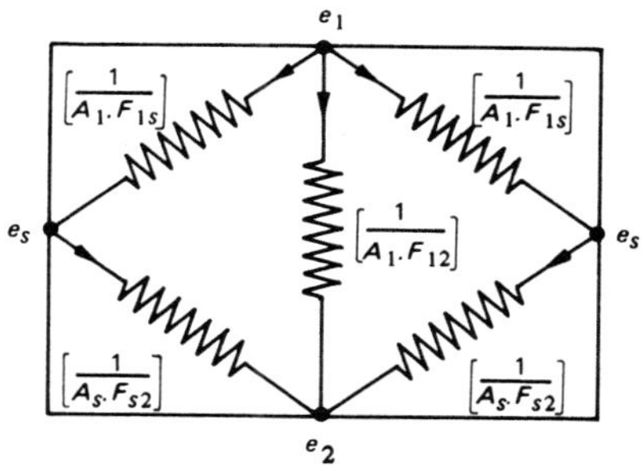


$$A_1 = 0.5\,\mathrm{m}^2, \quad A_s = 0.1\,\mathrm{m}^2 .$$

$F_{12}$ is obtained from Section D.1.5.

$$x/z = 1.0/0.2 = 5.0 .$$
$$y/z = 0.5/0.2 = 2.5 .$$
$$F_{12} = 0.55 .$$

$F_{1s}$ is obtained from Section D.1.7.

$$z/x = 0.2/0.5 = 0.4 .$$
$$y/x = 1.0/0.5 = 2.0 .$$
$$F_{1s} = 0.06 .$$

$F_{s2}$ is obtained from Section D.1.7.

$$z/x = 1.0/0.5 = 2.0 .$$
$$y/x = 0.2/0.5 = 0.4 .$$
$$F_{s2} = 0.34 .$$

$$\frac{1}{A_1 F_{1s}} = \frac{1}{0.5 \times 0.06} = 33.3 .$$

$$\frac{1}{A_s F_{s2}} = \frac{1}{0.1 \times 0.34} = 29.4 .$$

$$\frac{1}{A_1 F_{12}} = \frac{1}{0.5 \times 0.55} = 3.6 .$$

This gives:

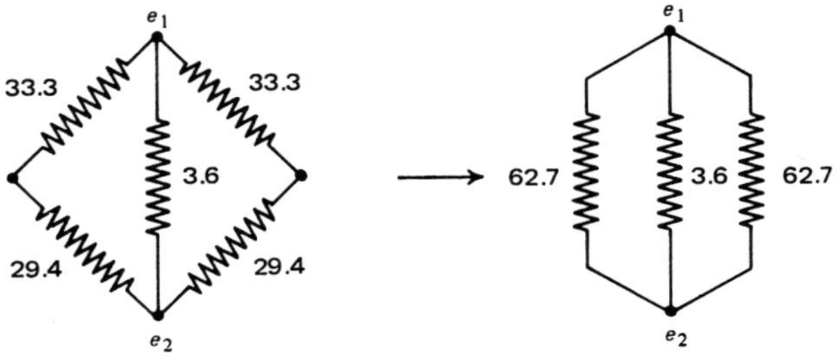


$$\frac{1}{R_T} = \frac{1}{62.7} + \frac{1}{3.6} + \frac{1}{62.7} = 0.310$$

therefore,

$$R = 3.2\,.$$

$$q = \frac{(e_1 - e_2)}{R} = \frac{56\,241}{3.2} = 17.6\,\text{kW}.$$

**(d)**

See Section 6.4.3.

$$R_{G2} = \left(\frac{1-\epsilon}{\epsilon A}\right) = \left(\frac{1-0.9}{0.9 \times 0.5}\right) = 0.2$$

$$R_T = 3.2$$

$$R_{G1} = \left(\frac{1-\epsilon}{\epsilon A}\right) = \left(\frac{1-0.8}{0.8 \times 0.5}\right) = 0.5\,.$$

New $R_T = 0.2 + 3.2 + 0.5 = 3.9$ .

$$q = \frac{56\,241}{3.9} = 14.4\,\text{kW}.$$

**(e)**

Case (a) represents the maximum possible rate of radiant heat transfer whilst case **(d)** is nearest to reality. In this practical case, the actual rate of radiation heat transfer is only $14.4/28.1 \times 100\% = 51\%$ of the maximum possible.

## CHAPTER 6: PROBLEM (3)

Section D.1.3 applies to this case, that is

$$q = \frac{(e_1 - e_2)}{\left(\frac{1}{A_G \epsilon_G}\right)} .$$

Griddle surface at 450 K:

$$e_1 = 5.67 \times 10^{-8} \times 450^4 = 2325 \text{ W/m}^2 .$$

Griddle surface at 500 K:

$$e_1 = 5.67 \times 10^{-8} \times 500^4 = 3544 \text{ W/m}^2 .$$

Surroundings at 20°C (293 K):

$$e_2 = 5.67 \times 10^{-8} \times 293^4 = 418 \text{ W/m}^2 .$$

For $\epsilon = 1.0, A = 1.0 \text{ m}^2$ (0% covered by food):

$$\frac{1}{A_G \epsilon_G} = \frac{1}{1.0 \times 1.0} = 1.0 .$$

For $\epsilon = 0.1, A = 1.0 \text{ m}^2$ (0% covered by food):

$$\frac{1}{A_G \epsilon_G} = \frac{1}{1.0 \times 0.1} = 10.0 .$$

Other areas left uncovered are calculated in simple proportion:

90% covered: $\epsilon = 1.0$ $R = 10$; $\epsilon = 0.1$ $R = 100.0$ .

50% covered: $\epsilon = 1.0$ $R = 2.0$; $\epsilon = 0.1$ $R = 20.0$ .

Heat losses (W):

| Griddle temperature K | | 450 | | 500 | |
|---|---|---|---|---|---|
| Emissivity | | 0.1 | 1.0 | 0.1 | 1.0 |
| Area covered | 90% | 19.1 | 190.7 | 31.3 | 312.6 |
| | 50% | 95.4 | 953.5 | 156.3 | 1563.0 |
| | 0% | 190.7 | 1907.0 | 312.6 | 3126.0 |

The major increases in heat losses from the surface are caused by the surface becoming blackened and less covered with food. Griddle-plate temperature over the 50°C range examined is relatively less important. There will also be a convection heat loss from the griddle surface which has not been considered in this calculation. Its effect is probably quite large.

## CHAPTER 7: PROBLEM (1)

As with the example in the text, the attenuation of microwave energy in calculated, but this time a value for $\alpha$ is based on the temperature in the appropriate slab at the end of the previous time interval.

## CHAPTER 7: PROBLEM (2)

A comparison of Table 7.1 with the calculations in Problem (2) opposite shows that, as the temperature of food is raised, it becomes less 'lossy' and therefore microwave energy can penetrate to a greater depth into the food. This means the calculated values in Table 7.1 overestimate the upper surface temperature and underestimate the lower surface temperature.

## CHAPTER 7: PROBLEM (3)

At the end of the microwave heating period the terms in the equations involving $q_i$ are eliminated. The equation for an internal slab simplifies to:

$$\theta_1^{n+1} = \tfrac{1}{2}(\theta_{i-1}^n + \theta_{i+1}^n)$$

and for a surface slab:

$$\theta_1^{n+1} = \frac{1}{2}\left[\frac{h\Delta x}{\lambda}(\theta_a - \theta_1^n) + \theta_1^n + \theta_2^n\right].$$

Ten minutes is equivalent to twelve time intervals.

| | 1 | 2 | 3 | 4 | 5 | 6 | 7 | 8 |
|---|---|---|---|---|---|---|---|---|
| $T_i^0$ | 114.3 | 101.2 | 100.6 | 91.2 | 70.1 | 65.0 | 54.5 | 51.0 |
| $T_i^1$ | 106.6 | 107.5 | 96.2 | 85.4 | 78.1 | 62.3 | 58.0 | 52.4 |
| $T_i^2$ | 106.0 | 101.4 | 96.5 | 87.2 | 73.9 | 68.1 | 57.4 | 54.8 |
| $T_i^3$ | 102.7 | 101.3 | 94.3 | 85.2 | 77.7 | 65.7 | 61.5 | 55.7 |
| $T_i^4$ | 101.0 | 98.5 | 93.3 | 89.8 | 75.5 | 69.6 | 60.7 | 58.2 |
| $T_i^5$ | 98.8 | 97.2 | 94.2 | 84.0 | 79.7 | 68.1 | 63.9 | 59.0 |
| $T_i^6$ | 97.1 | 96.5 | 90.6 | 87.0 | 81.9 | 71.8 | 63.6 | 61.0 |
| $T_i^7$ | 95.8 | 93.9 | 91.8 | 86.3 | 79.4 | 72.8 | 66.4 | 61.8 |
| $T_i^8$ | 94.0 | 93.8 | 90.1 | 85.6 | 79.6 | 76.1 | 67.3 | 63.6 |
| $T_i^9$ | 93.0 | 92.1 | 89.7 | 84.9 | 80.9 | 73.5 | 69.9 | 64.9 |
| $T_i^{10}$ | 91.7 | 91.4 | 88.8 | 85.3 | 79.2 | 75.4 | 69.2 | 66.9 |
| $T_i^{11}$ | 90.7 | 90.3 | 88.4 | 84.0 | 80.4 | 74.2 | 71.2 | 67.5 |
| $T_i^{12}$ | 89.7 | 89.6 | 87.2 | 84.4 | 79.1 | 75.8 | 70.9 | 68.8 |

| | 1 | 2 | 3 | 4 | 5 | 6 | 7 | 8 |
|---|---|---|---|---|---|---|---|---|
| Initial temp. | 0 | 0 | 0 | 0 | 0 | 0 | 0 | 0 |
| $P_i (P_o = 100\,000)$ | 53585 | 28713 | 15386 | 8245 | 4418 | 2367 | 1269 | 680 |
| $q_i$ | 46415 | 24872 | 13327 | 7141 | 3827 | 2051 | 1098 | 589 |
| $T_i$ (end of 1st period) | 108.7 | 58.1 | 31.2 | 16.7 | 8.9 | 4.8 | 2.6 | 1.6 |
| | $\alpha$ at 100 | $\alpha$ at 60 | $\alpha$ at 40 | $\alpha$ at 20 | $\alpha$ at 0 | $\alpha$ at 0 | $\alpha$ at 0 | $\alpha$ at 0 |
| $P_i (P_o = 100\,000)$ | 92969 | 80669 | 65462 | 46042 | 24679 | 13228 | 7090 | 3800 |
| $q_i$ | 7031 | 12300 | 15207 | 19420 | 21363 | 11451 | 6138 | 3290 |
| $T_i$ (end of 2nd period | 98.8 | 98.7 | 73.0 | 65.4 | 60.7 | 32.5 | 17.6 | 10.0 |
| | $\alpha$ at 100 | $\alpha$ at 100 | $\alpha$ at 80 | $\alpha$ at 60 | $\alpha$ at 60 | $\alpha$ at 40 | $\alpha$ at 20 | $\alpha$ at 0 |
| $P_i (P_o = 100\,000)$ | 92969 | 86432 | 78606 | 68207 | 59184 | 48026 | 33779 | 18101 |
| $q_i$ | 7031 | 6537 | 7826 | 10399 | 9023 | 11158 | 14247 | 15678 |
| $T_i$ (end of 3rd period) | 114.3 | 101.2 | 100.6 | 91.2 | 70.1 | 65.0 | 54.5 | 51.0 |

During the ten minutes holding time the temperature in the lower half of the food continues to rise by conduction from the hotter upper layers. The upper half cools, partly because of conduction and partly by convection from the surface. This results in a more uniform temperature at the end of the 10 minutes holding period.

**CHAPTER 8: PROBLEM (1)**

Refer to Section 8.3.

**(a)**

Since one surface is insulated, effective thickness $a = 2 \times 0.008\,\text{m} = 0.016\,\text{m}$

$$t = \frac{L\rho}{\Delta\theta}\left(P\frac{a}{h} + R\frac{a^2}{\lambda}\right)$$

$$= \frac{251 \times 10^3 \times 1040}{20-(-2)}\left(\frac{1}{2}\frac{0.016}{7.4} + \frac{1}{8}\frac{(0.016)^2}{0.48}\right)$$

$$= 11.9 \times 10^6\,(1.08 \times 10^{-3} + 0.07 \times 10^{-3})$$

$$= 13\,700\,\text{s}$$

$$= 3.8\,\text{hours (for 8 mm thickness)}.$$

**(b)**

For thickness of 16 mm (that is $a = 0.032$ m):

$$t = 11.9 \times 10^6\left(\frac{1}{2}\frac{0.032}{7.4} + \frac{1}{8}\frac{(0.032)^2}{0.48}\right)$$

$$= 11.9 \times 10^6\,(2.16 \times 10^{-3} + 0.27 \times 10^{-3})$$

$$= 28\,900\,\text{s}$$

$$= 8.0\,\text{hours (for 16 mm thickness)}.$$

For thickness of 24 mm (that is $a = 0.048$ m):

$$t = 11.9 \times 10^6\left(\frac{1}{2}\frac{0.048}{7.4} + \frac{1}{8}\frac{(0.048)^2}{0.48}\right)$$

$$= 11.9 \times 10^6\,(3.24 \times 10^{-3} + 0.60 \times 10^{-3})$$

$$= 45\,700\,\text{s}$$

$$= 12.7\,\text{hours (for 24 mm thickness)}.$$

**(c)**

Thickness has a substantial (but non-linear) effect on thawing time.

## CHAPTER 8: PROBLEM (2)

**(a)**

Refer to Section 8.3.

Calculate $h$ values for different equipment:

Still air:

$$\frac{1}{h} = \frac{1}{5.7} + \frac{0.001}{0.06} = 0.175 + 0.017 = 0.192$$

therefore

$$h = \frac{1}{0.192} = 5.2\ .$$

Air blast (2.5):

$$\frac{1}{h} = \frac{1}{17.0} + \frac{0.001}{0.06} = 0.059 + 0.017 = 0.076$$

therefore

$$h = \frac{1}{0.076} = 13.2\ .$$

Air blast (5.0):

$$\frac{1}{h} = \frac{1}{28.4} + \frac{0.001}{0.06} = 0.035 + 0.017 = 0.052$$

therefore

$$h = \frac{1}{0.052} = 19.2\ .$$

$$\beta_1 = \frac{100}{50} = 2.0\,, \quad \beta_2 = \frac{75}{50} = 1.5$$

therefore

$$P = 0.23\,, \quad R = 0.065\ .$$

Calculate $t$:

Still air:

$$t = \frac{L\rho}{\Delta\theta}\left(P\frac{a}{h} + R\frac{a^2}{\lambda}\right)$$

$$= \frac{256 \times 10^3 \times 1088}{-2-(-30)}\left(0.23 \times \frac{0.050}{5.2} + 0.065 \times \frac{(0.050)^2}{1.53}\right)$$

$$= 9.95 \times 10^6\,(2.21 \times 10^{-3} + 0.11 \times 10^{-3})$$

$$= 23\,084\ \mathrm{s}$$

$$= 6.4\ \mathrm{h}\ .$$

Air blast (2.5):

$$t = 9.95 \times 10^6 \left(0.23 \times \frac{0.050}{13.2} + 0.11 \times 10^{-3}\right)$$

$$= 9.95 \times 10^6 (0.87 \times 10^{-3} + 0.11 \times 10^{-3})$$

$$= 9751\,\text{s}$$

$$= 2.7\,\text{h}\,.$$

Air blast (5.0):

$$t = 9.95 \times 10^6 \left(0.23 \times \frac{0.050}{19.2} + 0.11 \times 10^{-3}\right)$$

$$= 9.95 \times 10^6 (0.60 \times 10^{-3} + 0.11 \times 10^{-3})$$

$$= 7064\,\text{s}$$

$$= 2.0\,\text{h}\,.$$

**(b)**
See Section 8.5.

**CHAPTER 10: PROBLEM (1)**

Section E.3 for flat plates indicates that the equation to use depends on the Reynolds number.

Calculate *Re* for the four cases:

| $u$ (m/s) | $T$(°C) | $Re = \frac{du\rho}{\mu} = \frac{du}{\nu}$ |
|---|---|---|
| 5 | 150 | $\frac{0.5 \times 5}{2.9 \times 10^{-5}} = 86\,000$ |
| 5 | 200 | $\frac{0.5 \times 5}{3.5 \times 10^{-5}} = 71000$ |
| 10 | 150 | $\frac{0.5 \times 10}{2.9 \times 10^{-5}} = 172\,000$ |
| 10 | 200 | $\frac{0.5 \times 10}{3.5 \times 10^{-5}} = 143\,000$ |

Notes: $d = 0.5$ m (length of plate is characteristic dimension ($d$)). Values of $\nu$ from Table I.9 (by interpolation).

In the above $Re > 80\,000$ but much less than $500\,000$; therefore we can use:

$$j_D = 0.664\, Re^{-0.5}$$

therefore

$$St\, Sc^{2/3} = 0.664\, Re^{-0.5}$$

therefore

$$\frac{Sh}{Re\, Sc} Sc^{2/3} = 0.664\, Re^{-0.5}$$

therefore

$$Sh = 0.664\, Re^{0.5}\, Sc^{0.33}$$

| $u$ (m/s) | $T$(°C) | $Re^{0.5}$ | $Sc$ | $Sc^{0.33}$ | $Sh$ | $D$ | $k$ | $R = \frac{1}{Ak}$ |
|---|---|---|---|---|---|---|---|---|
| 5 | 150 | 293 | 0.61 | 0.85 | 165 | $4.5 \times 10^{-5}$ | 0.015 | 667 |
| 5 | 200 | 267 | 0.61 | 0.85 | 151 | $5.3 \times 10^{-5}$ | 0.016 | 625 |
| 10 | 150 | 415 | 0.61 | 0.85 | 234 | $4.5 \times 10^{-5}$ | 0.021 | 476 |
| 10 | 200 | 378 | 0.61 | 0.85 | 213 | $5.3 \times 10^{-5}$ | 0.023 | 435 |

Notes: (i) $Sc$ from I.10 (by extrapolation),
(ii) $Sh = kl/D$,
(iii) $D$ from I.10 (by extrapolation),
(iv) $k = ShD/l$,
(v) $A = 0.5 \times 0.2 = 0.1\,\text{m}^2$,
(vi) $R$ = resistance to mass transfer ($\text{s/m}^3$).

## CHAPTER 10: PROBLEM (2)

Moisture contents of saturated air at food surface temperatures (from I.11):

| | |
|---|---|
| 20°C | 0.0149 kg water/kg dry air |
| 50°C | 0.0875 kg water/kg dry air |
| 80°C | 0.559 kg water/kg dry air . |

**(a)**

$$C_B = 0.0149 \times \frac{50}{100} = 0.007\,\text{kg water/kg dry air}$$

saturated
air at 20°C × $RH$.

**(b)**

$$1CH_4 + 2O_2 + 7.52\,N_2 = 1CO_2 + 2H_2O + 7.52\,N_2 \; .$$

Excess air:

$$1\,O_2 + 3.76\,N_2 = 3.76N_2 + 1\,O_2$$

Dry combustion products consist of:

$1\,O_2$
$11.28\,N_2$
$1\,CO_2$ .

The moisture formes from the combuston of methane is $2\,H_2O$.

The air used consists of:

$3\,O_2$
$11.28\,N_2$ .

Its mass is = (3 × 32) + (11.28 × 28) = 411.8 kg.

Moisture in the combustion air (at 20°C and $RH$ = 50%) = (411.8 × 0.007) kg = 2.88 kg. This moisture must be added to the moisture formed during combustion.

Water content of dry combustion products ($C_B$)

$$= \frac{(2 \times 18) + (2.88)}{(1 \times 44) + (11.28 \times 28) + (1 \times 32)} \frac{\text{kg}}{\text{kg}}$$

$$= \frac{38.88}{391.84}\,\text{kg/kg}$$

$$= 0.099\,\text{kg water/kg dry combustion products} \; .$$

**Electric oven**

| $u$ (m/s) | $T_{air}$ (°C) | $T_{surface}$ (°C) | $R$ | $C_S$ | $C_B$ | $m$ ($m^3$/s) | $M$(g/s) |
|---|---|---|---|---|---|---|---|
| 5 | 150 | 20 | 667 | 0.015 | 0.007 | $0.01 \times 10^{-3}$ | 0.01 |
| | | 50 | 667 | 0.088 | 0.007 | $0.12 \times 10^{-3}$ | 0.06 |
| | | 80 | 667 | 0.559 | 0.007 | $0.82 \times 10^{-3}$ | 0.43 |
| 5 | 200 | 20 | 625 | 0.015 | 0.007 | $0.01 \times 10^{-3}$ | – |
| | | 50 | 625 | 0.088 | 0.007 | $0.12 \times 10^{-3}$ | 0.06 |
| | | 80 | 625 | 0.559 | 0.007 | $0.88 \times 10^{-3}$ | 0.41 |
| 10 | 150 | 20 | 476 | 0.015 | 0.007 | $0.01 \times 10^{-3}$ | 0.01 |
| | | 50 | 476 | 0.088 | 0.007 | $0.17 \times 10^{-3}$ | 0.09 |
| | | 80 | 476 | 0.559 | 0.007 | $1.15 \times 10^{-3}$ | 0.60 |
| 10 | 200 | 20 | 435 | 0.015 | 0.007 | $0.01 \times 10^{-3}$ | – |
| | | 50 | 435 | 0.088 | 0.007 | $0.18 \times 10^{-3}$ | 0.08 |
| | | 80 | 435 | 0.559 | 0.007 | $1.26 \times 10^{-3}$ | 0.58 |

Notes: (i) $R$ from problem 1.

(ii) $m = \frac{(C_S - C_B)}{R}$ with units of $m$ as $m^3$/s since $C_S$ and $C_B$ are dimensionless.

(iii) To convert $m$ ($m^3$/s) to $M$(g/s) it is necessary to convert the volume of water vapour to a mass by using the density of water vapour (I.4) at the appropriate air/combustion products temperature ($\rho_{150°C}$ = 0.519, $\rho_{200°C}$ = 0.464).

**Gas oven**

| $u$(m/s) | $T_{air}$ (°C) | $T_{surface}$ (°C) | $R$ | $C_S$ | $C_B$ | $m$(m³/s) | $M$(g/s) |
|---|---|---|---|---|---|---|---|
| 5 | 150 | 20 | 667 | 0.015 | 0.099 | $-0.12 \times 10^{-3}$ | −0.06 |
| | | 50 | 667 | 0.088 | 0.099 | $-0.01 \times 10^{-3}$ | −0.01 |
| | | 80 | 667 | 0.559 | 0.099 | $0.68 \times 10^{-3}$ | 0.35 |
| 5 | 200 | 20 | 625 | 0.015 | 0.099 | $-0.13 \times 10^{-3}$ | −0.06 |
| | | 50 | 625 | 0.088 | 0.099 | $-0.02 \times 10^{-3}$ | −0.01 |
| | | 80 | 625 | 0.559 | 0.099 | $0.73 \times 10^{-3}$ | 0.34 |
| 10 | 150 | 20 | 476 | 0.015 | 0.099 | $-0.18 \times 10^{-3}$ | −0.09 |
| | | 50 | 476 | 0.088 | 0.099 | $-0.02 \times 10^{-3}$ | −0.01 |
| | | 80 | 476 | 0.559 | 0.099 | $0.96 \times 10^{-3}$ | 0.50 |
| 10 | 200 | 20 | 435 | 0.015 | 0.099 | $-0.19 \times 10^{-3}$ | −0.09 |
| | | 50 | 435 | 0.088 | 0.099 | $-0.02 \times 10^{-3}$ | −0.01 |
| | | 80 | 435 | 0.559 | 0.099 | $1.05 \times 10^{-3}$ | 0.49 |

Minus sign indicates food *gains* moisture.

Note that this assumes that the combustion products have the same properties as air, a reasonable assumption since they both contain predominantly nitrogen.

## CHAPTER 10: PROBLEM (3)

In this case:

$$Re = \frac{0.5 \times 10}{2.19 \times 10^{-5}} = 23\,000$$

$$Re^{0.5} = 152$$

$$Sc = 0.61$$

$$Sc^{0.33} = 0.85$$

$$Sh = 86$$

$$D = 3.6 \times 10^{-5}$$

$$k = 0.006$$

$$R = 1667$$

$$C_S = 0.559$$

$$C_B = 1.46 \times 0.95 = 1.387$$

therefore $m = -0.49 \times 10^{-3}\ \mathrm{m^3/s}$

therefore $M = -0.30 g/\mathrm{s}\ (\rho_{90°\mathrm{C}} = 0.605)$.

Thus there is a *gain* in moisture cf. a *loss* with other methods of heating at this high surface temperature.

The Rapidaire oven, by using moist air, avoids the moisture loss apparent with gas and electric ovens. There is a trend nowadays to provide facilities for injecting a fine water spray into the air stream in forced convection ovens for the same reason. See Section 10.3.

# Index